System Reliability
Evaluation and Prediction in Engineering

Alain Pagès & Michel Gondran

Studies and Research Division
Electricité de France

System Reliability
Evaluation and Prediction in Engineering

Translated by
E. Griffin

Springer-Verlag
New York Berlin Heidelberg Tokyo

Alain Pagès is Head of Department within the Research Division of Electricité de France and is responsible for metering, load management and voltage disturbance research.

Michel Gondran is in charge of the methodology and development of reliability programs in the Research Division.

Studies and Research Division
Electricité de France
1 avenue du Générale de Gaulle
92141 Clamart cedex
France

Revised and updated 1986

English edition first published 1986 by North Oxford Academic Publishers Ltd, a subsidiary of Kogan Page Ltd, 120 Pentonville Road, London N1 9JN

Sole distribution in the USA, Canada and Mexico
Springer-Verlag New York Inc.
175 Fifth Avenue
New York, NY 10010
USA

Library of Congress Cataloging-in-Publication Data
Pagès, Alain
System reliability: evaluation and prediction in engineering.
Includes index.
1. Reliability (Engineering) I. Gondran, Michel.
II. Title.
TA169.P34 1986 620'.00452 86-3766

ISBN 0-387-91276-2

Printed in Great Britain

CONTENTS

Chapter 2. Part I. — Representation of system logic

Chapter 2. Part II. — Representation of knowledge in reliability studies: FCP analysis

Chapter 3. — Direct analysis of simple systems

Chapter 4. — Evaluating system availability

Chapter 5. — Classical methods for evaluating the reliability of repairable systems

Chapter 6. — Reliability evaluation using the critical operating states method

Chapter 7. — Statistical treatment of reliability data

Chapter 8. — Optimization and maintenance

Chapter 9. — Case study (travelling bridge crane), by J.-F. Barbet

PREFACE

Once Man had invented tools he became dependent on their ability to work properly. The concept of reliability was born.

At the beginning of the industrial era, malfunctions or failures of system components or complex systems were not subjected to any kind of systematic examination. It was only gradually that this became standard practice, in the wake of several accidents and disasters. At that time, systems were essentially mechanical. Examples include ball bearings, whose service life has been the subject of much in-depth research for railway applications, and electricity distribution. In the latter case, as soon as the first distribution networks were set up, every effort was made to make this new energy source more reliable by arranging the generators, transformers and power lines in parallel.

Reliability entered a new era with the advent of electronics, subsonic and supersonic aircraft, missiles, nuclear energy applications and computers, as well as domestic applicances. However, systematic equipment reliability studies are of very recent origin.

There are three main reasons for the importance acquired by this discipline.

— The first is that it is vital that certain equipment or systems should function properly and perform a specified mission or task, e.g. where human lives are at stake in public transport systems or where society itself is affected – imagine the consequences of a bank being totally deprived on its processing facilities!

— The second reason is basically an economic one. If an item of equipment is to be selected on the basis of its total cost, the operating cost must be added to the purchase price. The cost of in-service failure is often high not only because of the price of replacement parts, but also in terms of the loss of use suffered by the user. By finding a compromise between purchase and operating costs, one can arrive at a minimum total cost, making it possible to evaluate the level of reliability which is economically desirable.

— The third reason concerns the need to rationalize various industrial activities due to the increased size of companies, the ever shorter lead times for new technologies and the complexity of the systems involved.

In this book, Alain Pagès and Michel Gondran present a synthesis of ten years' work on methods of evaluating and predicting system reliability. They have been closely involved with E.D.F. (Electricité de France) activities concerning power station reliability, transmission and interconnection networks. In particular, since 1974, they have collaborated on a probability analysis programme to assess the safety of nuclear power stations. Based on the example of the Fessenheim pressurized-water reactor, the main objectives of this programme are first to develop methods of examining the reliability of complex systems and to compile the necessary computer code, and secondly to apply these results to the optimization of safety systems for future nuclear plants.

The original feature of this book compared with other works on reliability engineering is that it is concerned with presenting methods of solving practical problems. It also introduces a new method: the ***critical operating states method****, which is based on the simple idea of the various operating states of a system, some of which are more 'dangerous' than others. For example, it is clear that in a chain of servomechanisms, the failure of a redundant component mounted in parallel with its counterpart initiates a 'critical period' which will last until the faulty component is repaired, as a new failure of the second component during this period would immobilize the servomechanism.*

This idea developed into the critical operating states method which also provides an automatic method of calculating the reliability of large repairable systems, a calculation which is impossible using conventional methods. It also allows immediate interpretation of the Markov diagram used to model the behaviour of a small system.

This last point would seem to be of particular interest to the engineer, as the method allows an analytical expression for the availability or reliability of numerous sytems to be found quickly. He will thus be able to find a large number of expressions for solving his problem, in addition to all the conventional formulas.

The expectations of theoreticians anxious for applications and of practicians avid for rigour will not be disappointed by this book which has every likelihood of becoming a reliability classic. I wish it well.

Maurice Magnien

ACKNOWLEDGMENTS

We wish to thank Mr Maurice Magnien, Research and Development Director of Electricite de France, for writing the preface to this book.

Our thanks are also due to Bernard Gachot, whose skill and enthusiasm inspired the 'Reliability methodology' working party which we found a useful source of contacts and exchanges.

We are extremely grateful to our colleagues in the Atomic Energy Commissariat (D.S.N.) and in Electricite de France (S.P.T., SEPTEN, DER) who participated in this working party and whose help and encouragement made this book possible.

We would like to give special thanks to Michel Llory and to all the engineers in his 'Probabilistic Studies' division for their constructive comments and criticisms at the manuscript preparation stage, and in particular Jean-François Barbet for writing chapter 9.

CHAPTER 1

INTRODUCTION TO RELIABILITY: BASIC MATHEMATICS

1. DEFINITIONS

1.1. The term 'reliability'

The science of reliability has a terminology of its own and this section will be devoted to introducing a few definitions. One of the most important concepts in 'reliability science' is that of 'reliability' itself, which the International Electrotechnical Commission defines as follows [I.E.C., 1974]:

"The ability of an item to perform a required function, under stated conditions, for a stated period of time. The term reliability is also used as a reliability characteristic denoting a probability of success or a success ratio."

A cursory examination of this definition indicates just how important it can be to evaluate an item's reliability:

—at the design stage, it is essential to have methods of predicting reliability in order to meet a specification, to achieve consistency and to realize an objective at minimum cost (improved reliability increases production costs but reduces operating expenses) (cf. Ch. 8);

—at the manufacturing stage, it will be necessary to check that the components manufactured or used comply fully with the design characteristics. Manufacturing control methods will not be discussed in this book, as they involve specialized techniques (see, for example [Beliav *et al.*, 1972, Ch. 7]);

—at the operational stage, it is useful, for many reasons, to check the accuracy of reliability predictions: not only to verify that the system is behaving normally and, if not, to locate the cause of the trouble, but also to obtain better reliability data (cf. Ch. 7). In addition, reliability constraints may affect equipment operation through the introduction of certain test procedures or the need for stock control of standby components (cf. Ch. 8).

For a long time the engineer's judgement, based on experience, was sufficient as far as equipment design and operation were concerned. Although this judgement is still vital, it is useful for the engineer to be able to back it up with a set of qualitative and quantitative methods for evaluating reliability, in view of increased system complexity and the associated risks (large-capacity aircraft, rockets, nuclear power stations, computers, etc.).

Kline and coworkers [Kline *et al.*, 1978] give an interesting analysis of the many articles and books devoted to different aspects of reliability with respect to various criteria, notably for areas of application: air transport and communications in the 1950s and space exploration in the 1960s generated an enormous amount of published material. From the 1970s onwards, electricity generation seems to have come to the fore with the advent of nuclear power.

1.2. Examination of the I.E.C. definition

From a more detailed examination of the I.E.C. definition, the following important concepts emerge.

(i) The concept of an 'item having to perform a required function under stated conditions'. In the following chapters, we will refer to the various items whose reliability is to be evaluated as *systems.* A system is defined as a set of interacting components. The *components* of the system are capable of failing (i.e. of being unable to perform their mission or function). Failure of a component can be *sudden* or *gradual, partial* or *total.* In this book, we will only consider sudden and total failures which are termed *catastrophic.* Types of failure are classified in terms of *failure modes*: operating, standby, under load, etc. The components may also be *repairable* or *non-repairable.* A component will be termed repairable if it can be restored to its original condition after it has failed, without affecting system operation. The 'repairable' character of a component may be related to the nature of the mission assigned to the system. A system may also perform several missions; when its mission is to prevent the occurrence of a failure with disastrous consequences, reliability becomes synonymous with safety.

(ii) The concept of *probability.* Evaluating system reliability therefore requires probability calculations. Consequently, a large part of this first chapter will be devoted to summarizing probability theory.

(iii) The concept of mission *duration.* Reliability is therefore seen to be a function of time.

(iv) The concept of *operating condition.* This refers not only to the entire *physical environment* in which the system must perform its intended function (temperature, humidity, pressure, etc.) but also to the system's *operating* and *maintenance modes.* The system may in fact be continuously operating or normally on standby and required to operate under given conditions. In the first case, component failures are generally detected immediately, which is not so in the second case.

Maintenance will obviously be implemented following component failure, but can also take the form of preventive action, which is particularly necessary in the case of systems normally on standby.

In short, we can formulate a mathematical expression for the reliability $R(t)$ of a system S required to *perform a mission under specified conditions*:

$$R(t) = \text{Probability } (S \text{ will be operable during } [0, t]) \tag{1}$$

Consequently, $R(t)$ is a *non-increasing* function varying between 1 and 0 over $[0, \infty[$.

1.3. Other definitions

Reliability is not the only probability of interest associated with system operation.

We are also interested in the two following probabilities:

—*availability* $A(t)$, which is the probability of the system S being operable at instant t. Note that in the case of non-repairable systems, the definition of $A(t)$ is equivalent to that of reliability (expression (1)), which explains the fact that reliability and availability are often confused.

$$A(t) = \text{Probability } (S \text{ will be operable at time } t) \tag{2}$$

—*maintainability* $M(t)$, which is the complement to 1 of the probability of the system not being repaired during interval $[0, t]$, given that it failed at instant $t = 0$.

$$M(t) = 1 - \text{Probability } (S \text{ will not be repaired during } [0, t])$$

This concept only applies to repairable systems. $M(t)$ is a *non-decreasing* function varying between 0 and 1 over $[0, +\infty[$.

It can be shown that under certain very general conditions the availability of a continuously operating repairable system tends to a non-zero limit as t tends to infinity and that this limit is equal to the proportion of the time during which the system is in an operable state. In practice, this value is reached very quickly (a few times the mean repair time of the system).

On the other hand, the reliability and maintainability of all systems tend to the same limit (0 for reliability and 1 for maintainability) as t tends to infinity. The system is characterized by the manner in which these limits are reached.

One way of characterizing them is by the *Mean Time To Failure* (MTTF) and the *Mean Time To Repair* (MTTR) (cf. section 1.4). It will be shown in section 6 that these values can be expressed as follows:

$$\text{MTTF} = \int_0^\infty R(t)\,\mathrm{d}t$$

$$\text{MTTR} = \int_0^\infty [1 - M(t)]\,\mathrm{d}t$$

A second method consists of introducing the instantaneous failure rate $\Lambda(t)$ and instantaneous repair rate $\mathcal{M}(t)$ of the system, given respectively by:

$$\Lambda(t) = \lim_{\Delta t \to 0} \frac{1}{\Delta t} \text{Probability } (S \text{ will fail between } t \text{ and } t + \Delta t, \text{ given that no failure has occurred during } [0, t]) \tag{3}$$

$$\mathcal{M}(t) = \lim_{\Delta t \to 0} \frac{1}{\Delta t} \text{Probability } (S \text{ will be repaired between } t \text{ and } t + \Delta t \text{ given that it has failed during } [0, t]) \tag{4}$$

Under certain conditions, these expressions have limits when t tends to infinity and $R(t)$ and $M(t)$ behave as $\exp(-\Lambda(\infty)t)$ and $1 - \exp(-\mathcal{M}(\infty)t)$ respectively (cf. Ch. 6).

1.4. Meaning of certain abbreviations

A number of abbreviations denoting mean times are used in reliability studies and unfortunately their definitions sometimes cause confusion. We intend to use the following notations:

MTTF: Mean Time To (first) Failure
MTTR: Mean Time To Repair

These two values naturally depend on the initial state of the system (cf. Ch. 6).

MUT: Mean Up Time (after repair)

The MUT is different from the MTTF because when the system is restored to an operable condition following a failure, not all the defective components have necessarily been repaired. However, we shall see in Ch. 6 that for many systems these two quantities are very similar.

MDT: Mean Down Time

This includes fault detection, delay time, repair time and service restoration time.

MTBF: Mean Time Between Failures (repairable system)

Naturally, MTBF = MUT + MDT.

Some authors use MTBF for the 'mean operating time', thereby assigning it the meaning which we are giving to MTTF or MUT. However, for many systems, the MDT is negligible compared with the MUT and the difference between the MTTF and the MTBF is minimal.

2. CONCEPT OF PROBABILITY

The following sections do not provide a comprehensive treatment of probability but merely summarize the main concepts and formulae of probability theory used for reliability calculations. For an in-depth treatment of these concepts the reader is referred to [Feller, 1966; Reniy, 1966; Neveu, 1972].

2.1. Algebra of events

Given a *set of observables* Ω and a trial ω for extracting an element from this set, we call a subset of Ω *event* A. The set of events is therefore included (in the broad sense) in the set of subsets of Ω.

Two events A and B are identical if, for each trial, they both occur or do not occur. The event denoted by $\bar{A}$ which occurs when A does not occur and vice versa, is called the *negation* of A. The event which never occurs is the *impossible event* corresponding to the empty set in Ω. The *certain event* is the event which always occurs, corresponding to the complete set Ω.

The *simultaneous* realization of two events A and B is also an event denoted by A and B (or $A \cap B$); the associated subset C of Ω is $C = A \cap B$. Two events A and B are termed *mutually exclusive* when $A \cap B = \varnothing$. The realization of *at least one* of events A and B is also an event denoted by A or B (or $A \cup B$); associated subset C of Ω is $C = A \cup B$. When set Ω is finite, the set of events $\mathscr{A}$ has a Boolean algebra structure for the operations 'negation', 'and', 'or'. If set Ω is countable, the structure is called a σ-algebra.

2.2. Conditional events

Let us assume that we know that an event X of $\mathscr{A}$ has occurred; the result of the trials is modified by this knowledge. Given an event A of $\mathscr{A}$, we will call the subset $A \cap X$ a *conditional event* at X. The set of conditional events at X possesses an algebraic structure, denoted by $\mathscr{A}|X$.

2.3. Probability spaces

Let Ω be a non-empty set, $\mathscr{A}$ a σ-algebra of the events during Ω and a function $\mathscr{P}$ of $\mathscr{A}$ in R^+ having the following properties:

(a) $\mathscr{P}(\Omega) = 1$.

(b) $A_1, A_2, \ldots, A_n$ being a finite or denumerable sequence of pairwise disjoint sets of $\mathscr{A}$.

$$\mathscr{P}(A_1 \cup A_2 \ldots \cup A_n \cup \ldots) = \mathscr{P}(A_1) + \mathscr{P}(A_2) \ldots + \mathscr{P}(A_n) \ldots$$

Function $\mathscr{P}$ is called the *probability function* and the triplet $(\Omega, \mathscr{A}, \mathscr{P})$ is termed the *probability space.*

Function $\mathscr{P}$ has the following properties:

$$\mathscr{P}(\overline{A}) = 1 - \mathscr{P}(A)$$
$$A \subset B \Rightarrow \mathscr{P}(A) \leq \mathscr{P}(B)$$
$$0 \leq \mathscr{P}(A) \leq 1$$
$$\mathscr{P}(A \cup B) = \mathscr{P}(A) + \mathscr{P}(B) - \mathscr{P}(A \cap B).$$

This last property can be generalized in a form known as Poincaré's theorem:

$$\mathscr{P}\left(\bigcup_{i=1}^{n} A_i\right) = \sum_{j=1}^{n} \mathscr{P}(A_i) - \sum_{j=2}^{n} \sum_{i=1}^{j-1} \mathscr{P}(A_i \cap A_j) +$$

$$+ \sum_{j=3}^{n} \sum_{k=2}^{j-1} \sum_{i=1}^{k-1} \mathscr{P}(A_i \cap A_j \cap A_k) - \cdots + (-1)^n \mathscr{P}\left(\bigcap_{i=1}^{n} A_i\right). \tag{5}$$

Given an event X of $\mathscr{A}$, let us consider the application $\mathscr{P}(.|X)$ of $\mathscr{A}|X$ in R^+ defined by:

$$\mathscr{P}(A/X) = \frac{\mathscr{P}(A \cap X)}{\mathscr{P}(X)} \tag{6}$$

It is easy to verify that:

$$\mathscr{P}(A/X) \geq 0$$
$$\mathscr{P}(X/X) = 1$$
$$A \cap B = 0 \Rightarrow \mathscr{P}(A/X \cup B/X) = \mathscr{P}(A/X) + \mathscr{P}(B/X).$$

It follows that $\mathscr{P}(.|X)$ is a probability and that Eqn (6) should be regarded as defining the *conditional probabilities.*

Equation (6) may also be written:

$$\mathcal{P}(A \cap X) = \mathcal{P}(A/X)\,\mathcal{P}(X)$$

and is generalized in the form:

$$\mathcal{P}(A_1 \cap A_2 \ldots \cap A_n) =$$
$$= \mathcal{P}(A_1/(A_2 \cap \ldots \cap A_n))\,\mathcal{P}(A_2/(A_3 \ldots A_n)) \ldots \mathcal{P}(A_{n-1}/A_n)\,\mathcal{P}(A_n)\,. \quad (7)$$

Returning to the definitions of $\Lambda(t)$ and $\mathcal{M}(t)$ given by Eqns (3) and (4) respectively, it can be seen that conditional probabilities are involved.

Using Eqn (6), we obtain:

$$\Lambda(t) = \lim_{\Delta t \to 0} \frac{1}{\Delta t} \frac{\mathcal{P}(S \text{ will fail between } t \text{ and } t + \Delta t \text{ and } S \text{ will be operable during } [0, t])}{\mathcal{P}(S \text{ will be operable during } [0, t])}$$

or using Eqn (1):

$$\Lambda(t) = \lim_{\Delta t \to 0} \frac{1}{\Delta t} \cdot \frac{1}{R(t)} (\mathcal{P}(S \text{ will fail during } [0, t + \Delta t]) - \mathcal{P}(S \text{ will fail during } [0, t]))$$

$$\Lambda(t) = \lim_{\Delta t \to 0} \frac{R(t) - R(t + \Delta t)}{\Delta t\, R(t)}\,.$$

If $R(t)$ is differentiable, we have:

$$\Lambda(t) = \frac{-\dfrac{\mathrm{d}R}{\mathrm{d}t}(t)}{R(t)} \geq 0 \qquad (8)$$

and similarly:

$$\mathcal{M}(t) = \frac{\dfrac{\mathrm{d}M}{\mathrm{d}t}(t)}{1 - M(t)} \geq 0\,. \qquad (9)$$

Two events A and B are termed *independent* if and only if

$$\mathcal{P}(A/B) = \mathcal{P}(A)$$

i.e., if and only if:

$$\mathcal{P}(A \cap B) = \mathcal{P}(A)\,.\,\mathcal{P}(B)\,.$$

2.4. Total probability theorem

A denumerable set A_i of pairwise mutually exclusive events such as

$$\mathcal{P}\left(\bigcup_i A_i\right) = \sum_i \mathcal{P}(A_i) = 1\,. \qquad (10)$$

is known as a complete system of events of the probability space $(\Omega, \mathscr{A}, \mathscr{P})$.

For any element B of $\mathscr{A}$, we can therefore write:

$$\mathscr{P}(B) = \sum_i \mathscr{P}(B/A_i)\,\mathscr{P}(A_i) \qquad (11)$$

which expresses the *total probability theorem.*

This theorem will be used for direct calculation of the availability of systems which cannot be reduced to a series-parallel structure (cf. Ch. 3).

2.5. Bayes' theorem

Given an event B of non-zero probability and a set of complete events A_i, Bayes' theorem states that:

$$\mathscr{P}(A_i/B) = \frac{\mathscr{P}(B/A_i)\,\mathscr{P}(A_i)}{\sum_i \mathscr{P}(B/A_i)\,\mathscr{P}(A_i)}\,. \qquad (12)$$

If the *a priori* probabilities $\mathscr{P}(A_i)$ are known, Bayes' theorem enables the *a posteriori* probabilities $\mathscr{P}(A_i|B)$ to be calculated, hence the alternative name: 'theorem concerning the probability of causes'.

Example: Let us consider a computer whose failures are due 80% to hardware and 20% to software; half of the hardware failures and 90% of the software failures result in computer failure. Given that the computer is down (event B), what is the probability of the failure being due to hardware?

By calling event A_1 'hardware failure of system' and event A_2 'software failure of system', Eqn (12) can be written as:

$$\mathscr{P}(A_1/B) = \frac{\mathscr{P}(B/A_1)\,\mathscr{P}(A_1)}{\mathscr{P}(B/A_1)\,\mathscr{P}(A_1) + \mathscr{P}(B/A_2)\,\mathscr{P}(A_2)}$$

$$= \frac{0.5 \times 0.8}{0.5 \times 0.8 + 0.9 \times 0.2} \simeq 0.69\,.$$

This theorem has aroused much controversy; moreover, as the $\mathscr{P}(A_i)$ are not generally known, one is often tempted to make unverifiable assumptions, thereby misusing Bayes' theorem. An application of this theorem is given in Ch. 7 (reliability data).

3. RANDOM VARIABLES

3.1. Definition

Let $(\Omega, \mathscr{A}, \mathscr{P})$ be a probability space and v an application of Ω in R. v is said to be a *random variable* if:

$$\forall x \in R,\; v^{-1}(]-\infty, x[) \in \mathscr{A}\,.$$

This condition can always be verified if $\mathscr{A}$ is the set of subsets of Ω.

Let us assume that

$$A_x = v^{-1}(]-\infty, x]),\ x = v(\omega),\ \omega \in \Omega$$

and $\mathscr{B}$ = σ-algebra produced by the disjoint sums of intervals of R which are closed on the left and open on the right.

One then defines a probability $\mathscr{P}'$ during $(R, \mathscr{B})$ by the expression:

$$\mathscr{P}'(X \leq x) = \mathscr{P}(A_x)\,.$$

This probability $\mathscr{P}'$ is called the *distribution function F* of the random variable v.

In the following chapters we will make the usual comparison between the function v and the range x of any element ω of Ω.

We will therefore say that F is the distribution function of X. It can easily be shown that F has the following properties:

(a) $0 \leq F(x) \leq 1$; $F(\infty) = 1$.

(b) F is non-decreasing.

(c) $\mathscr{P}(X = x) = F(x^+) - F(x^-)$.

(d) The various classes of distribution functions are linear combinations of step functions (corresponding to discrete random variables) and of continuous functions (corresponding to continuous random variables for which $\mathscr{P}(X = x)$ is zero).

When the distribution function F is differentiable, its derivative f is called the *probability density* of the random variable X and:

$$F(x) = \int_{-\infty}^{x} f(u)\,\mathrm{d}u \tag{13}$$

since $F(\infty) = 1$, it follows that

$$\int_{-\infty}^{+\infty} f(u)\,\mathrm{d}u = 1\,. \tag{14}$$

When F is differentiable virtually everywhere and has points of discontinuity, it is necessary to regard it as a distribution in order to define the density; the jumps then appear as Dirac distributions in the density expression and the integral of Eqn (13) is a Stieljes integral (cf. [Schwartz, 1961]).

3.2. Moments

The real value, when it exists, defined by:

$$E[X^k] = \int_{-\infty}^{+\infty} x^k\,\mathrm{d}F(x) \qquad \text{also denoted by } \mu'_k \tag{15}$$

is called the kth *moment* of the random variable X.

In the case of a discrete random variable, Eqn (15) can be written as:

$$E[X^k] = \sum_i x_i^k\,\mathscr{P}(x = x_i)$$

and in the case where f exists, we obtain:

$$E[X^k] = \int_{-\infty}^{\infty} x^k f(x)\,\mathrm{d}x\,.$$

$E[X]$ is called the *mean value* or mathematical expectation.

The real value μ_k, when it exists, defined by:

$$\mu_k = \int_{-\infty}^{+\infty} (x - E[X])^k \, \mathrm{d}F(x) \tag{16}$$

is called the kth *central moment.* μ_2 is called the *variance* and $\sqrt{\mu_2}$ the *standard deviation*, which is denoted by σ.

3.3. Random vectors

By replacing R by R^n in the definition of a random variable, we obtain the definition of a *random vector of dimension n.*

We are interested here in the case $n = 2$.

As with a random variable, a random vector is completely defined by its distribution function $F(.,.)$:

$$F(x, y) = \mathcal{P}(X \leqslant x \text{ et } Y \leqslant y) . \tag{17}$$

$F(x, \infty)$ is called the marginal distribution function of x and $F(\infty, y)$ the marginal distribution function of y.

If F has second order differentials in x and y, the probability density $f(.,.)$ is given by:

$$f(x, y) = \frac{\partial^2 F(x, y)}{\partial x \, \partial y} . \tag{18}$$

$f(x, \infty)$ is called the marginal probability density of x and $f(\infty, y)$ the marginal probability density of y.

$$f(x, \infty) = \int_{-\infty}^{+\infty} f(x, y) \, \mathrm{d}y$$

$$f(\infty, y) = \int_{-\infty}^{+\infty} f(x, y) \, \mathrm{d}x .$$

The *conditional probability density* of x with respect to y is defined as:

$$p(x/y) = \frac{f(x, y)}{f(\infty, y)} \tag{19}$$

and the conditional probability density of y with respect to x is given by:

$$q(y/x) = \frac{f(x, y)}{f(x, \infty)} . \tag{20}$$

The random vector defined by $F(.,.)$ may be regarded as being formed from two random variables X and Y. The two random variables are termed independent if, irrespective of the subsets I and J of R, the two probabilities $\mathcal{P}(X \in I)$ and $\mathcal{P}(Y \in J)$ are independent. Consequently:

$$F(x, y) = F(x, \infty) . F(\infty, y) \tag{21}$$

and if $f(x, y)$ exists:

$$f(x, y) = f(x, \infty) \,.\, f(\infty, y) \,. \tag{22}$$

The real number $\mathrm{Cov}(X, Y)$, if it exists, defined by:

$$\mathrm{Cov}(X, Y) = E[(X - E[X])(Y - E[Y])] \tag{23}$$

is called the *covariance* between the two random variables X and Y. Note that $\mathrm{Cov}(X, Y) = \mathrm{Cov}(Y, X)$.

The ratio ρ_{xy}, if it exists, defined by:

$$\rho_{xy} = \frac{\mathrm{Cov}(x, y)}{\sigma(x)\,\sigma(y)} \,. \tag{24}$$

is called the *linear correlation coefficient*.

It can be shown that $|\rho_{xy}| \leq 1$.

If the linear correlation coefficient is zero, the two variables are termed *non-correlated*.

Using Eqn (15), the covariance can be written as:

$$\mathrm{Cov}(X, Y) = E[XY] - E[X]\,E[Y] \,. \tag{25}$$

The symmetrical matrix Θ defined by:

$$\Theta = \begin{bmatrix} \mu_2(X) & \mathrm{Cov}(X, Y) \\ \mathrm{Cov}(X, Y) & \mu_2(Y) \end{bmatrix} . \tag{26}$$

is called the matrix of variance-covariance between X and Y.

When the random variables X and Y are independent, using Eqns (15) and (21) we obtain:

$$E[XY] = E[X] \,.\, E[Y]$$

hence

$$\mathrm{Cov}(X, Y) = 0 \,.$$

Independent random variables are therefore non-correlated, but the converse is not true; one only has to consider the two random variables $X = \cos u$ and $Y = \sin u$ where u is a uniformly distributed random variable on $[0, 2\pi]$, $E(XY) = 0$ even though X and Y are related by the equation

$$X^2 + Y^2 = 1 \,.$$

3.4. Transformations of random variables

Given a random variable X with probability density f_x and a function g, can one determine the probability density, if it exists, of $Y = g(X)$?

Let us first assume that g is strictly monotonic and continuously differentiable. If F_x and F_y are the respective distribution functions of X and Y, we can write:

$$F_y(y + \Delta y) - F_y(y) = |\, F_x(g^{-1}(y + \Delta y)) - F_x(g^{-1}(y)) \,|$$

therefore

$$f_y(y) = f_x(g^{-1}(y)) \left| \frac{\mathrm{d}g^{-1}(y)}{\mathrm{d}y} \right|$$

where $g^{-1}(y)$ is the solution of $g(x) = y$.

If g is not monotonic, the calculation is more complicated. In the following case, for example, we obtain:

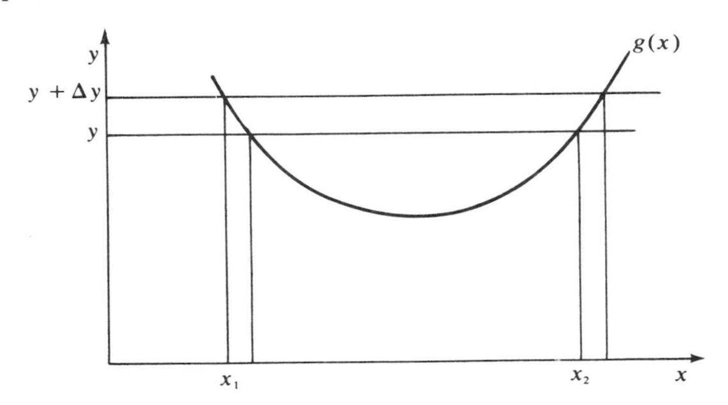

Fig. 1.

$$f_y(y) = f(x_1)\left|\frac{dx}{dy}(x_1)\right| + f(x_2)\left|\frac{dx}{dy}(x_2)\right|.$$

In the case of a function of several variables, it is necessary to replace $\left|\frac{dg^{-1}(y)}{dy}\right|$ by the absolute value of the Jacobian of the transformation.

3.5. Algebra of random variables

From the definition of central moments (Eqn (16)) and covariance (Eqn (23)) it is easy to derive the following relations which are true for any finite constant a:

$$E[aX] = aE[X] \tag{27}$$
$$E[X + a] = E[X] + a \tag{28}$$
$$\sigma^2[aX] = a^2\sigma^2[X] \tag{29}$$
$$\sigma^2[X + a] = \sigma^2[X] \tag{30}$$
$$\sigma^2[X] = E[X^2] - (E[X])^2 \tag{31}$$
$$E(X + Y) = E[X] + E[Y] \tag{32}$$
$$\sigma^2[X + Y] = \sigma^2[X] + \sigma^2[Y] + 2\,\mathrm{Cov}(X, Y) \tag{33}$$

hence, when X and Y are non-correlated:

$$\sigma^2[X + Y] = \sigma^2[X] + \sigma^2[Y]) \tag{34}$$
$$E[X - Y] = E[X] - E[Y] \tag{35}$$
$$\sigma^2[X - Y] = \sigma^2[X] + \sigma^2[Y] - 2\,\mathrm{Cov}(X, Y) \tag{36}$$
$$E[XY] = E[X]\,E[Y] + \mathrm{Cov}(X, Y). \tag{37}$$

If X and Y are *independent*:

$$\sigma^2[XY] = E^2[X]\,\sigma^2[Y] + E^2[Y]\,\sigma^2[X] + \sigma^2[X]\,\sigma^2[Y]. \tag{38}$$

4. MAIN PROBABILITY DISTRIBUTIONS USED IN RELIABILITY

This section will present a few properties of the main distributions used in reliability. If the reader wishes to carry out a more detailed study, he is referred to [Johnson and Kotz, 1970; Kendall and Stuart, 1963].

4.1. Discrete distributions

4.1.1. *Binomial distribution*

Let us consider a distribution for which the set of events reduces to $\{A, \bar{A}, \varphi, \Omega\}$. Let p be the probability of A and $(1 - p)$ that of $\bar{A}$. The discrete random variable X representing the number of occurrences of event A in the course of n experiments is binomially distributed with parameters (p, n) such that:

$$\mathcal{P}(X = k) = C_n^k p^k (1 - p)^{n-k} \quad 0 \leq k \leq n \quad 0 \leq p \leq 1 \tag{39}$$

giving the distribution function:

$$F(k) = \mathcal{P}(x \leq k) = \sum_{i=0}^{k} C_n^i p^i (1 - p)^{n-i} \tag{40}$$

then:

$$E[X] = np$$
$$\sigma^2[X] = np(1 - p) .$$

This distribution will be encountered in reliability calculations when an item has a probability γ of failure under load. After n loadings, the number of failures is distributed according to a binomial law $B(\gamma, n)$.

When $n \to \infty$, and np remains constant, the binomial distribution tends to a Poisson distribution (cf. section 4.1.2) with parameter $m = np$. The approximation holds good as soon as $n > 10$ and $p < 0.1$.

When $n \to \infty$, the binomial distribution tends to a normal law (cf. section 4.2.2) of mean $m = np$ and variance $\sigma^2 = np(1 - p)$. The approximation is good for $p = 1/2$ and bad when $p < 1/(n + 1)$ or $p > n/(n + 1)$ and outside the interval $[m - 3\sigma, m + 3\sigma]$.

4.1.2. *Poisson distribution*

This distribution has one positive parameter m and is defined by:

$$\mathcal{P}(X = k) = e^{-m} \frac{m^k}{k!}$$

giving the distribution function:

$$F(k) = \sum_{i=0}^{k} e^{-m} \frac{m^i}{i!} = 1 - \frac{\Gamma(k + 1, m)}{k!} \tag{41}$$

with $\Gamma(k+1, m) = \int_0^m t^k e^{-t} \, dt$ (incomplete gamma function)

hence

$$E[X] = m \; ; \quad \sigma^2[X] = m .$$

4.2. Continuous distributions

In this section we will briefly consider six of the main distributions used in reliability. A fuller statistical study will be undertaken in Ch. 7. Table 1 contains the main characteristics of these distributions, $Y(t)$ representing the Heaviside function defined by:

$$Y(t) = 0 \quad t < 0$$
$$Y(t) = 1 \quad t \geq 0$$

Table 1.

Distribution	Probability density		Mean	Variance
Exponential	$\lambda \, e^{-\lambda t} \, Y(t)$	$\lambda > 0$	$\frac{1}{\lambda}$	$\frac{1}{\lambda^2}$
Normal	$\frac{1}{\sigma \sqrt{2\pi}} \exp\left(-\frac{1}{2}\left(\frac{t-m}{\sigma}\right)^2\right)$	$\sigma > 0$	m	σ^2
Lognormal	$\frac{1}{tb\sqrt{2\pi}} \exp\left(-\frac{1}{2}\left(\frac{\log(t-a)}{b}\right)^2\right) Y(t)$	$b > 0$	$\exp\left(a + \frac{b^2}{2}\right)$	$e^{2a+b^2}(e^{b^2} - 1)$
Weibull	$\frac{\beta}{\eta}\left(\frac{t-\gamma}{\eta}\right)^{\beta-1} \exp\left(-\left(\frac{t-\gamma}{\eta}\right)^{\beta}\right) Y(t-\gamma)$	$\eta > 0$ $\beta > 0$	$\eta\Gamma\left(1+\frac{1}{\beta}\right)+\gamma$	$\eta^2\left[\Gamma\left(\frac{2}{\beta}+1\right) - \Gamma^2\left(1+\frac{1}{\beta}\right)\right]$
Gamma	$\frac{1}{\Gamma(\beta)} \lambda^{\beta} t^{\beta-1} e^{-\lambda t} \, Y(t)$	$\alpha > 0$ $\beta > 0$	$\frac{\beta}{\lambda}$	$\frac{\beta}{\lambda^2}$
Chi-squared	$\frac{1}{2^{\nu/2}\Gamma(\nu/2)} t^{(\nu-2)/2} \exp\left(-\frac{t}{2}\right) Y(t)$	ν positive integer	ν	2ν

4.2.1. *Exponential distribution* (see figure 2)

The exponential distribution is very frequently used in reliability studies as it is one of the few which lend themselves to computation. The failure rate of a component whose distribution of operating times is exponential, is constant and equal to λ (cf. Ch. 2); this assumption is quite often found to be realistic.

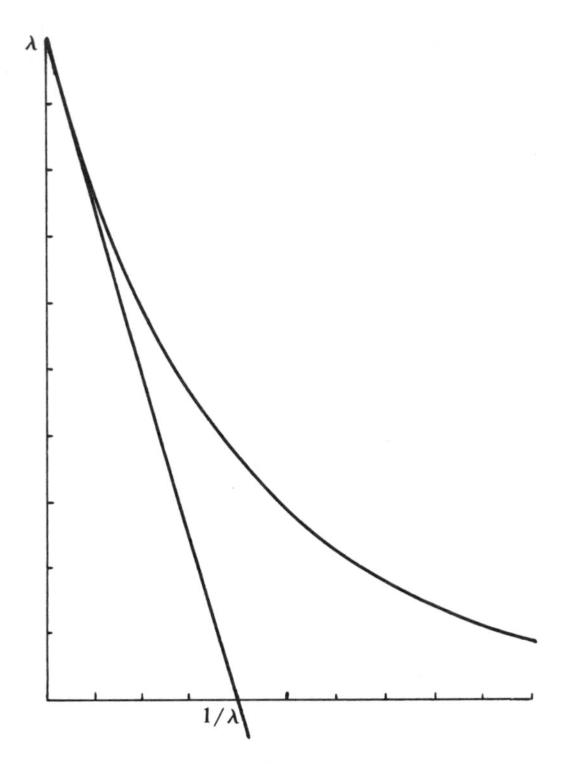

Fig. 2.
Exponential Distribution

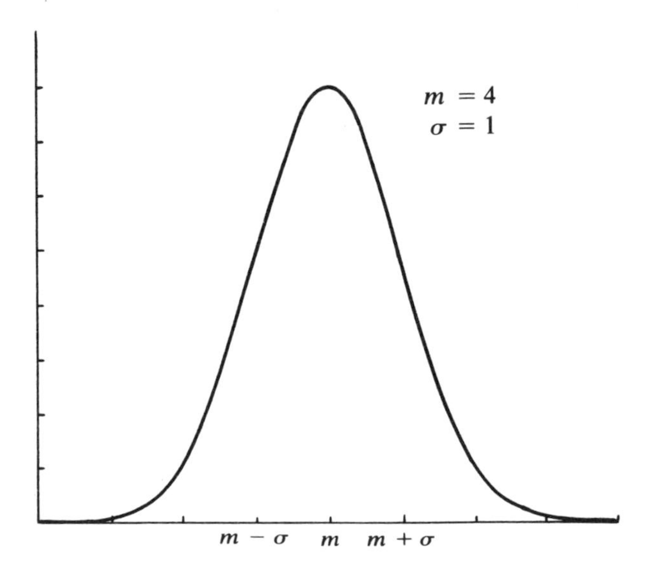

Fig. 3.
Normal Distribution

4.2.2. ***Normal distribution*** (see figure 3)

The normal or Gaussian distribution is symmetrical about the mean m. A normal distribution of mean m and standard deviation σ will be denoted by $N(m, \sigma)$. The $N(0, 1)$ distribution is called a standard normal distribution.

If X obeys a normal law $N(m, \sigma)$, $Y = (X - m)/\sigma$ has a standard normal distribution.

The normal distribution is often encountered in physics because of the following property: (*Central limit theorem*).

Given n independent random variables X_i of identical distribution with mean m and of finite standard deviation σ, the random variable X'_n defined by:

$$X'_n = \frac{1}{n} \sum_{i=1}^{n} X_i$$

tends to a normal distribution $N\left(m, \frac{\sigma}{\sqrt{n}}\right)$ as n tends to infinity.

The sum of two independent normal random variables $X_1(m_1, \sigma_1)$ and $X_2(m_2, \sigma_2)$ is a normal random variable $N(m_1 + m_2, \sqrt{\sigma_1^2 + \sigma_2^2})$. Conversely, if two independent random variables are such that their sum is normal, they too are normal.

The standard normal distribution is tabulated in the form:

$$\alpha = \int_{u_\alpha}^{\infty} \frac{1}{\sqrt{2\pi}} e^{-\frac{x^2}{2}} \, dx.$$

u_α can also be computed using a program giving an approximate analytical expression. For example, the values of u_α given in Appendix 1 were calculated using the XFROMP program [Cunningham, 1959]. Note the following special values:

$$\alpha = 0.025 \qquad u_\alpha = 1.96$$

$$\alpha = 0.05 \qquad u_\alpha = 1.64$$

$$\alpha = 0.10 \qquad u_\alpha = 1.28.$$

4.2.3. ***Lognormal distribution*** **[Aitchison and Brown, 1957]** (see figure 4)

A random variable is said to be lognormally distributed if its logarithm is distributed according to a normal law $N(a, b)$.

As the sum of two normally distributed independent variables is normal, the product of two lognormally distributed independent variables is lognormally distributed. For small values of b, the distribution is virtually normal. See Exercise 2 for complements.

This distribution is often used to deal with uncertainty problems concerning data (cf. Ch. 7).

4.2.4. ***Weibull distribution*** **[Weibull, 1954]** (see figure 5)

γ is the location parameter and it is generally zero. When $\alpha = 0$ and $\beta = 1$, an exponential distribution is found. The Weibull distribution is very useful as, by

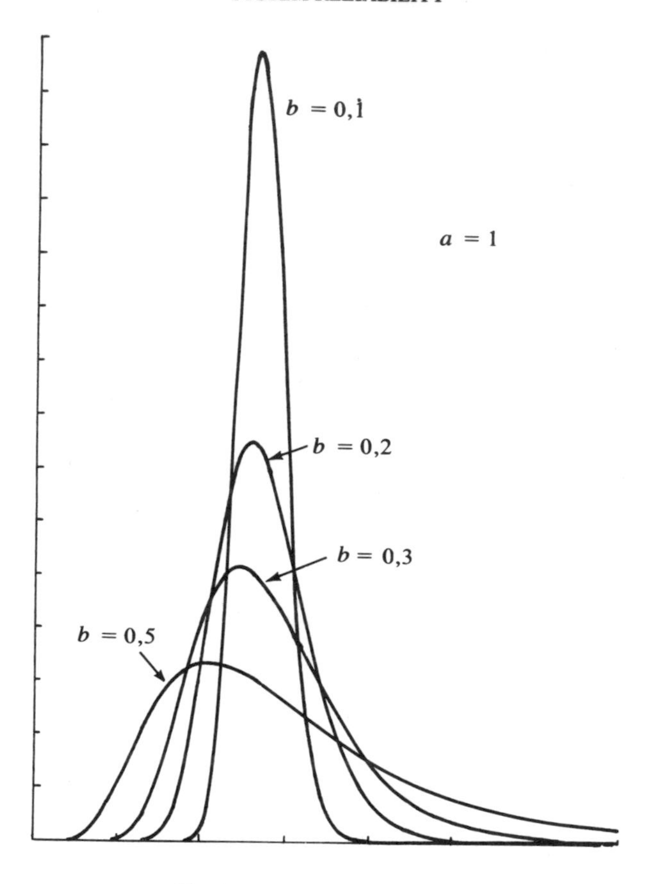

Fig. 4. *Lognormal Distribution*

varying the parameters, it is possible to represent a large number of experimental distributions.

4.2.5. *Gamma distribution* (see figures 6 and 7)

In accordance with Table 1, the distribution function of the gamma distribution is given by:

$$F(t) = \int_0^t \frac{\lambda^\beta x^{\beta-1} e^{-\lambda x} \, dx}{\Gamma(\beta)} = \frac{1}{\Gamma(\beta)} \Gamma(\beta, \lambda t) .$$

When parameter β is an integer, this is known as the Erlangian distribution and

$$F(t) = 1 - e^{-\lambda t} \sum_{k=1}^{\beta} \frac{(\lambda t)^{k-1}}{(k-1)!} .$$

When $\beta = 1$, the distribution is exponential.

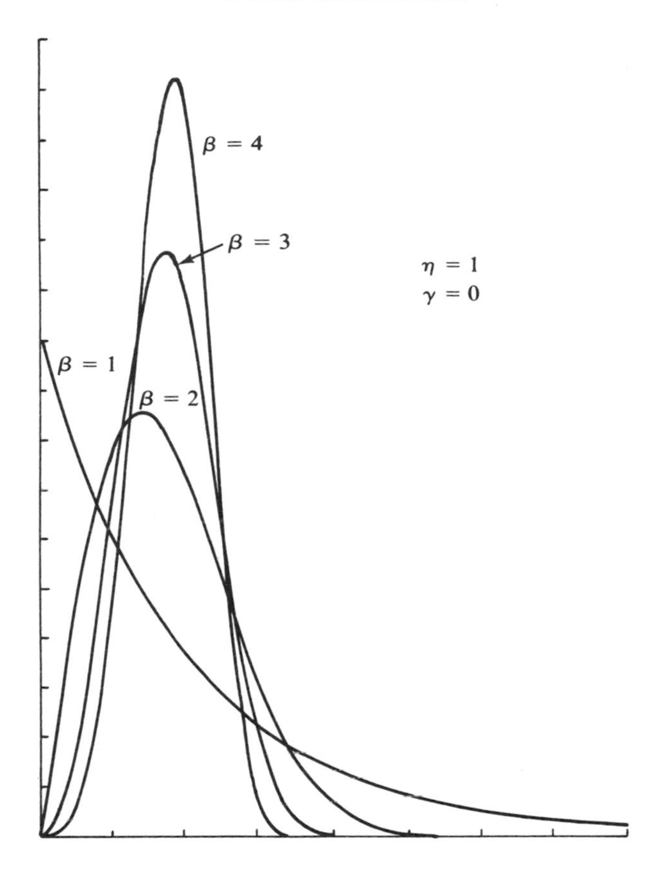

Fig. 5. *Weibull Distribution*

4.2.6. *Chi-squared distribution* (see figure 8)

Let us consider ν X_i-independent standard normal variables. It can then be shown (cf. Exercise 1) that $X_\nu = \sum_{i=1}^{\nu} X_i^2$ has a chi-squared distribution with parameter ν. Consequently, $X_{\nu 1} + X_{\nu 2}$ has a chi-squared distribution with parameter $\nu_1 + \nu_2$ if all the X_i random variables are independent.

If there are p linear relations between the variables, ν must be replaced by $\nu - p$. It is for this reason that ν is called a degree of freedom. The chi-squared distribution is related to the gamma distribution. In fact, putting $k = (\nu/2) - 1$ and $y = t/2$, we obtain the density of a gamma distribution

$$f(y) = \frac{y^k \, \mathrm{e}^{-y}}{\Gamma(k+1)}.$$

The distribution function of the chi-squared law is suitably tabulated to give the values $\chi_\alpha^2(\nu)$ such that:

$$\int_{\chi_\alpha^2(\nu)}^{\infty} \frac{t^{\frac{\nu-2}{2}} \, \mathrm{e}^{-t/2}}{2^{\nu/2} \, \Gamma(\nu/2)} \, \mathrm{d}t = \alpha \tag{42}$$

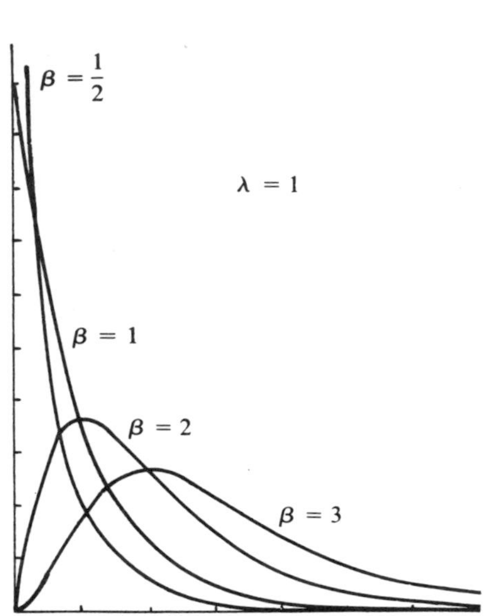

Fig. 6. *Gamma Distribution*

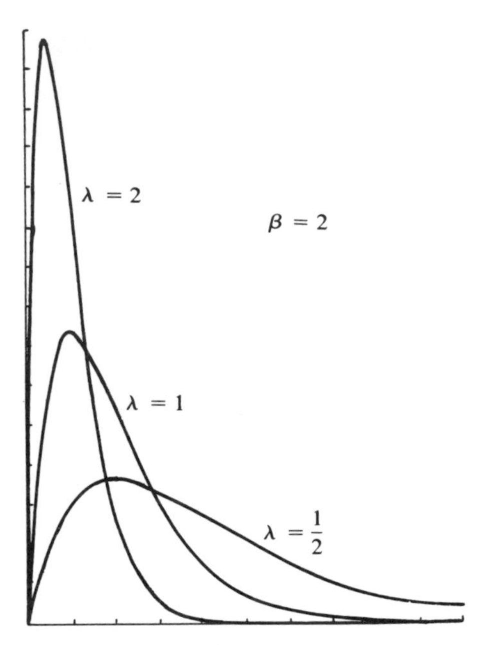

Fig. 7.

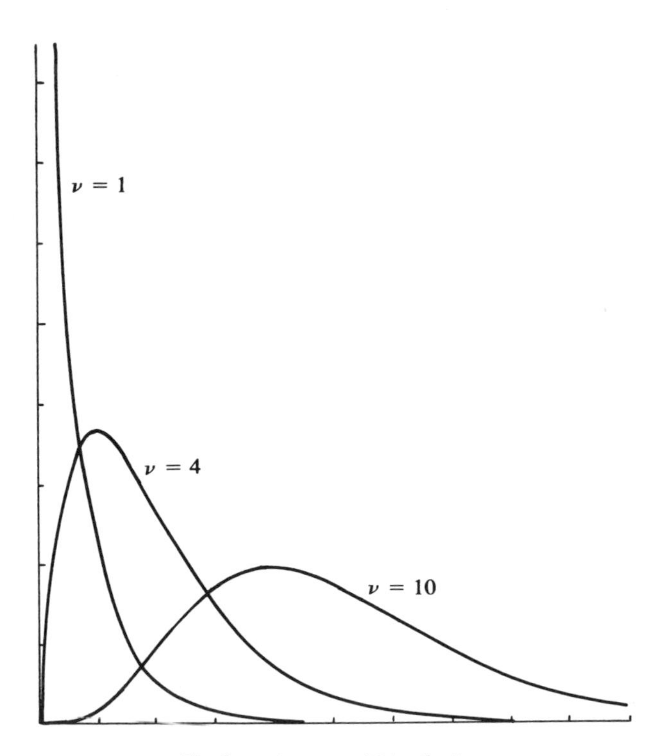

Fig. 8. *Chi-squared Distribution*

$\chi^2_\alpha(\nu)$ is therefore a decreasing function of α and an increasing function of ν.
For $\nu \leqslant 30$, we will use the table in Appendix 2.

For $\nu > 30$, we will use the fact that $\sqrt{2\chi^2} - \sqrt{2\nu - 1}$ approximates to a standard normal distribution, i.e.

$$\chi^2_\alpha(\nu) \simeq \frac{1}{2}(u_\alpha + \sqrt{2\nu - 1})^2 .$$

5. CHARACTERISTIC FUNCTION AND LAPLACE TRANSFORM

5.1. Characteristic function

The complex function of the real-number variable u defined by:

$$\varphi(u) = E[\mathrm{e}^{iuX}] \tag{43}$$

is called the *characteristic function* of the random variable X. φ has the following properties:

(a) $\varphi(0) = 1$.
(b) $|\varphi(u)| \leqslant 1$.
(c) $\varphi(u)$ is continuous.
(d) φ is the Fourier transform $\mathscr{F}(f)$ (cf. [Schwartz, 1961]) of the probability density f of X when the latter exists:

$$\varphi = \mathscr{F}(f) \text{ therefore } f = \mathscr{F}^{-1}(\varphi)$$

(e) When φ can be expanded as a MacLaurin series in the neighbourhood of 0:

$$\varphi(u) = \sum_{k=0}^{\infty} \frac{u^k}{k!} \varphi^{(k)}(0) .$$

On the other hand, by taking the mathematical expectation of the expansion of e^{iuX} as a series when it is meaningful to do so, we obtain:

$$\varphi(u) = \sum_{k=0}^{\infty} \frac{i^k u^k E[X^k]}{k!}$$

hence:

$$E[X^k] = \frac{\varphi^{(k)}(0)}{i^k} = \text{coefficient of} \frac{(iu)^k}{k!} \text{ in the expansion of } \varphi(u).$$

(f) When X_1 and X_2 are two independent random variables:

$$\varphi_{X_1+X_2}(u) = \varphi_{X_1}(u) \cdot \varphi_{X_2}(u) .$$

5.2. Laplace transform

Sometimes functions other than the characteristic function are used. For example, the second characteristic function $\psi(u) = \log \varphi(u)$ (cf. Exercise 1).

5.2.1. *Definition*

When the random variables only assume *positive values*, the *Laplace transform* is used. This is defined by:

$$\mathscr{L}[f(t), s] = \bar{f}(s) = \int_0^\infty f(t)\,\mathrm{e}^{-st}\,\mathrm{d}t \qquad (44)$$

s being a complex number.

It can be shown that there is a real number a of any sign such that for $R(s) > a$, the integral (Eqn (44)) is summable and for $R(s) < a$ it is not summable (cf. [Schwartz, 1961]). a is called the *convergence abscissa.* If $a = +\infty$, the integral is not summable and the Laplace transform of $f(t)$ does not exist.

When X has a density, the integral is summable for $s = 0$, hence $a \leqslant 0$. Equation (44) can be extended when f is a distribution (cf. [Schwartz, 1961]).

5.2.2. *Some properties*

When all the integrals are summable, it is easy to establish the following results:

$$\mathscr{L}\left(\frac{\mathrm{d}f}{\mathrm{d}t}(t), s\right) = s\bar{f}(s) - f(0) \qquad (45)$$

$$\mathscr{L}\left(\frac{d^k f(t)}{\mathrm{d}t^k}, s\right) = s^k \bar{f}(s) - s^{k-1} f(0) - s^{k-2}\frac{\mathrm{d}f}{\mathrm{d}t}(0) \cdots - \frac{d^{k-1}f}{\mathrm{d}t^{k-1}}(0)$$

$$\mathscr{L}\left(\int_0^t f(u)\,\mathrm{d}u, s\right) = \frac{\bar{f}(s)}{s} \qquad R(s) > 0 \qquad (46)$$

$$\mathscr{L}\left(\int_0^\infty f(u)\,\mathrm{d}u, s\right) = \frac{\bar{f}(0)}{s} \qquad R(s) > 0 \qquad (47)$$

$$\mathscr{L}(f(\lambda t), s) = \frac{1}{\lambda}\bar{f}\left(\frac{s}{\lambda}\right) \qquad \lambda = \text{positive constant} \qquad (48)$$

$$\mathscr{L}\left(\sum_i k_i f_i(t), s\right) = \sum_i k_i \bar{f}_i(s) \qquad k_i = \text{constant} \qquad (49)$$

$$\mathscr{L}(f(t - t_0), s) = \bar{f}(s)\,\mathrm{e}^{-st_0} \qquad t_0 = \text{positive constant} \qquad (50)$$

$$\mathscr{L}((-t)^k f(t), s) = \frac{d^k \bar{f}(s)}{\mathrm{d}s^k} \qquad \text{the convergence abscissa is that of } \bar{f}(s) \qquad (51)$$

$$\mathscr{L}(\mathrm{e}^{-\lambda t} f(t), s) = \bar{f}(s + \lambda) \qquad \lambda = \text{constant} \qquad (52)$$

$$\bar{f}(s) \equiv \bar{g}(s) \Leftrightarrow f(t) = g(t) \text{ almost everywhere} \qquad (53)$$

Note that Eqns (45) to (48) make the Laplace transform useful for solving integro-differential equations.

5.2.3. *Laplace transforms of some functions and distributions*

Table 2.

Function or distribution		Laplace transform	Convergence abscissa
Heaviside	$Y(t)$	$1/s$	0
Dirac	$\delta(t)$	1	$-\infty$
Integer-valued power	$\dfrac{t^{n-1}\,Y(t)}{(n-1)!}\quad(n \geqslant 1)$	$1/s^n$	0
Gamma	$\dfrac{\lambda^{\beta}\,t^{\beta-1}\,e^{-\lambda t}}{\Gamma(\beta)}\,Y(t)$	$\dfrac{\lambda^{\beta}}{(s+\lambda)^{\beta}}$	$-\lambda$
Exponential	$\lambda\,e^{-\lambda t}\,Y(t)$	$\dfrac{\lambda}{s+\lambda}$	$-\lambda$
Sine	$\operatorname{Sin}(\omega t)\,Y(t)$	$\dfrac{\omega}{s^2+\omega^2}$	0
Cosine	$\operatorname{Cos}(\omega t)\,Y(t)$	$\dfrac{s}{s^2+\omega^2}$	0

A number of Laplace transforms are given in [Erdelyi, 1954].

5.2.4. *Inverse Laplace transform*

Given $\bar{f}(s)$, is it possible to determine $f(t)$?
Under certain conditions (cf. [Schwartz, 1961]), it can be shown that:

$$f(t) = \frac{1}{2\,i\pi}\int_{c-i\infty}^{c+i\infty} \bar{f}(s)\,e^{st}\,ds\,.$$

All the singularities of $\bar{f}(s)$ must be on the left of the integration contour. This being the case, the above integral is independent of c. This integral is generally calculated using the residue theorem (cf. for example [Schwartz, 1961]).

In certain simple cases (e.g. rational fractions in s) it is possible to calculate the inverse using the table of Laplace transforms (see, for example [Erdelyi, 1954]).

For example:

• $\bar{f}(s) = \dfrac{\overline{Q}(s)}{\overline{P}(s)}$;with $\overline{P}(s) = (s-\alpha_1)(s-\alpha_2)\ldots(s-\alpha_n)$

and $d^0 Q < d^0 P \quad \alpha_i \neq \alpha_j \quad$ for $\quad i \neq j$

then

$$f(t) = \sum_{m=1}^{n} \frac{Q(\alpha_m)}{P_m(\alpha_m)}\,e^{\alpha_m t} \quad \text{with} \quad \overline{P}_m(s) = \frac{\overline{P}(s)}{s-\alpha_m}$$

- $\overline{f}(s) = \dfrac{\overline{Q}(s)}{\overline{P}(s)}$ with $\overline{P}(s) = (s - \alpha_1)^{m_1}(s - \alpha_2)^{m_2} \dots (s - \alpha_n)^{m_n}$

and $d^0 Q < d^0 P \quad \alpha_i \neq \alpha_j \quad$ for $\quad i \neq j$

then

$$f(t) = \sum_{k=1}^{n} \sum_{l=1}^{m_k} \frac{\overline{Q}_{kl}(\alpha_k)}{(m_k - l)!(l-1)!} t^{m_k - l} e^{\alpha_k t} \tag{54}$$

with $\overline{Q}_{kl}(s) = \dfrac{d^{l-1}}{ds^{l-1}}\left(\dfrac{\overline{Q}(s)}{\overline{P}_k(s)}\right)$, $\overline{P}_k(s) = \dfrac{\overline{P}_{(s)}}{(s - \alpha_k)^{mk}}$.

There are also numerical inversion methods.

5.2.5. *Asymptotic expansions*

When it is not possible to invert $\overline{f}(s)$ simply, one naturally looks for asymptotic behaviour of $f(t)$, generally when $t \to \infty$ or $t \to 0$.

The theorems for deducing the limiting behaviour of $f(t)$ from that of $\overline{f}(s)$ are known as Tauber's theorems. These are difficult to manipulate as they require certain assumptions concerning $f(t)$ (e.g. no rapid oscillations when $t \to \infty$). As these assumptions are generally satisfied by the functions which will be encountered, we shall use the following results:

$$\begin{array}{l} \overline{f}(s) \sim \dfrac{A}{s^{\gamma+1}} \quad \text{when} \quad s \to 0 \quad (\gamma \geqslant 0) \\ \Updownarrow \\ f(t) \sim \dfrac{At^{\gamma}}{\Gamma(\gamma+1)} \quad \text{when} \quad t \to \infty . \end{array} \tag{55}$$

If $A = 0$, $\dfrac{A}{s^{\gamma}}$ must be replaced by $0(s^{-\gamma})$ and $\dfrac{At^{\nu}}{\Gamma(\gamma+1)}$ by $0(t^{\gamma})$; when $\gamma = 1$, Eqn (55) is generally written as

$$\lim_{s \to 0} s\overline{f}(s) = \lim_{t \to \infty} f(t) ;$$

Likewise, when $s \to \infty$, we obtain

$$\begin{array}{l} \overline{f}(s) \sim \dfrac{A}{s^{\gamma+1}} \quad \text{when} \quad s \to \infty \quad (\gamma \geq 0) \\ \Updownarrow \\ f(t) \sim \dfrac{At^{\gamma}}{\Gamma(\gamma+1)} \quad \text{when} \quad t \to 0 ; \end{array}$$

When $\gamma = 1$, we generally write

$$\lim_{s \to \infty} s\overline{f}(s) = \lim_{t \to 0} f(t) .$$

Returning to the validity conditions of Tauber's theorems, the Abel theorem shows us that if the convergence abscissa of $\mathscr{L}(f(t), s)$ is zero and $\int_0^\infty f(t)\,\mathrm{d}t$ converges, then

$$\lim_{s \to 0^+} \bar{f}(s) = \int_0^\infty f(t)\,\mathrm{d}t .$$

However, the converse is not true. Taking the cosine function as an example

$$\bar{f}(s) = \frac{s}{s^2+1} = \int_0^\infty \mathrm{e}^{-st} \cos t \,\mathrm{d}t$$

$\lim_{s \to 0^+} \bar{f}(s) = 0$ even though $\int_0^\infty \cos t\,\mathrm{d}t$ does not converge.

In 1897, Tauber gave a condition under which the converse would be true (cf. [Widder, 1946]):

Given a function $f(t)$ integrable on $[0, R]\forall R$ such that the convergence abscissa of $\mathscr{L}(f(t), s)$ is negative or zero and $\lim_{s \to 0^+} \bar{f}(s) = A$, then

$$f(t) \underset{t \to \infty}{\sim} 0\left(\frac{1}{t}\right) \Rightarrow \bar{f}(0^+) = \int_0^\infty f(t)\,\mathrm{d}t = A .$$

A whole chapter in [Widder, 1946] is devoted to Tauber's theorems.

The functions which we shall use in this book are such that Tauber's theorems will apply whenever necessary.

5.2.6. *Application of the Laplace transform to random variables*

Let $f(t)$ be the probability density of a non-negative random variable and $\bar{f}(s)$ its Laplace transform. Assuming that $a < 0$, Eqn (44) can be written as:

$$\bar{f}(s) = E[\mathrm{e}^{-sX}]$$

hence, if the moments exist

$$\bar{f}(s) = \sum_{k=0}^{\infty} \frac{(-s)^k E[X^k]}{k!} .$$

We find a property similar to that of the characteristic function, which is normal because

$$\varphi(u) = E[\mathrm{e}^{iuX}] = \bar{f}(-iu) . \tag{56}$$

Let X and Y be two *independent* random variables with distribution functions F_x and F_y respectively, and Z the random variable $X + Y$.

Its distribution function is given by:

$$F(Z) = \int_{-\infty}^{+\infty} F_y(z-u)\,\mathrm{d}F_x(u) = \int_{-\infty}^{+\infty} F_x(z-u)\,\mathrm{d}F_y(u)\,. \tag{57}$$

If X and Y have densities f_x and f_y respectively, Z has a density f given by:

$$f(z) = \int_{-\infty}^{+\infty} f_x(z-u)\,f_y(u)\,\mathrm{d}u = \int_{-\infty}^{+\infty} f_x(u)\,f_y(z-u)\,\mathrm{d}u \tag{58}$$

Function f is called the *convolution* of densities f_x and f_y, and is denoted by $f = f_x * f_y$.

It can therefore be shown that in the case of non-negative random variables:

$$\bar{f}(s) = \bar{f}_x(s)\,.\,\bar{f}_y(s)\,. \tag{59}$$

These formulae can be generalized to the case of n random variables:

$$\bar{f}(s) = \prod_{j=1}^{n} \bar{f}_j(s) \Leftrightarrow f(t) = \int_0^{\infty} \therefore \int_0^{\infty} f_1(u_1)\,f_2(u_2-u_1)\ldots f_n(t-u_{n-1})\,\mathrm{d}u_1\ldots\mathrm{d}u_n\,.$$

The Laplace transform is therefore well suited to dealing with the sums of random variables. Note that the same applies to the characteristic function by reason of Eqn (56).

Detailed examinations of the Laplace transform can be found in [Schwartz, 1961; Widder, 1946].

6. STOCHASTIC PROCESSES

6.1. Definition

Given a set of indices T and a probability space $(\Omega, \mathscr{A}, \mathscr{P})$, an application of T in the set of random variables defined on $(\Omega, \mathscr{A}, \mathscr{P})$ is called a *stochastic process*. We will often use the parameter t, denoting time, as elements of the set T. The sets T and Ω can be discrete or continuous, which gives rise to four types of process.

For each element ω of Ω, the application of T in R which associates the real number $X_t(\omega)$ with t is called a *trajectory* (or realization) of the process. When we come to use stochastic processes in reliability theory, Ω will represent the set of realizations of the system under examination in its various states (cf. Ch. 5).

Let $X_t(t \in T)$ be a stochastic process such that for any finite sequence $t_1 < t_2 < \ldots < t_n < t$ of elements of T and for any finite sequence $A_1, A_2, \ldots A_n$ of elements of $\mathscr{A}$, we obtain:

$$\mathscr{P}[X_t \in A \mid X_{t_n} \in A_n, X_{t_{n-1}} \in A_{n-1} \ldots X_{t_1} \in A_1] = \mathscr{P}[X_t \in A \mid X_{t_n} \in A_n]\,.$$

This is known as a *Markov* process. Knowledge of the state at instant t_n encapsulates the entire history of the system. When the set Ω of the states is discrete, this process is called a *Markov chain* (of the discrete parameter type if T is discrete, and of the continuous parameter type if T is continuous).

6.2. Poisson process

Let us now assume that T is continuous and we wish to determine the number of events X_{t_i} occurring up to instants t_i ($t_i \in T$). It is also assumed that the random variables representing the numbers of events occurring between two disjoint intervals are independent; i.e. if

$$t_0 < t_1 \ldots < t_n, (X_{t_1} - X_{t_0}), (X_{t_2} - X_{t_1}), \ldots, (X_{t_n} - X_{t_{n-1}})$$

are mutually independent random variables, X_t is termed an *independent increment* process.

A *Poisson process* is an independent increment process such that:

(a) the random variable $X_{t_0+t} - X_{t_0}$ only depends on t,
(b) $\mathcal{P}(X_{t_0+\Delta t} - X_{t_0} \geq 1) = \lambda \, \Delta t + 0(\Delta t)$,
(c) $\mathcal{P}(X_{t_0+\Delta t} - X_{t_0} > 1) = 0(\Delta t)$,

where $0(\Delta t)$ is a function of Δt such that $\lim_{\Delta t \to 0} \frac{0(\Delta t)}{\Delta t} = 0$.

Property (c) indicates that two events cannot occur simultaneously.

It can then be shown (cf. [Carton, 1975]) that:

$$\mathcal{P}(X_t = m) = \frac{(\lambda t)^m \, e^{-\lambda t}}{m\,!} \tag{60}$$

(Poisson distribution of mean λt).

It can also be shown (cf. [Carton, 1975]) that:

—At whatever instant, the interval between that instant and the next event is an exponentially distributed random variable with parameter λ.

—The intervals separating consecutive events are random variables with the same exponential distribution.

6.3. Renewal process

Let us consider a set of components whose lifetime is a continuous random variable F with a probability density function f. At time $t = 0$, the first component is put into service and is replaced by the succeeding one when it fails at time F_1. If F_r is the lifetime of the rth component in service, its failure will occur at time k_r defined by:

$$k_r = F_1 + F_2 + \cdots + F_r .$$

If this process involves independent increments, it is called a *simple renewal process*. If f is an exponential distribution, we again have the Poisson process.

The *renewal function* is the mean value of the number of renewals $N(t)$ having occurred during $(0, t)$, the introduction of the first component at time $t = 0$ not being counted as a renewal.

$$H(t) = E[N(t)] .$$

The derivative of $H(t)$ is called the *renewal density* $h(t)$. $h(t)\Delta t$ is equal to the mean number of renewals in the interval $[t, t + \Delta t]$ when Δt tends to zero.

A *modified renewal process* is the term given to a renewal process for which the random variable F_1 has a different density from the other random variables F_i.

The *residual lifetime* V_t is the random variable representing the remaining service life of the component at time t.

In [Cox, 1966], a renewal theory is given with proofs for the following important results concerning the modified processes (in order to simplify the notations, the distribution function of a random variable is denoted by the same upper-case letter, and the probability density by the corresponding lower-case letter):

$$\bar{k}_r(s) = \bar{f}_1(s)\, \bar{f}^{r-1}(s) \tag{61}$$

$$\mathcal{P}(N(t) = r) = k_r(t) - k_{r+1}(t) \quad \text{with} \quad k_0(t) = Y(t)$$

$$\bar{H}(s) = \frac{\bar{f}_1(s)}{s(1 - \bar{f}(s))} \tag{62}$$

$$\bar{h}(s) = \frac{\bar{f}_1(s)}{1 - \bar{f}(s)} \tag{63}$$

$h(t)$ is therefore the solution to the Volterra equation

$$h(t) = f_1(t) + \int_0^t h(t-u)\, f(u)\, \mathrm{d}u \tag{64}$$

$$\lim_{t \to \infty} h(t) = \frac{1}{\mu} \quad \text{with} \quad \mu = \int_0^\infty t f(t)\, \mathrm{d}t \tag{65}$$

$$v_t(x) = f_1(t+x) + \int_0^t h(t-u)\, f(u+x)\, \mathrm{d}u \tag{66}$$

If $f_1 \to 0$ when $t \to \infty$, we obtain

$$\lim_{t \to \infty} v_t(x) = \frac{1 - F(x)}{\mu}. \tag{67}$$

For a simple renewal process, it is sufficient to replace f_1 by f. When $f_1 = \dfrac{1-F}{\mu}$, the process is called a *stationary renewal process*. In this case we obtain:

$$\bar{f}_1(s) = \frac{1 - \bar{f}(s)}{s\mu}$$

hence

$$\bar{H}(s) = \frac{1}{s^2 \mu}$$

$$\bar{h}(s) = \frac{1}{s\mu} \qquad \text{i.e.} \qquad h(t) = \frac{1}{\mu} Y(t)$$

$$v_t(x) = \frac{1 - F(x)}{\mu} Y(x).$$

A stationary renewal process can be regarded as a simple renewal process which has been operating for a long time prior to the initial date $t = 0$.

When the random variables F_{2n-1} have a density f_1 and the random variables F_{2n} have a density f_2 (different from f_1), all these random variables being independent, the process is called an *alternating renewal process*. In reliability terms, f_1 might represent a failure probability density and f_2 a repair probability density.

Calling H_1 the renewal function of the type 1 components and H_2 the renewal function of the type 2 components, we obtain (cf. [Cox, 1966]).

$$\bar{H}_1(s) = \frac{\bar{f}_1(s)}{s(1 - \bar{f}_1(s)\,\bar{f}_2(s))} \tag{68}$$

$$\bar{H}_2(s) = \frac{\bar{f}_1(s)\,\bar{f}_2(s)}{s(1 - \bar{f}_1(s)\,\bar{f}_2(s))} \tag{69}$$

It would be useful to know the probability $\Pi(t)$ of a component of a given type being in service at time t.

If a type 1 component is in service at time $t = 0$

$$\Pi_1(t) = R_1(t) + \int_0^t h_2(u)\,R_1(t-u)\,\mathrm{d}u \tag{70}$$

where $R_1(t) = 1 - F_1(t)$

hence
$$\bar{\Pi}_1(s) = \frac{1 - \bar{f}_1(s)}{s(1 - \bar{f}_1(s)\,\bar{f}_2(s))} = \bar{H}_2(s) - \bar{H}_1(s) + \frac{1}{s} \tag{71}$$

and
$$\Pi_1(t) = H_2(t) - H_1(t) + 1\,.$$

Putting $\mu_i = \int_0^\infty t f_i(t)\,\mathrm{d}t$ (mcan lifetime), it can be shown that:

$$\lim_{t \to \infty} \Pi_1(t) = \frac{\mu_1}{\mu_1 + \mu_2}\,. \tag{72}$$

When f_1 represents a failure density and f_2 a repair density, $\Pi_1(t)$ is in fact the system availability.

From the expression

$$\Pi_1(t) + \Pi_2(t) = 1$$

we obtain:

$$\bar{\Pi}_2(s) = \frac{1}{s} - \bar{\Pi}_1(s) = \bar{H}_1(s) - \bar{H}_2(s) = \frac{\bar{f}_1(s)(1 - \bar{f}_2(s))}{s(1 - \bar{f}_1(s)\,\bar{f}_2(s))}\,. \tag{73}$$

The renewal process will be used in Ch. 3 to examine the reliability of a renewed component and in Ch. 5 to examine non-Markovian systems, particularly the renewal density of the type 1 component, $h_1(t)$, whose Laplace transform is given by:

$$\bar{h}_1(s) = s\bar{H}_1(s) = \frac{\bar{f}_1(s)}{(1 - \bar{f}_1(s)\,\bar{f}_2(s))}\,.$$

A more detailed treatment of stochastic processes will be found in [Neveu, 1972; Carton, 1975; Cox, 1966; Cox and Miller, 1945].

7. BASIC RELATIONSHIPS

7.1. Relationship between MTTF and reliability

Let T be the random variable measuring the operating time of the system.
The reliability definition (expression (1)) can therefore be written as:

$$R(t) = \mathcal{P}(T > t) .$$

The distribution function U of the random variable T can then be written as:

$$U(t) = \mathcal{P}(T \leq t) = 1 - R(t)$$

$U(t)$ representing the probability of failure during $[0, t]$.
The failure density $u(t)$, if it exists, is therefore given by:

$$u(t) = \frac{\mathrm{d}U(t)}{\mathrm{d}t} = -\frac{\mathrm{d}R(t)}{\mathrm{d}t} .$$

It follows that, the mean time to failure, if it exists, is given by:

$$\mathrm{MTTF} = \int_0^\infty tu(t)\,\mathrm{d}t = -\int_0^\infty t\frac{\mathrm{d}R}{\mathrm{d}t}(t)\,\mathrm{d}t \tag{74}$$

hence, integrating by parts

$$\mathrm{MTTF} = \int_0^\infty R(t)\,\mathrm{d}t - [tR(t)]_0^\infty .$$

For $t = 0$, $tR(t) = 0$.
Let us evaluate $tR(t)$ for $t \to \infty$.
If $R(t)$ does not tend sufficiently rapidly to zero, $t(R)$ will tend to infinity as $t \to \infty$.
This is the case, for example, when $R(t) = (t + 1)^{-1/2}$.

However, $\mathrm{MTTF} = \int_0^\infty \frac{t\,\mathrm{d}t}{2(t+1)^{3/2}}$ is not defined as the integral is not convergent.

We will assume that the integral (Eqn (74)) is convergent and that Tauber's theorems apply. In the case where $tR(t)$ tends to zero when t tends to infinity, $\int_0^\infty R(T)\,\mathrm{d}t$ converges and integration by parts is justified.

Hence, as the $\mathrm{MTTF} < +\infty$:

$$\mathrm{MTTF} = \int_0^\infty [R(t)]\,\mathrm{d}t \tag{75}$$

and similarly:

$$\mathrm{MTTR} = \int_0^\infty [1 - M(t)]\,\mathrm{d}t . \tag{76}$$

Given that:

$$\mathcal{L}\left(\int_0^t R(u)\,\mathrm{d}u, s\right) = \frac{\overline{R}(s)}{s}$$

Eqn (75) can be rewritten as:

$$\mathrm{MTTF} = \lim_{t\to\infty} \int_0^t R(u)\,\mathrm{d}u = \lim_{s\to 0} \bar{R}(s)\,. \tag{77}$$

Similarly, in accordance with Eqns (47) and (48):

$$\mathscr{L}\left(\int_0^t [1 - M(u)]\,\mathrm{d}u,\ s\right) = \left(\frac{1}{s} - \bar{M}(s)\right)\frac{1}{s}$$

hence:

$$\mathrm{MTTR} = \lim_{t\to\infty} \int_0^t [1 - M(u)]\,\mathrm{d}u = \lim_{s\to 0} \left(\frac{1}{s} - \bar{M}(s)\right). \tag{78}$$

7.2. Instantaneous failure rate and instantaneous repair rate

We have shown that expressions (3) and (4) were equivalent to:

$$\Lambda(t) = \frac{-\dfrac{\mathrm{d}R}{\mathrm{d}t}(t)}{R(t)}$$

$$\mathcal{M}(t) = \frac{\dfrac{\mathrm{d}M}{\mathrm{d}t}(t)}{1 - M(t)}$$

and consequently:

$$R(t) = \exp\left(-\int_0^t \Lambda(u)\,\mathrm{d}u\right) \tag{79}$$

$$M(t) = 1 - \exp\left(-\int_0^t \mathcal{M}(u)\,\mathrm{d}u\right). \tag{80}$$

The probability density $u(t)$ can therefore be written as:

$$u(t) = -\frac{\mathrm{d}R}{\mathrm{d}t}(t) = \Lambda(t)\,R(t) = \Lambda(t)\exp\left(-\int_0^t \Lambda(u)\,\mathrm{d}u\right). \tag{81}$$

Similarly, the repair probability density $m(t)$ is written as:

$$m(t) = \frac{\mathrm{d}M}{\mathrm{d}t}(t) = m(t)[1 - M(t)] = \mathcal{M}(t)\exp\left(-\int_0^t \mathcal{M}(u)\,\mathrm{d}u\right). \tag{82}$$

The failure rate can be interpreted in a different way.

Consider N identical non-repairable systems operating at the initial time $t = 0$ and of reliability $R(t)$.

The number of systems $N(t)$ operable at time t has a binomial distribution with parameters $(R(t), N)$, hence:

$$\mathcal{P}(N(t) = k) = C_n^k [R(t)]^k [1 - R(t)]^{N-k}$$

and $$E[N(t)] = NR(t).$$

It follows that:

$$R(t) = \frac{E[N(t)]}{N}.$$

The reliability may be interpreted as the mean number of systems operable at time t.

The density $u(t)$ of the mean number of failed systems in t and $t + \Delta t$ is:

$$u(t) = \frac{E[N(t)] - E[N(t + \Delta t)]}{N \Delta t}.$$

By definition, the failure rate is:

$$\Lambda(t) = \lim_{t \to 0} \frac{R(t) - R(t + \Delta t)}{\Delta t R(t)} = \frac{E[N(t)] - E[N(t + \Delta t)]}{\Delta t E[N(t)]}$$

hence

$$\Lambda(t) = \frac{Nu(t)}{E[N(t)]}.$$

$\Lambda(t)\Delta t$ is the mean ratio of the number of systems failing within the interval $[t, t + \Delta t]$ to the number of systems operable at time t.

7.3. Failure rate values for some distributions of the random variable *T*

Table 3 shows the characteristics of the failure rates of certain distributions frequently used in reliability studies. As the times to failure generally vary between zero and infinity, the function $Y(t)$ has been omitted to simplify the notations.

The exponential distribution is particularly simple since the failure rate is constant. Note that in certain cases the failure rate is monotonic (increasing or decreasing). It is then possible (see, for example [Barlow and Proschan, 1967]) to determine reliability bounds.

8. SUMMARY

Most of the sections in this chapter cover basic mathematics. As they are themselves summaries, we will merely refer the reader to the references given in the course of the text.

As far as reliability proper is concerned, the following definitions and formulas should be noted particularly:

Table 3.

Distribution of T	Failure density $u(t)$	Failure rate $\lambda(t)$		MTTF
Exponential	$\lambda\ e^{-\lambda t}$	λ		$\frac{1}{\lambda}$
Lognormal	$\frac{1}{tb\sqrt{2\pi}}\exp\left(-\frac{1}{2}\left(\frac{\log(t-a)}{b}\right)^2\right)$	$\frac{\exp\left(-\frac{1}{2}\left(\frac{\log(t-a)}{b}\right)^2\right)}{t\int_t^\infty b\sqrt{2\pi}\,u(t)\cdot dt}$ It can be shown that $\lambda(t)$ increases then decreases, tending to zero.	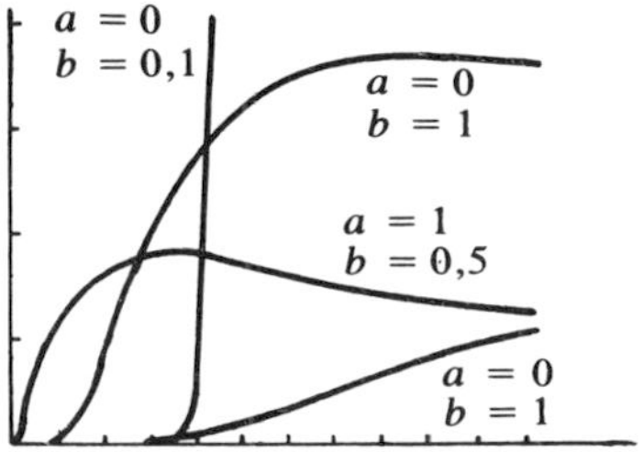 $a=0$, $b=0,1$; $a=0$, $b=1$; $a=1$, $b=0,5$; $a=0$, $b=1$	$\exp\left(a+\frac{b^2}{2}\right)$
Gamma	$\frac{\lambda^\beta t^{\beta-1} e^{-\lambda t}}{\Gamma(\beta)}$	$\frac{\lambda^\beta t^{\beta-1} e^{-\lambda t}}{\int_t^\infty \Gamma(\beta)\,u(t)\,dt}$ $\lambda(t)$ increasing if $\beta>1$ decreasing if $\beta<1$ constant if $\beta=1$	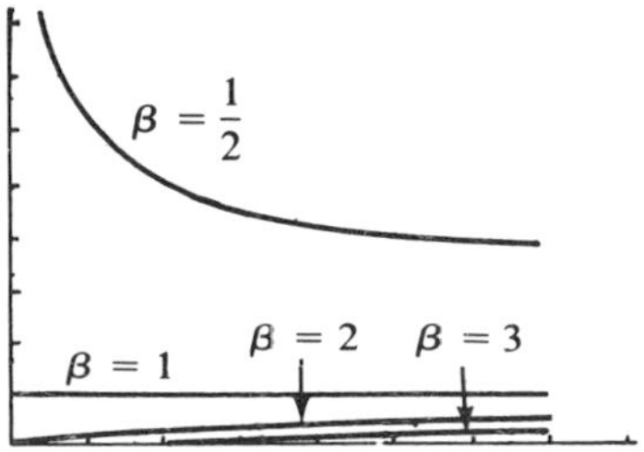 $\beta=\frac{1}{2}$; $\beta=1$; $\beta=2$; $\beta=3$	$\frac{\beta}{\lambda}$
Weibull	$\frac{\beta}{\eta}\left(\frac{t-\gamma}{\eta}\right)^{\beta-1}\exp\left(-\left(\frac{t-\gamma}{\eta}\right)^\beta\right)Y(t-\gamma)$	$\frac{\beta}{\eta}\left(\frac{t-\gamma}{\eta}\right)^{\beta-1}Y(t-\gamma)$ $\lambda(t)$ increasing if $\beta>1$ decreasing if $\beta<1$ constant if $\beta=1$ increasing linearly if $\beta=2$	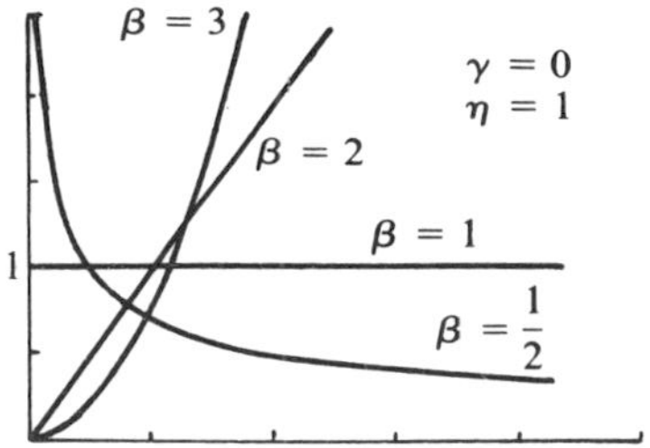 $\beta=3$; $\beta=2$; $\gamma=0$, $\eta=1$; $\beta=1$; 1; $\beta=\frac{1}{2}$	$\eta\Gamma\left(1+\frac{1}{\beta}\right)+\gamma$

• The reliability $R(t)$ of a system is the probability of no failure occurring during the interval $[0, t]$ under stated conditions.

• The availability $A(t)$ of a system is the probability of its not being in a failed state at time t under stated conditions.

• The failure rate $\Lambda(t)$ of a system is defined by

$$\Lambda(t) = \lim_{\Delta t \to 0} \frac{1}{\Delta t} \mathscr{P}\begin{pmatrix} \text{the system will fail between } t \text{ and } t + \Delta t \\ \text{given that there has been no failure during } [0, t] \end{pmatrix}.$$

Knowing the function $\Lambda(t)$ is equivalent to knowing the function $R(t)$. Both of these functions are in fact related by the two following equivalent expressions:

$$R(t) = \exp\left(-\int_0^t \Lambda(u)\,du\right)$$

$$\Lambda(t) = \frac{-\dfrac{dR}{dt}(t)}{R(t)}.$$

• Of the main lifetime (or repair time) distributions discussed in section 4, there is one which is fundamental to reliability. This is the exponential, for which the function $\Lambda(t)$ reduces to a constant usually denoted by λ. In this case, $R(t) = \exp(-\lambda t)$.

The full importance of this distribution will become clear in the following chapters.

• Lastly, there is a formula involving the Laplace transform which enables us to evaluate the mean time to failure (MTTF) of a system

$$\text{MTTF} = \lim_{s \to 0} \bar{R}(s) = \int_0^\infty R(t)\,dt.$$

Laplace transforms are difficult to invert, but the calculation for $s = 0$ is generally much more simple. The MTTF gives an idea of the behaviour of the system but provides much less information than the function $\Lambda(t)$, especially for low values of the time t.

EXERCISES

Exercise 1 — Characteristic function and moments

(a) Let X be a random variable with moments up to the rth order. Show that:

$$\mu_r = \sum_{j=0}^{r} C_r^j E[X^{r-j}](-m)^j$$

where $m = E[X]$.

Then calculate μ_2, μ_3, μ_4.

(b) 'Second characteristic function' is the term applied to the function ψ defined by

$$\psi(u) = \log \varphi(u)$$

as $\varphi(u)$ is complex, it is necessary to select a branch of $\log \varphi(u)$; the principal branch is generally used.

By analogy with the expression giving the moments from the characteristic function, the coefficients k_r of $(it)^r/r!$ in the expansion of $\psi(u)$ are called cumulants.

From the expression $\left(\sum_{j=1}^{\infty} k_j \frac{(it)^j}{j!}\right) = 1 + \sum_{j=1}^{\infty} E[X^j]\frac{(it)^j}{j!}$, deduce the relations existing between k_1, k_2, k_3, k_4 and $\mu_1, \mu_2, \mu_3, \mu_4$.

(c) Determine the characteristic function of the chi-squared distribution. Deduce the cumulants, moments and central moments.

(d) Show that if n independent variables X_i are identically distributed and if $\sum_{i=1}^{n} X_i^2$ has a chi-squared distribution with n degrees of freedom, the distribution of the X_i obeys a standard normal law.

Exercise 2—Lognormal distribution

(a) A random variable X is said to be lognormally distributed if its logarithm has a normal distribution $N(a, b)$. Knowing the density of the normal distribution, deduce the density $f(x)$ of the lognormal distribution using the method described in section 3.4.

(b) Calculate the following characteristics of the lognormal distribution: mean (m), variance (σ^2), median (med) defined by $\int_{med}^{\infty} f(x)\,dx = \frac{1}{2}$, mode (value of x giving maximum probability density) and the variation coefficient (σ/m).

(c) Show that the product of two independent lognormal variables is distributed lognormally. Deduce from this that the mean and median of the product are equal to the product of the means and the product of the medians respectively.

(d) Let x_+ and x_- defined by $(0 < \alpha < 1)$ be:

$$\int_{x_+}^{\infty} f(x)\,dx = \alpha$$

$$\int_{-\infty}^{x_-} f(x)\,dx = \alpha\,.$$

Show that $\frac{x_+}{\text{med}} = \frac{\text{med}}{x_-} = e^{u_\alpha b}$. We shall call this ratio f_e (u_α being defined in section 4.2.3).

The ratio f_e is called the error factor. If the median (or the mean) and the error factor are given, the lognormal distribution can be completely determined.

(e) Determine the parameters a and b when the following are known:

—the mean m and the standard deviation σ,
—the mean m and the error factor f_e,
—the median med and the error factor f_e,
—the median med and the standard deviation σ.

Exercise 3—Extreme value distribution

Let $X_1, X_2, \ldots, X_n$ be n independent random variables with the same distribution function $F(X)$ and X'_n defined by

$$X'_n = \text{Sup}\,(X_1, X_2, \ldots, X_n)\,.$$

(a) Show that

$$F'(x) = \text{Prob}\,(X'_n \leq x) = [F(x)]^n\,.$$

Deduce from this that:

$$\lim_{n\to\infty} F'(x) = 0 \qquad \text{if} \quad F(x) < 1$$

$$\lim_{n\to\infty} F'(x) = 1 \qquad \text{if} \quad F(x) = 1\,.$$

(b) As these limits are trivial, we are only interested in the random variables of the type $a_n X'_n$

$+ b_n$ and their limit distributions with distribution function G (where a_n and b_n are sequences which have to be determined).

Let us consider the $n \times N$ random variables

$$X_1, X_2, \ldots, X_n, X_{n+1}, X_{n+2}, \ldots, X_{2n}, \ldots, X_{jn+1}, X_{jn+2}, \ldots, X_{(j+1)n}, \ldots, X_{nN}.$$

Putting $X^j = \text{Sup}\,(X_{jn+1}, X_{jn+2}, \ldots, X_{(j+1)n})$, we obtain:

$$X'_{nN} = \text{Sup}\,(X^0, X^1, \ldots, X^{n-1}).$$

We will assume that the limit distribution function G (when $n \to \infty$) of a variable $a_n X'_n + b_n$ satisfies the equation:

$$[G(x)]^N = G(a_N x + b_N). \tag{1}$$

This equation introduced by Frechet is known as the Stability Postulate. Show that

$$a_{NM} = a_N\, a_M$$
$$b_{NM} = a_M\, b_N + b_M$$

then

$$a_N = N^k.$$

(c) Consider the case $k = 0$, i.e. $a_N = 1$.
Show that $b_N = -\theta \log N$.
Determine the function $h(x) = \log(-\log G(x))$ using Eqn (1).
Deduce from this that the limit distribution of $F'(x)$ is the distribution function distribution:

$$G(x) = \exp(-\mathrm{e}^{-(x-\xi)/\theta}) \qquad (\theta > 0) \quad (\text{type I}).$$

This is known as the Gumbel distribution.

(d) Now consider the case $k \neq 0$.
Show that:

$$b_N = -\xi(N^k - 1).$$

Putting $h(x) = -\log G(x)$, show that

$$Nh(x) = h[(x - \xi)N^k + \xi] \quad \text{et} \quad h'(x) \leq 0. \tag{2}$$

Verify that when $k > 0$

$$h_1(x) = \frac{(\xi - x)^{\frac{1}{k}}}{\theta}\, Y(\xi - x) \qquad \theta > 0$$

is the solution of Eqn (2).

Verify that when $k < 0$

$$h_2(x) = \frac{(x - \xi)^{\frac{1}{k}}}{\theta}\, Y(x - \xi) \qquad \theta > 0$$

is the solution of Eqn (2).

Deduce from this the two following limit distributions of distribution functions:

$$k > 0 \quad G(x) = \exp\left(-\frac{(\xi - x)^{\frac{1}{k}}}{\theta}\right) Y(\xi - x) \qquad \theta > 0 \quad (\text{Type III})$$

$$k < 0 \quad G(x) = \exp\left(-\frac{(x - \xi)^{\frac{1}{k}}}{\theta}\right) Y(x - \xi) \qquad (\text{Type II})$$

The distributions of $(-x)$ are also called extreme value distributions. In this case, the type III distribution is a Weibull distribution.

Gnedenko has shown that there are no other limit distributions. In addition, if the tail of the distribution of $F(x)$ is exponential, the limit distribution is the Gumbel distribution and if the tail of the distribution varies as x^{-k}, the limit distribution is of type II or III.

(e) Study of the Gumbel distribution.

Let $Y = \dfrac{x - \xi}{\theta}$, known as Gumbel's standard variable.

Calculate the probability density $f(y)$ of Y.
Calculate the characteristic function $\varphi(u)$ of Y.
Deduce the mean and the variance.
Plot the graph of $f(y)$.

(f) Using the extreme value distributions in reliability studies.
These distributions are used in the following cases:

—failures caused by corrosion: the item fails when the maximum corrosion depth reaches a certain threshold;

—for studying phenomena of exceptional intensity: floods, earthquakes, etc.

An extensive bibliography can be found in [Johnson and Kotz, 1970].

SOLUTIONS TO EXERCISES

Exercise 1

(a) The equation is derived directly from:

$$(x - m)^r = \sum_{j=0}^{r} C_r^j x^{r-j} (-m)^j \tag{1}$$

hence

$$\begin{aligned}
\mu_2 &= E[X^2] - m^2 \\
\mu_3 &= E[X^3] - 3\, m E[X^2] + 2\, m^3 \\
\mu_4 &= E[X^4] - 4\, m E[X^3] + 6\, m^2 E[X^2] - 3\, m^4 .
\end{aligned}$$

(b) Putting $\mu'_k = E[X^k]$

$$1 + \mu'_1 t + \mu'_2 \frac{t^2}{2!} + \cdots = \exp(k_1 t) \exp\left(k_2 \frac{t^2}{2!}\right) \exp\left(k_3 \frac{t^3}{3!}\right) \exp \ldots \tag{2}$$

$$= \prod_{i=1}^{\infty} \left(1 + \frac{k_i t}{1!} + \frac{k_i^2 t^2}{2!} + \cdots\right)$$

hence:

$$\mu'_r = \sum_{n=0}^{r} \sum \left(\frac{k_{\rho_1}}{\rho_1 !}\right)^{\Pi_1} \left(\frac{k_{\rho_2}}{\rho_2 !}\right)^{\Pi_2} \cdots \left(\frac{k_{\rho_m}}{\rho_m !}\right)^{\Pi_m} \frac{r!}{\Pi_1 ! \, \Pi_2 ! \ldots \Pi_m !}$$

where the second summation sign extends to all the sets $(\rho_1, \rho_2, \ldots, \rho_m)$ such that

$$\rho_1 \Pi_1 + \rho_2 \Pi_2 + \cdots + \rho_m \Pi_m = r$$

hence

$$\begin{aligned}
\mu'_1 &= k_1 \\
\mu'_2 &= k_2 + k_1^2 \\
\mu'_3 &= k_3 + 3\, k_2 k_1 + k_1^3 \\
\mu'_4 &= k_4 + 4\, k_3 k_1 + 3\, k_2^2 + 6\, k_2 k_1^2 + k_1^4
\end{aligned}$$

therefore, in accordance with (a)

$$\mu_2 = k_2, \quad \mu_3 = k_3, \quad \mu_4 = k_4 + 3\, k_2^2 .$$

Note that by differentiating Eqn (2), we obtain

$$\frac{t^j}{j\,!}\left(1+\mu'_1 t+\cdots+\mu'_r\frac{t^r}{r\,!}+\cdots\right)=\frac{\partial\mu'_1}{\partial k_j}t+\cdots+\frac{t^r}{r\,!}\frac{\partial\mu'_r}{\partial k_j}+\cdots$$

hence

$$\frac{\partial\mu'_r}{\partial k_j}=C^j_r\,\mu'_{r-j}$$

and in particular

$$\frac{\partial'\mu_r}{\partial k_r}=r\mu'_{r-1}\,.$$

Given that $\mu'_1=k_1$ and that the coefficient of k_r in μ'_r is 1, it is easy to deduce the μ'_i.

(c) $$\varphi(u)=\int_0^\infty\frac{e^{iu\chi^2}(\chi^2)^{(\nu-2)/2}\exp\left(-\frac{\chi^2}{2}\right)}{2^{\nu/2}\,\Gamma(\nu/2)}\,d\chi^2\,;$$

putting $\chi^2=\dfrac{y}{\frac{1}{2}-iu}$, we obtain

$$\varphi(u)=\frac{1}{(1-2\,iu)^{\nu/2}}\qquad\Psi(u)=-\frac{\nu}{2}\log(1-2\,iu)$$

hence

$$\frac{\partial\varphi^n}{\partial u}=\frac{i^n\,\nu(\nu+2)(\nu+4)\ldots(\nu+2\,n-2)}{(1-2\,iu)^{-\nu/2-n}}\qquad\frac{\partial\psi^n}{\partial u}=\frac{i^n\,\nu\,2^{n-1}(n-1)\,!}{(1-2\,iu)^n}\,.$$

It follows that:

$$\mu'_n=\nu(\nu+2)(\nu+4)\ldots(\nu+2\,n-2)$$
$$K_n=\nu\,2^{n-1}(n-1)\,!$$

then using (a):

$$\mu_2=2\,\nu$$
$$\mu_3=8\,\nu$$
$$\mu_4=48\,\nu+12\,\nu^2\,.$$

(d) As the variables are independent, by denoting the characteristic function of the common distribution by $\theta(u)$, we have:

$$\theta^\nu(u)=\varphi(u)$$

hence

$$\theta(u)=\frac{1}{\sqrt{1-2\,iu}}\,.$$

Consequently the common distribution is a standard normal distribution.

Exercise 2

(a) $$f(y)=\frac{1}{by\sqrt{2\,\pi}}\exp\left(-\frac{(\log y-a)^2}{2\,b^2}\right)$$

(b)

- $$m=\int_0^\infty\frac{1}{b\sqrt{2\,\pi}}\exp\left(-\frac{(\log y-a)^2}{2\,b^2}\right)dy=\frac{\exp\left(\frac{b^2}{2}+a\right)}{\sqrt{\pi}}\int_{-\infty}^{+\infty}e^{-y^2}\,dy=\exp\left(\frac{b^2}{2}+a\right)$$

similarly:

- $E[X^2] = e^{2b^2+2a}$

hence

$$\sigma^2 = e^{b^2+2a}(e^{b^2} - 1) .$$

- As the median of $N(a, b)$ is equal to a, consequently:

$$\text{med} = e^a$$
$$\text{mod} = e^{a-b^2}$$
$$\frac{\sigma}{m} = \sqrt{e^{b^2} - 1} .$$

(c) The direct consequence of the fact that the sum of two normal independent random variables is a normal random variable.

(d) $\mathscr{P}(X > x_+) = \alpha$

$$\Updownarrow$$

$\mathscr{P}(\log X > \log x_+) = \alpha = \alpha$

$$\Updownarrow$$

$\log x_+ = a + u_\alpha b$

hence

$$x_+ = e^{a+u_\alpha b}$$

and

$$x_- = e^{a-u_\alpha b}$$

hence the equations giving f_e.

(e) • $a = \log \dfrac{m^2}{\sqrt{m^2 + \sigma^2}}$, $\quad b = \sqrt{\log \dfrac{\sigma^2 + m^2}{m^2}}$

• $a = \log m - \left(\dfrac{\log f_e}{u_\alpha}\right)^2$, $\quad b = \dfrac{\log f_e}{u_\alpha}$

• $a = \log \text{med}$, $\quad b = \dfrac{\log f_e}{u_\alpha}$

• $a = \log \text{med}$, $\quad b = \sqrt{\log \dfrac{1 + \sqrt{1 + 4\left(\dfrac{\sigma}{\text{med}}\right)^2}}{2}}$.

Exercise 3

(a) $\mathscr{P}(X'_n \leq x) = \mathscr{P}(X_1 \leq x \text{ et } X_2 \leq x, \ldots, \text{et } X_n \leq x)$

$$= \prod_{i=1}^{n} \mathscr{P}(X_i \leq x) = [F(x)]^n$$

since the random variables are independent.

(b) $[G(x)]^{NM} = G(a_{NM}\,x + b_{NM}) = [G(a_N\,x + b_N)]^M$
$$= G(a_M(a_N\,x + b_N) + b_M)$$

hence

$$a_{NM} = a_N\,a_M \Rightarrow a_N = N^k$$
$$b_{NM} = a_M\,b_N + b_M = b_{MN} = a_N\,b_M + b_N .$$

(c) $k = 0 \Rightarrow a_N = 1 \Rightarrow b_{NM} = b_N + b_M \Rightarrow b_N = -\theta \log N.$

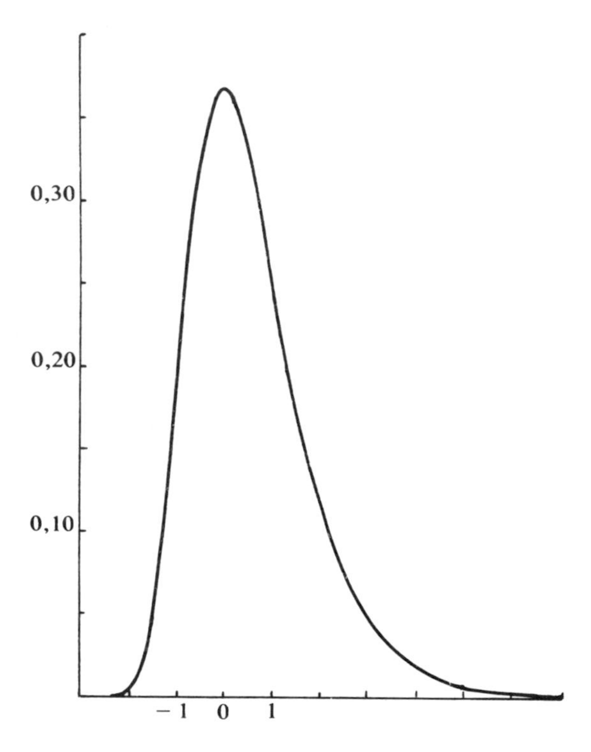

Fig. 9. *Density of the Gumbel distribution.*

Taking the logarithm of Eqn (1) two times, we obtain

$$+\log N + h(x) = h(x - \theta \log N)$$

hence

$$h(x) = -\frac{x-\xi}{\theta}$$

$$-\log G(x) = \exp\left(-\frac{x-\xi}{\theta}\right)$$

and

$$G(x) = \exp(-e^{-(x-\xi)/\theta})$$

as $h(x)$ is a decreasing function, θ is consequently positive.

(d) It follows from (b) that

$$b_N(a_M - 1) = b_M(a_N - 1)$$

therefore, as a_N and a_M differ from 1 ($k \neq 0$):

$$\frac{b_N}{a_N - 1} = \frac{b_M}{a_M - 1} = -\xi$$

hence

$$b_N = -\xi(N^k - 1).$$

Equation (2) can then be derived from Eqn (1) by substituting the values of a_N and b_N.
Verification of the fact that $h_1(x)$ and $h_2(x)$ satisfy Eqn (2) is immediate.

(e) $f(y) = e^{-y} e^{-e^{-y}}$

$\varphi(u) = \Gamma(1 - iu)$

$m = -\Gamma'(1) = \gamma$ (Euler's constant = 0.577 216)

$\sigma^2 = \Gamma''(1) - \Gamma'^2(1) = \frac{\pi^2}{6}$ (cf. [Schwartz, 1961]).

REFERENCES

AITCHISON J. and BROWN J. A. C. (1957): *The Lognormal Distribution*; Cambridge University Press, Cambridge.

BARLOW R. E. and PROSCHAN F. (1967): *Mathematical Theory of Reliability*; Wiley, Chichester.

BELIAV Y., GNEDENKO B. and SOLOVIEV A. (1972): *Méthodes mathématiques en théorie de la fiabilité*; Edition MIR, Moscow.

CARTON D. (1975): *Processus aléatoires utilisés en recherche opérationnelle*; Masson, Paris.

COX D. R. (1966): *Théorie du renouvellement*; Dunod, Paris.

COX D. R. and MILLER H. D. (1945): *The Theory of Stochastic Processes*; Methuen, London.

CUNNINGHAM S. (1959): From normal integral to deviate. Algorithm AS 24; *Applied Statistics*, **18**, no. 3.

ERDELYI A. (1954): *Tables of Integral Transforms*; Vol. 1, McGraw-Hill, New York.

FELLER W. (1966): *An Introduction to Probability Theory and its Applications*; Wiley, New York.

I.E.C. (1974): List of basic terms, definitions and related mathematics for reliability; Publication no. 271, 2nd edn, International Electrotechnical Commission.

JOHNSON N. L. and KOTZ S. (1970): *Distributions in Statistics*; Vol. 1: Discrete distributions, Vol 2: Continuous univariate distributions, Houghton Mifflin, Boston.

KENDALL M. G. and STUART A. (1963): *The Advanced Theory of Statistics*; Vol. 1, Charles Griffin, London.

KLINE M. B., MASTEN R. L., DI PASQUALE J. and HAMILTON T. A. (1978): An analysis of the evaluation of the reliability and maintainability disciplines; *Colloque international sur la fiabilité et la maintenabilité*, Paris.

NEVEU J. (1972): *Bases mathématiques du calcul des probabilitiés*; Dunod, Paris.

PEYRACHE G. (1968): Article sur la fiabilité; *Encyclopaedia Universalis*, Paris, Vol. 6, p. 1059.

RENIY A. (1966): *Calcul des probabilités*; Dunod, Paris.

SCHWARTZ L. (1961): *Méthodes mathématiques pour les sciences physiques*; Hermann, Paris.

WEIBULL, W. (1954): *A Statistical Representation of Fatigue Failure in Solids*; Royal Institute of Technology, Stockholm.

WIDDER D. V. (1946): *The Laplace Transform*; Princeton University Press, London.

CHAPTER 2 PART I

REPRESENTATION OF SYSTEM LOGIC

1. INTRODUCTION

1.1. Concept of logic representation

The first problem encountered by the reliability engineer when analysing the reliability or availability of a system is how to describe that system. In this chapter we will approach the problem by attempting to find the *simplest representation* irrespective of the method of computing the reliability or availability. In fact, it will then often be possible to progress automatically (via a program) to another representation lending itself more readily to computation.

Let us therefore consider a system composed of n interacting components and required to perform a given function (*). We will assume that each component has only a finite number of states.

Generally speaking, the two following cases arise:

Case 1. The component is normally operating in the system and has two possible states: the operating state and the failed state.

Case 2. The component is normally non-operating in the system and only begins to operate if the main component fails (standby redundancy).

It therefore has four possible states: operating, under repair, non-operating in an operable state and, lastly, non-operating in a failed state (note that there is not always sufficient information to discriminate between these last two states).

In order to take account of external effects on the system (natural disasters, human errors, etc.) we introduce a certain number of *events* which are regarded as *components* of the system. This special type of component type will also generally assume a finite number of states, i.e. a particular human error has or has not occurred and a particular disaster has or has not occurred, etc. Finally, as each component has a finite number of states, the overall system has a finite number of states.

Now, the number of system states increases exponentially with the number of components. In other words, if each component has two states, the system has 2^n states.

(*) The concepts 'system', 'components' and 'functions' will subsequently be discussed in the context of the environment, human failures, common modes, etc.

Representing the logic of a system means representing all the operating (*) *and non-operating states of the system and the connections between these various states.*

In many cases, the connections between the various states of a system are simple and can be deduced from knowledge of the component states constituting the system states. As the connections between the various system states are in these cases implicit, it is not necessary to represent them. The simplest representation is therefore the list of successful operating states (or failed states) known as the truth table (cf. section 5). However, this representation is generally excessively cumbersome (of the order of $n2^n$) and can only be used for systems with very few components. Other more manageable representations are therefore used (cf. section 5).

In sections 2, 3 and 4 we will describe the three most commonly used methods of representing the logic of a system: the reliability block diagram (cf. section 2) which represents the successful operating states of the system, the fault tree (cf. section 3) and the minimal cut sets (cf. section 4) which represent the failed states of the system. The cut sets method can only be used for coherent systems.

1.2. Coherent systems

A system is termed *coherent* [Birnbaum et al., 1961] if:

— when the system has failed, no failure will restore the system to a successful state,

— when the system is operating successfully, no repair will cause the system to fail,

— failure of all the components causes the system to fail,

— when all the components are working, the system is successful.

Most systems discussed will have this property.

Note that the majority of these representations are equivalent and that it is often possible to move automatically from one to another (e.g. using a translation program). When we want to represent the operating logic of a system (whose interstate connections are implicit), we will therefore try to find the *simplest and surest representation*; cf. section 7 for a discussion of selecting the representation.

When the connections between the various states of the system are not implicit, merely knowing the successful operating states of the system is not sufficient, and the system logic is represented by its state transition diagram; cf. section 6.

Note that *it is not necessary to have the same logic representation for all parts of the system.* It is often very useful to have a different representation for the subsystems constituting the overall system.

Finally, in Chapter 2, Part II, we describe a new method of modelling system failures which we regard as a crucial step towards computer-aided reliability analysis.

(*) If the system performs several functions, there will be a different *a priori* system logic for each of these functions.

2. RELIABILITY BLOCK DIAGRAM

2.1. Definition

This is the most natural way of representing the operating logic of a system, as it is often similar to the functional diagram.

In this representation, the *blocks representing components (equipments or events) or functions* whose failure (*) causes the system to fail, are placed in series, and those whose failure does not result in system failure except in conjunction with other blocks, are arranged in parallel with the latter.

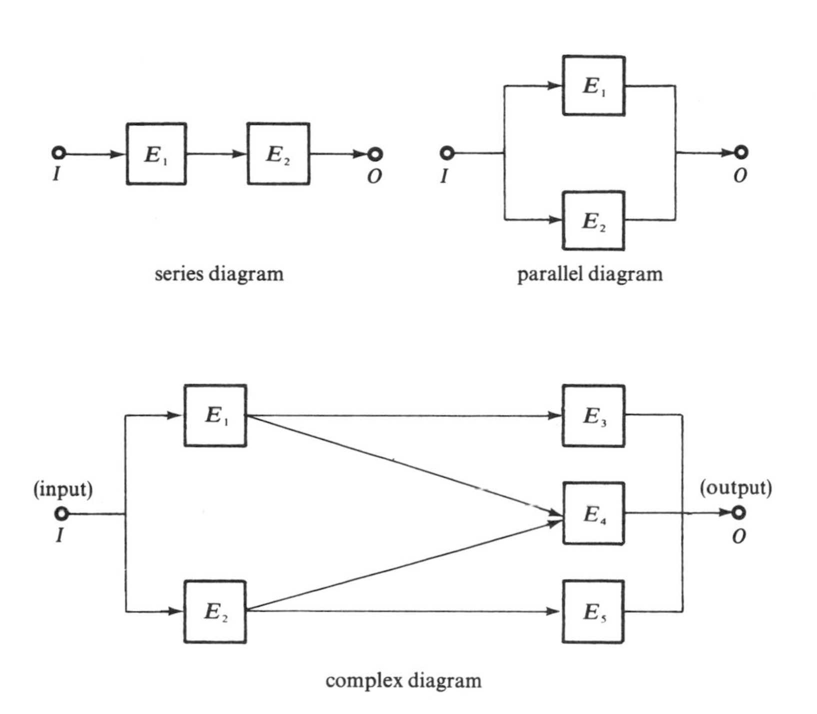

series diagram parallel diagram

complex diagram

The reliability block diagram is therefore a circuitless diagram [Gondran and Minoux, 1979] with an input and an output whose vertices (called blocks) represent the components of the system and whose arcs describe the relationships between the various components.

The system operates if there is a *successful path* between the input and output of the reliability block diagram. The list of successful paths therefore enables us to represent all the operating states of the system.

In the previous example, both components of one of the *successful paths* E_1, E_3, $E_1\,E_4$, $E_2\,E_4$, $E_2\,E_5$ must be working for the system to be functioning.

(*) Although a misuse of the term, we will refer to the occurrence of an undesirable event as a 'failure'.

The (logic) reliability diagram can represent more complex cases by introducing the following symbols:

Table 1. — *Redundancy symbols.*

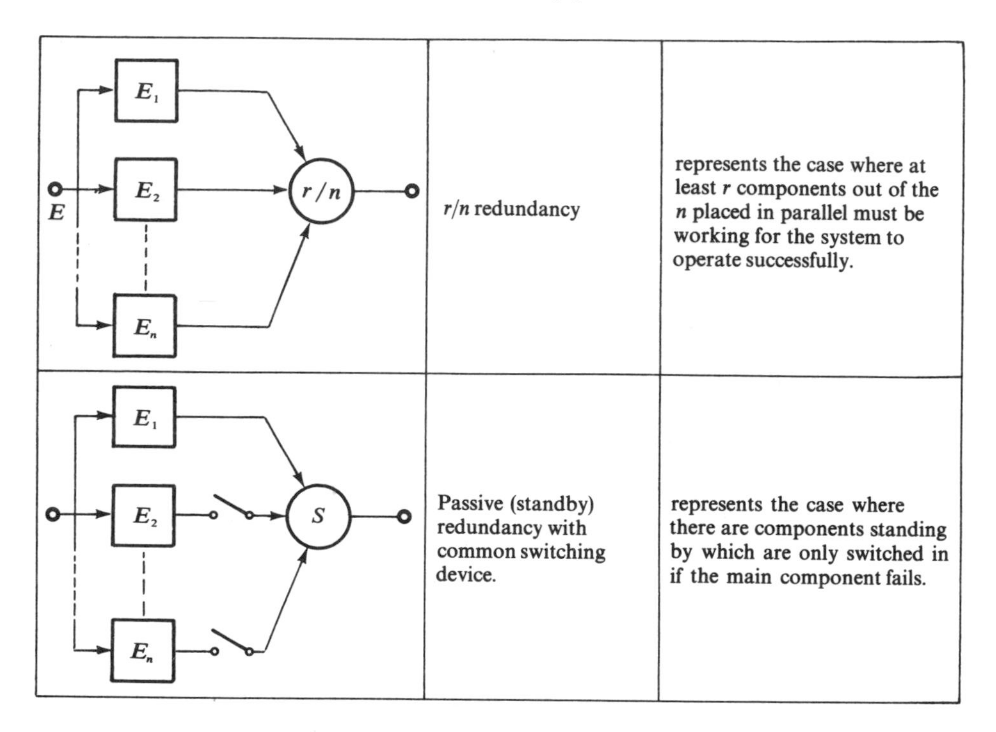

	r/n redundancy	represents the case where at least r components out of the n placed in parallel must be working for the system to operate successfully.
	Passive (standby) redundancy with common switching device.	represents the case where there are components standing by which are only switched in if the main component fails.

The blocks generally represent components. It is often useful to combine a number of components to form a single block; such a block is often termed a *supercomponent.*

We will now give a typical representation of a system's operating logic.

2.2. Example

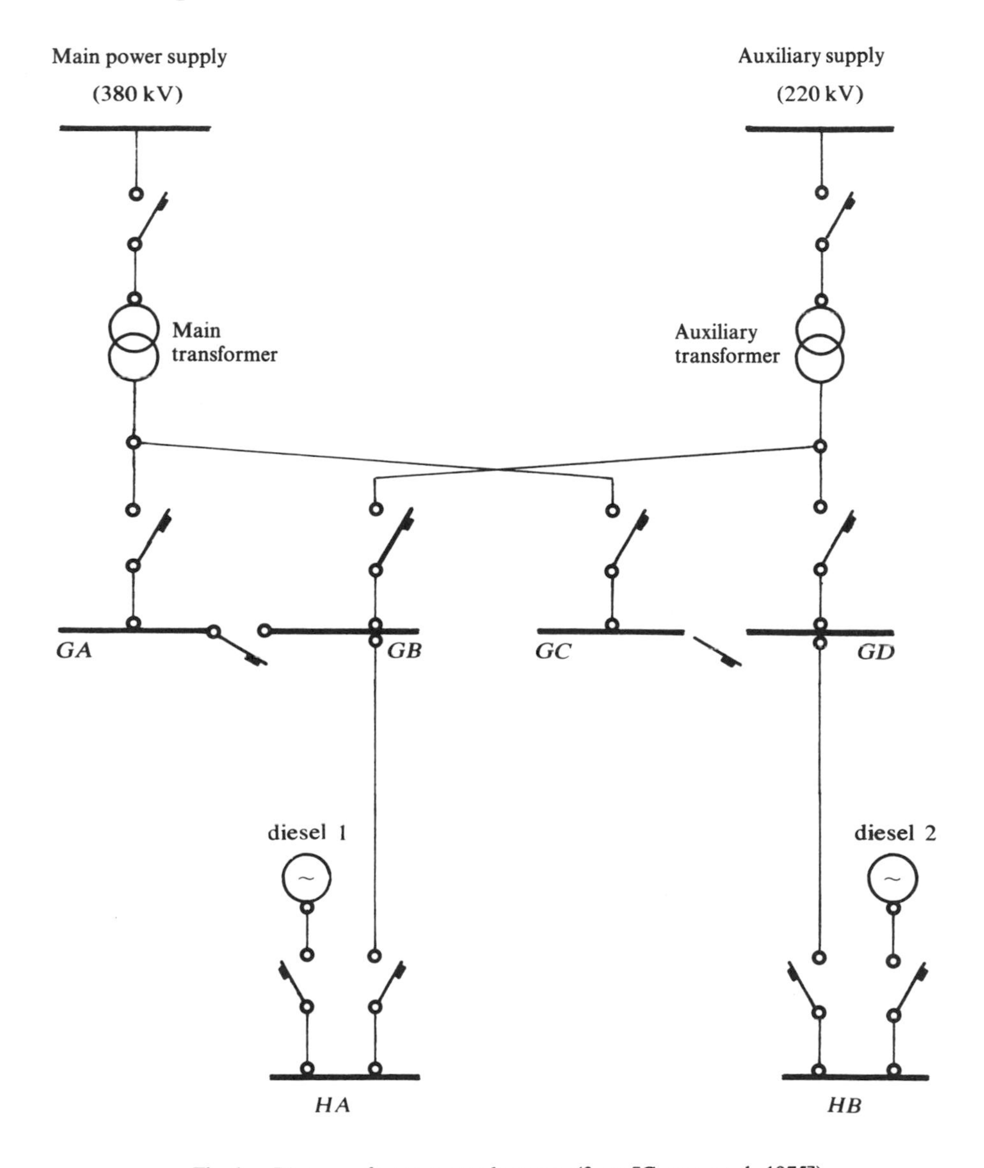

Fig. 1. *Diagram of a power supply system* (from [Greppo et al., 1975]).

Figure 1 shows the (simplified) diagram of a power supply system.

If the main (380 kV) supply fails, the auxiliary network (220 kV) and the two diesels take over.

The power supply function is said to be operative if voltage is present across either of the busbars *HA* or *HB*.

Let us now consider the following supercomponents:

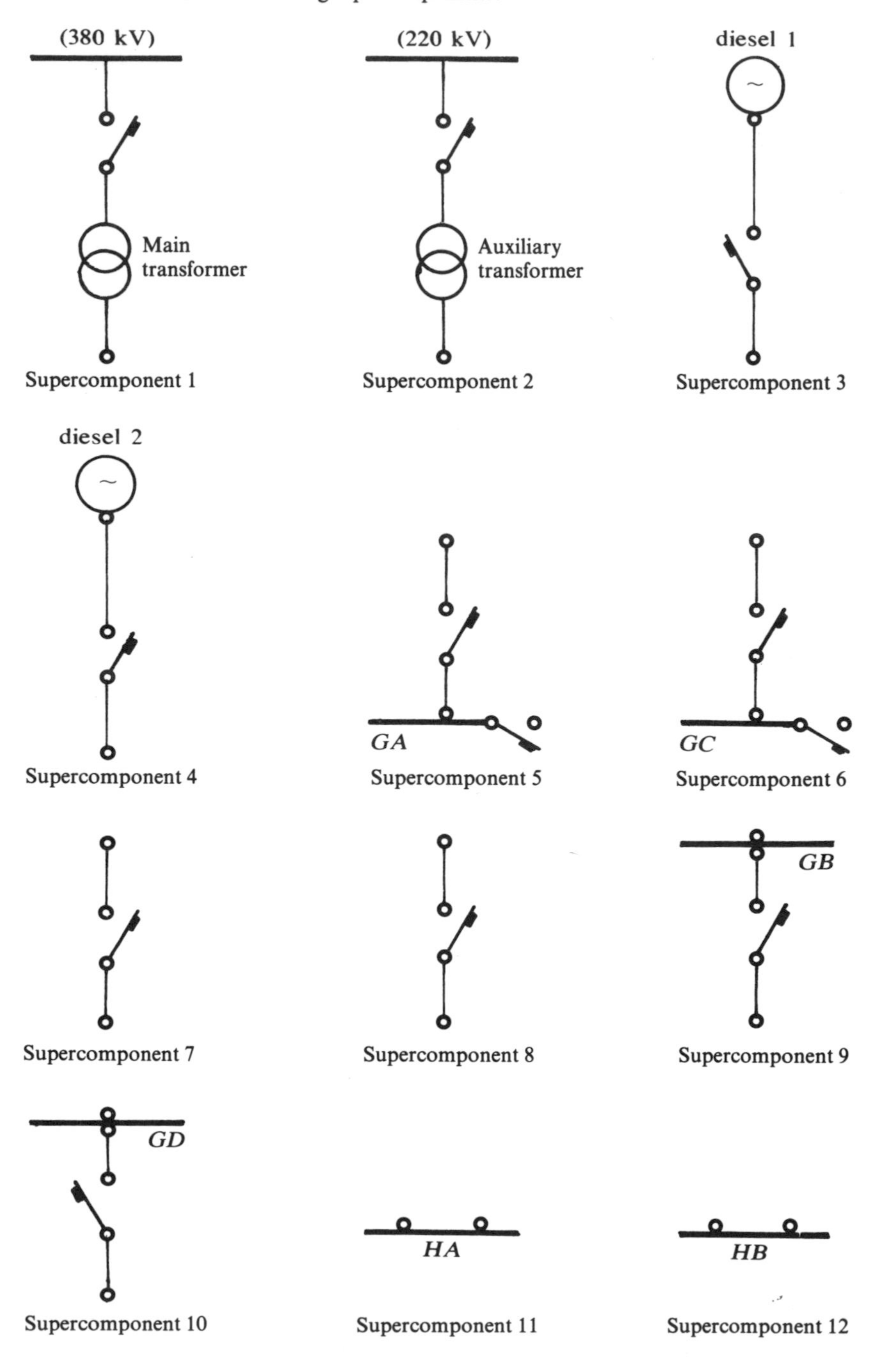

The reliability block diagram is therefore given by Fig. 2.

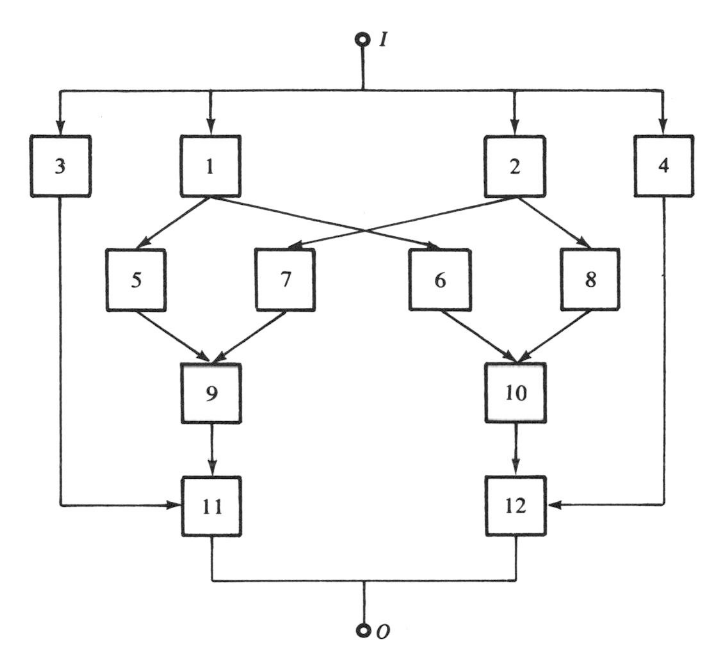

Fig. 2. *Reliability block diagram of power supply.*

Let us now determine the successful paths for this example. The following paths can be found:
3, 11
1, 5, 9, 11
1, 6, 10, 12
2, 7, 9, 11
2, 8, 10, 12
4, 12
In this example, these paths are minimal in the set-inclusion sense: there is no successful path strictly containing one of them.

In more complex cases, this condition may not be satisfied; however, in practice it is the minimal successful paths which will be of interest, as they alone provide complete information concerning the operation of the system: the system is successful $\Leftrightarrow$ one of the paths is successful $\Leftrightarrow$ one of the minimal paths is successful.

2.3. Some extensions

In order to take account of certain more complex conditions, some extensions will have to be defined.

— We will first examine the case where two components of the system exhibit a certain dependence.

Consequently, in the above example, the 380 and 220 kV networks are not independent; they can fail due to a common cause, e.g. a violent storm. The event capable of causing common failures (common mode failures) can therefore be *represented by an imaginary component* (here, component 13) placed in series with supercomponents 1 and 2. The reliability block diagram is then as shown in Fig. 3.

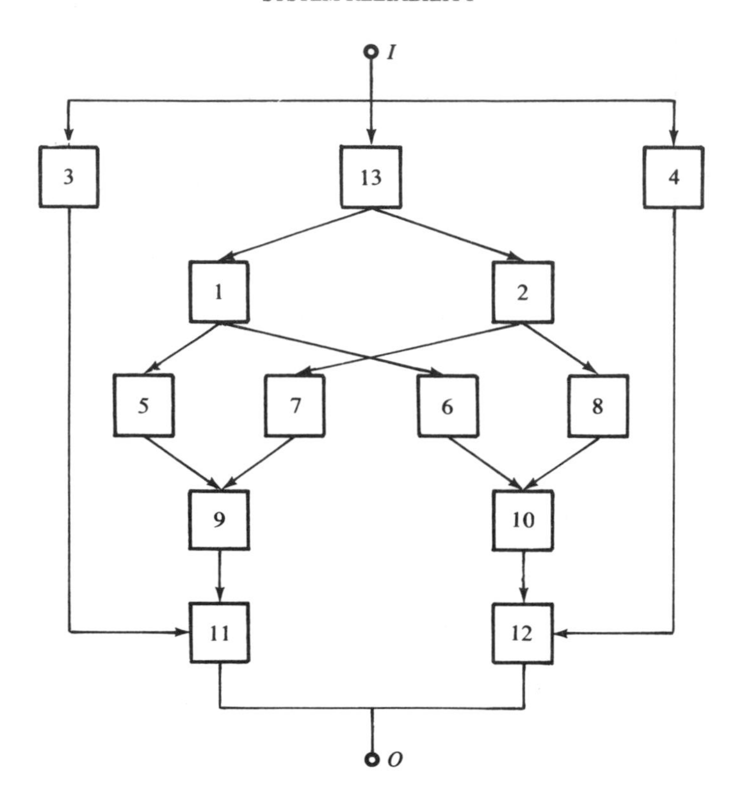

Fig. 3. *Reliability block diagram with common mode for 1 and 2.*

In this case, the minimal successful paths become: (3, 11), (13, 1, 5, 9, 11), (13, 1, 6, 10, 12), (13, 2, 7, 9, 10), (13, 2, 8, 10, 12) and (4, 12).

— Other dependences can be taken into account:

Thus, in the above power supply example, the supervisory control facility provides a functional connection between the system components. In fact, there are two supervisory control units, one supplied from *HA* and monitoring the left-hand side, the other supplied from *HB* and monitoring the right-hand side.

Now, the loss of *HA* causes the power supply of the left-side supervisory control unit to fail, which in turn causes failure of supercomponent 1.

In short, the dependence introduced is as follows: failure of supercomponent 1 if supercomponent 11 is down.

This can be represented in two ways:

— leaving the reliability block diagram unchanged but adding the condition: supercomponent 1 is down if supercomponent 11 is down;

— *replicating* block 11 on the diagram in order to incorporate this condition. This is the most frequently adopted solution.

Thus, in the power supply example, we now have the reliability block diagram shown in Fig. 4.

Supercomponent 11 has been replicated and inserted ahead of supercomponent 1.

The minimal successful paths are now: (3, 11), (1, 5, 9, 11), (1, 6, 10, 11, 12), (2, 7, 9, 11), (2, 8, 10, 12) and (4, 12).

Note that the above condition (supercomponent 1 down if supercomponent 11 failed) merely amounts to replacing successful path (1, 6, 10, 12) by (1, 6, 10, 11, 12).

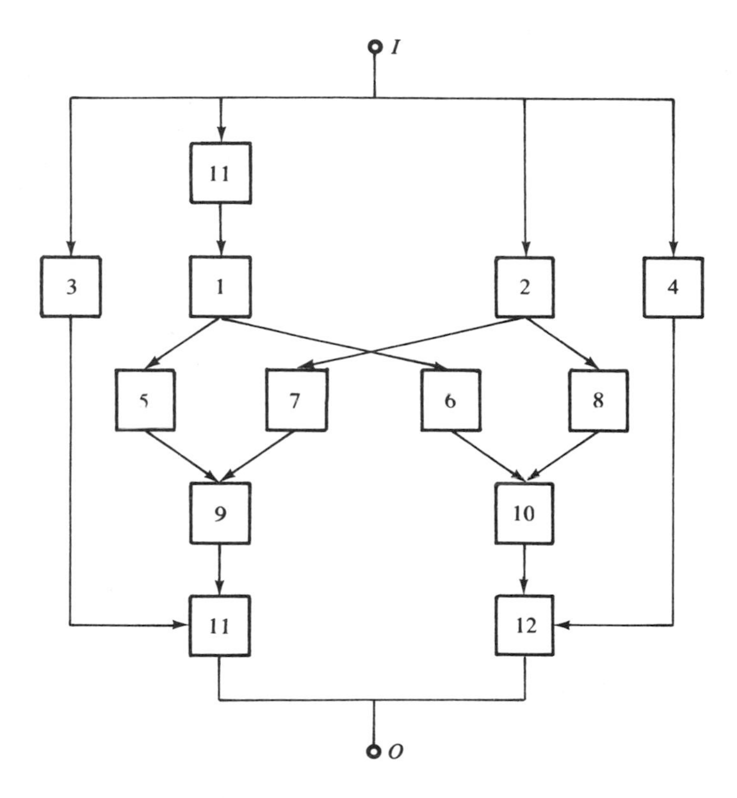

Fig. 4. *Reliability block diagram with replication of 11.*

In short, the two extensions described above only involve:

—introducing imaginary blocks to take account of certain dependences (e.g. common mode),
—replicating a particular block in the reliability block diagram.

Exercise 1 and Ch. 9 provide two other actual examples of system logic representation (cf. also Exercise 1 in Ch. 4).

3. FAULT TREE ANALYSIS

3.1. Definition

One of the most commonly used representations of system logic is the *fault tree*. This deductive method first appeared in 1962 in the Bell laboratories and was used to eliminate several weak points in the MINUTEMAN project.

It has been used extensively in nuclear reactor safety calculations (Rasmussen Report [WASH 1400, 1974]). The starting point is *a single and well-defined undesirable event*. In the case of system availability and reliability studies, this undesirable event is the non-operation of the system. Where system safety is concerned, this undesirable event is an event with serious repercussions.

The fault tree provides a diagrammatic representation of the *event combinations resulting in the occurrence of this undesirable event*. It is made up of *successive levels* such that each event is generated out of lower-level events via various logic operators (or gates).

This *deductive process* continues until one arrives at *basic events* which are mutually independent and probabilizable (even if the probabilities cannot be evaluated with absolute certainty). These basic events can be failures, human errors, external conditions, etc.

Let us consider, for example, the system represented by the following reliability block diagram:

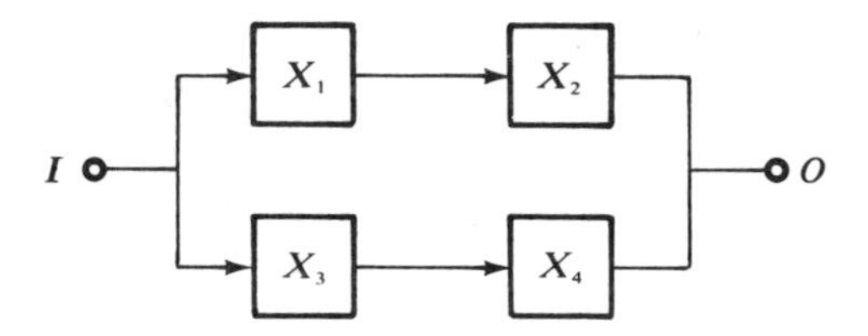

System failure occurs if both lines are down.

A fault tree for this system is therefore:

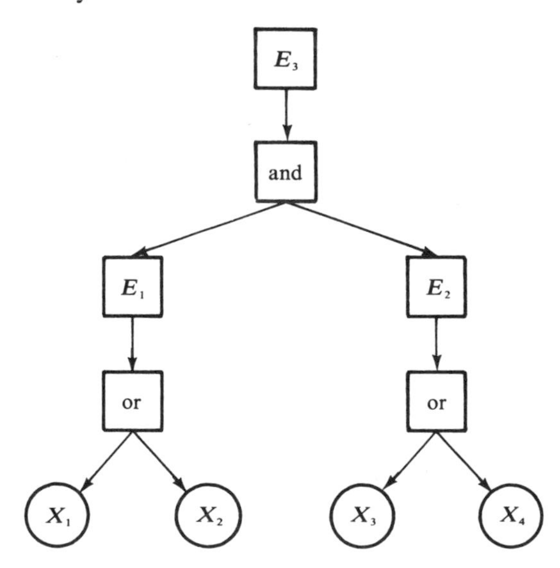

A fault tree is described using the symbols listed below.

3.2. Representation of events

Events are represented by the symbols given in Table 2.

Table 2. *Event representation.*

Symbol	Name of symbol	Meaning of symbol
	Circle	*Basic failure.* Representation of an event where the failure probability can be derived from empirical data.
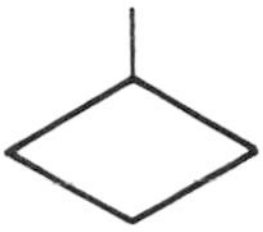	Diamond	*Fault assumed basic.* Representation of an event which could be subdivided into basic events but where this is not done through lack of information or usefulness.
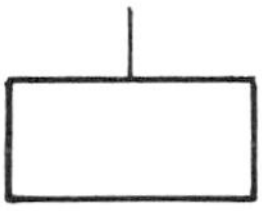	Rectangle	*Resultant event.* Representation of an event (intermediate event) resulting from the combination of other events via a logic gate.
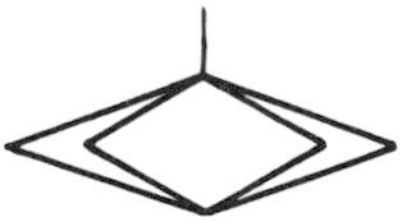	Double diamond	Representation of an event whose causes will not be specified until later.
	House	Representation of a *basic event* which is a normal occurrence while the system is operating.

The triangles in Table 3 enable us to split up a fault tree into several smaller trees (cf. section 3.3).

Table 3. *Representation of subtree transfers.*

Symbol	Name of symbol	Meaning of symbol
	Triangle	The part of the fault tree which follows the symbol  i is found at the point indicated by the symbol  i but is not shown in order to avoid repetition.
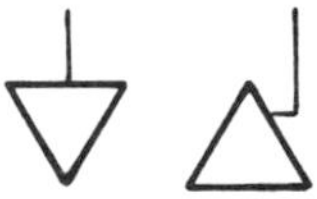	Inverted triangle	A part *similar* but not identical to that which follows the symbol  i is found at the point indicated by the symbol  i. It is not shown to avoid repetition.

3.3. Representation of logic gates (operators)

The simplest logic gates are represented by the symbols given in Table 4.

Table 4. *Representation of conventional logic gates.*

Symbol	Name of symbol	Meaning of symbol
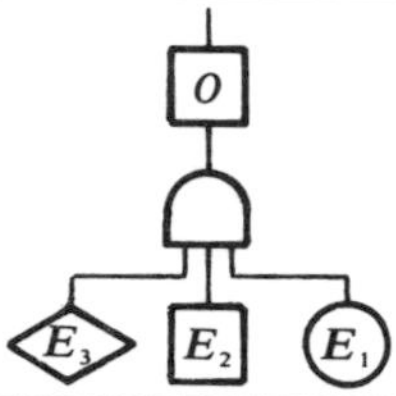 O; E_3 E_2 E_1	AND gate	The 'output' (here, O) event of the AND gate is generated if all the gate 'inputs' (here, the events E_1, E_2 and E_3) are present.
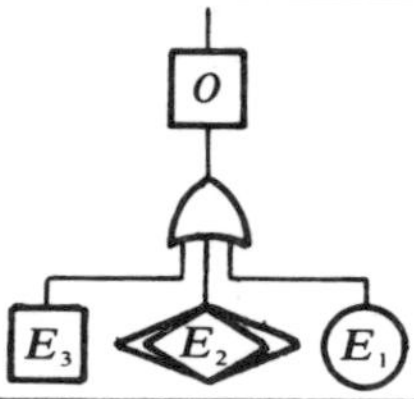 O; E_3 E_2 E_1	OR gate	The 'output' (O) event of the OR gate is generated if at least one of the gate 'inputs' (here, the events E_1, E_2 and E_3) is realized.
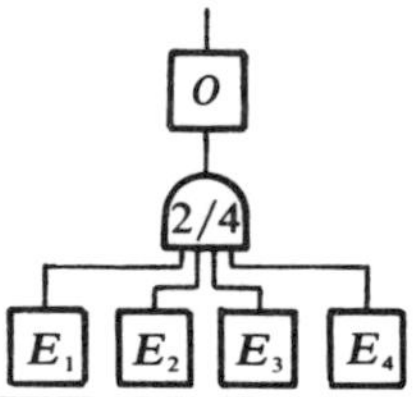 O; 2/4; E_1 E_2 E_3 E_4	r/n combination gate	The 'output' (O) event occurs if r of the n input events are realized. (Here, it is sufficient that two of the events E_1, E_2, E_3 and E_4 are realized).

A number of other, less common logic gates are represented by the symbols in Table 5.

Table 5. *Representation of complex logic gates* (cf. [Lievens, 1976]).

Symbol	Name of symbol	Meaning of symbol
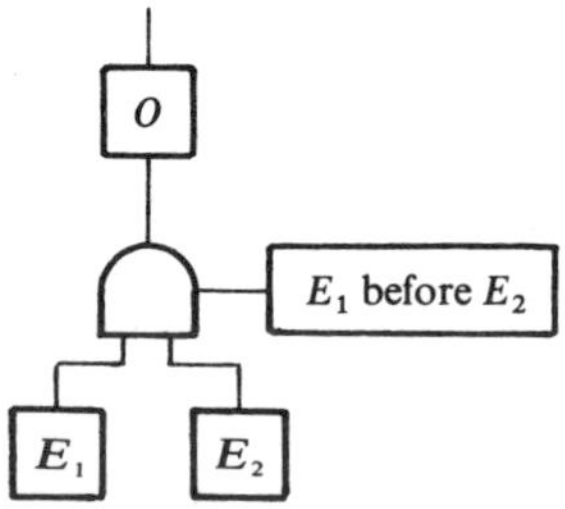 O; E_1 before E_2; E_1 E_2	AND gate with condition	The 'output' event it generated if all the gate 'inputs' (here, E_1 and E_2) are present and if the condition is satisfied (here, E_1 generated before E_2).

Symbol	Name of symbol	Meaning of symbol
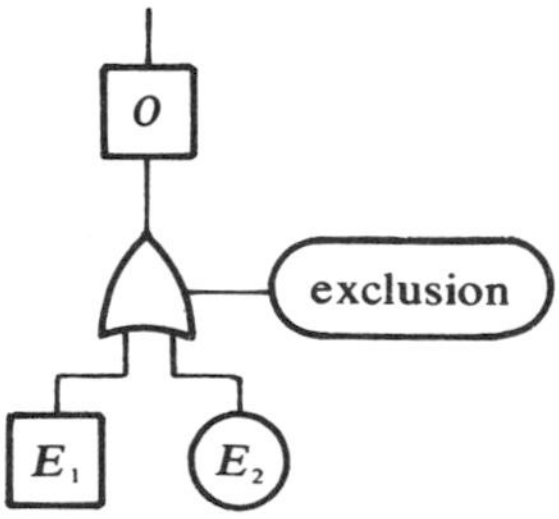 O; exclusion; E_1; E_2	OR gate with condition	The 'output' (O) event is generated if at least one of the gate 'inputs' (here, E_1 or E_2) is present and if the condition is satisfied (in this case, E_1 and E_2 must not occur simultaneously).
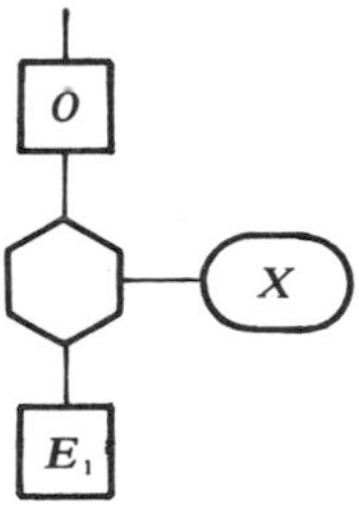 O; X; E_1	IF gate	The 'output' event is generated if the 'input' event is present and condition X is satisfied.
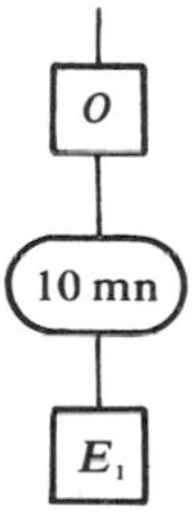 O; 10 mn; E_1	Delay gate	The 'output' event is generated with a given delay (here 10 min) after the appearance of the 'input event' (here, E_1) provided the latter has not disappeared in the meantime.
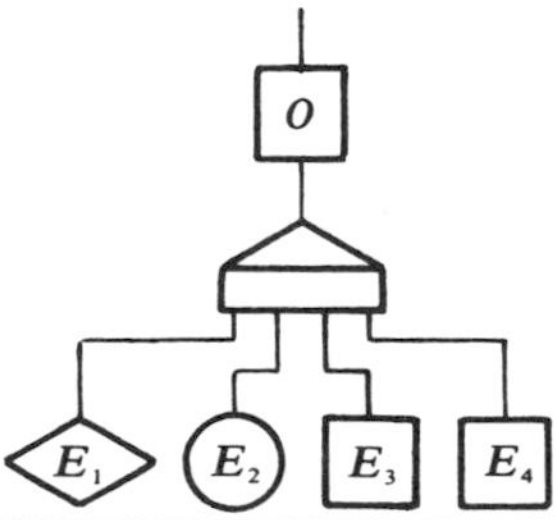 O; E_1; E_2; E_3; E_4	Matrix gate	The 'output' event is generated as a function of certain combinations of 'inputs' which are not specified at the time.
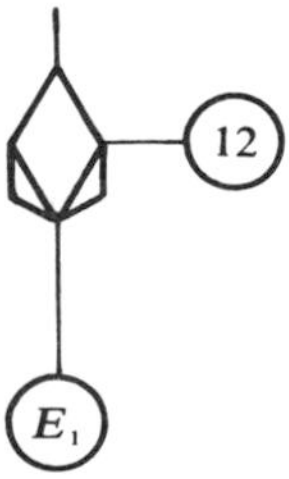 12; E_1	Quantification gate	A quantification gate makes a numerical value (here, 12) correspond to an event (here, E_1) and constitutes one of the 'inputs' of a summation.

Symbol	Name of symbol	Meaning of symbol
	Summation gate	A summation gate adds together the values from the various quantification gates and provides the 'input' to a comparison gate.
	Comparison gate	The 'output' event (O) of the comparison gate is generated if the value supplied from the summation gate satisfies a certain inequality (here, it is greater than or equal to 20).

3.4. A simple example

Let us consider the system shown below:

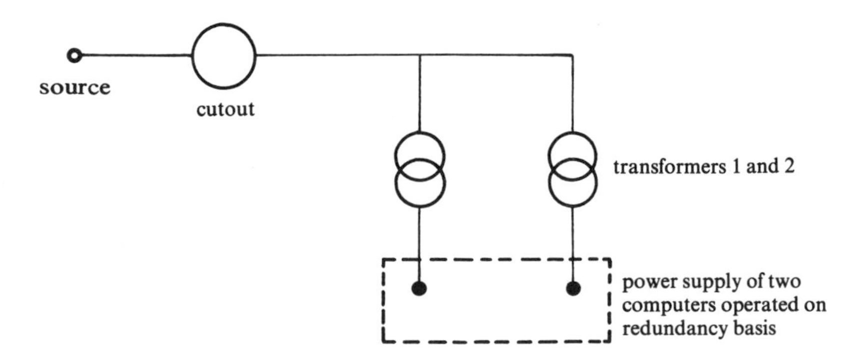

supplying power to two computers operated on a redundancy basis. The undesirable event is: simultaneous supply failure to both computers.

For the moment we shall consider the four events: source failure, cutout failure, failure of transformer 1 and failure of transformer 2 as basic components.

The fault tree is then as shown in Fig. 5.

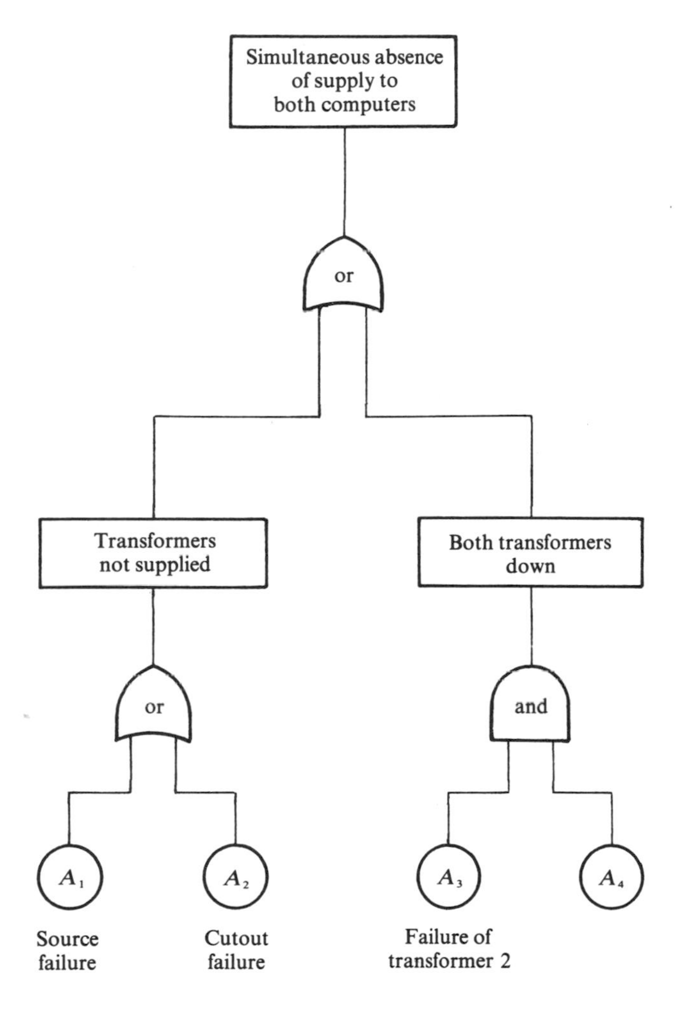

Fig. 5. *A fault tree.*

Let us now consider the case where source failure has to be subsequently specified and where transformer failures, although not having to be given separately, cannot be regarded as single events.

We get the same tree as in Fig. 5 by replacing

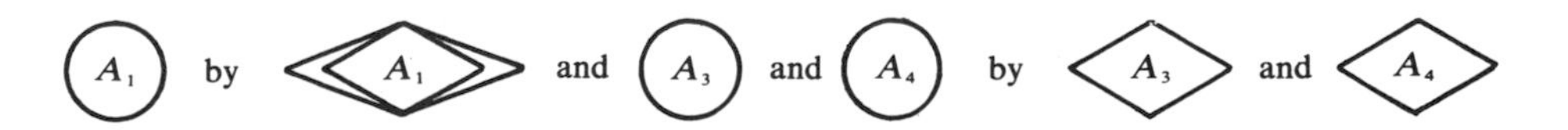

If there is a second power supply system whose reliability is 0.99, it is necessary to introduce an IF gate, the undesirable event being generated with a probability of 0.01 when failure of the first system occurs. The fault tree is then as shown in Fig. 6.

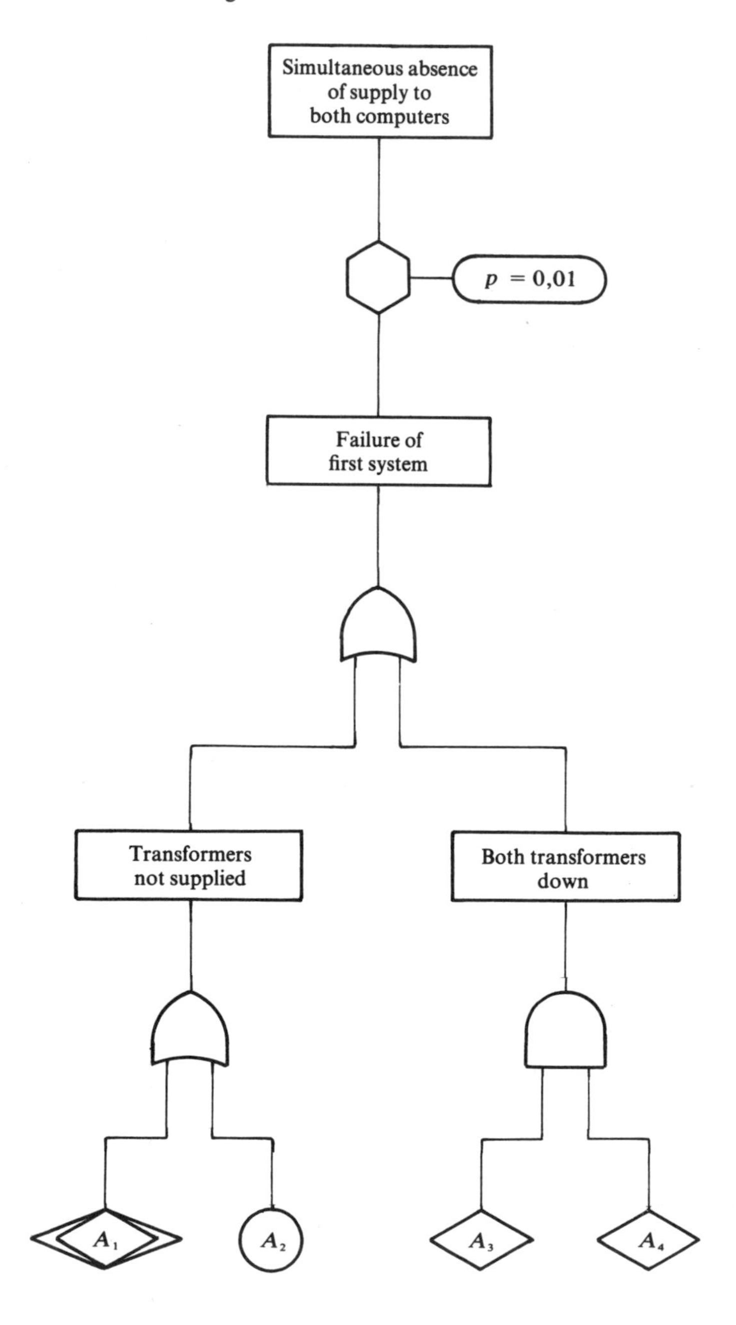

Fig. 6. *A fault tree with IF gate.*

If one now assumes that the second system is similar to the first and that there is true redundancy of these two systems, the fault tree can be drawn up as follows, using inverted triangle 1:

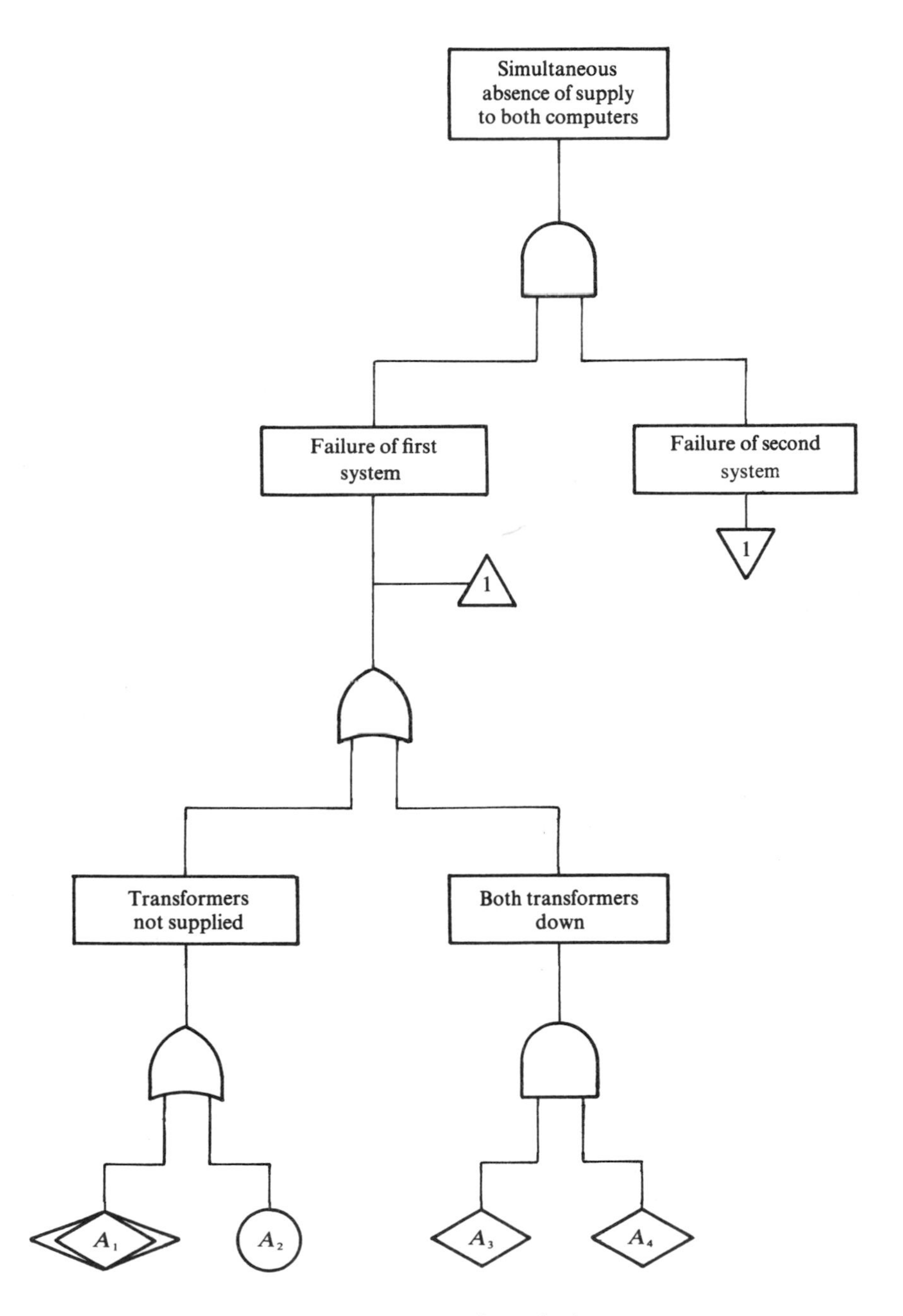

Fig. 7. *Use of inverted triangle in redundancy case.*

4. MINIMAL CUT SETS

4.1. Definition

A *cut set* is a set of components whose failure causes the system to fail. A minimal cut set is a cut set not containing any other cut set.

The series diagram in section 2 has two cut sets $\{E_1\}$ and $\{E_2\}$. The parallel diagram of section 2 has one cut set $\{E_1, E_2\}$. The complex diagram of section 2 has four minimal cut sets $\{E_1, E_2\}$, $\{E_1, E_4, E_5\}$, $\{E_2, E_3, E_4\}$ and $\{E_3, E_4, E_5\}$.

Notice that a cut set can be interpreted as a set of components intersecting each successful path.

Enumerating the minimal cut sets is an essential part of system reliability and availability analysis. In fact, each minimal cut set corresponds to a *significant combination of failures* at system level.

Interpretation of these minimal cut sets therefore gives a number of *qualitative* results such as the *weak points* of the system, *false redundancies* or *the effect of a given component* on system reliability. In addition, as we shall see in Chs 4 and 5, minimal cut sets are essential for *computing* the availability and reliability.

In practice, there are three methods of determining minimal cut sets.

—The first is inductive. One looks for the minimal cuts directly by intercombining 'significant failures' of system components. (See [Lievens, 1976, pp. 139–156] for a discussion of this approach.)

—The second proceeds automatically from the fault tree; cf. section 4.2.

—The third proceeds automatically from the reliability block diagram or successful paths; cf. section 4.3.

4.2. Fault tree approach

Minimal cut sets are determined by converting the fault tree to a Boolean expression.

—a Boolean variable is assigned to each basic event.

Thus, in the example given in Fig. 5, 'source failure' is assigned variable A_1, 'cutout failure' is assigned variable A_2, 'transformer 1 failure' is assigned variable A_3 and 'transformer 2 failure' is assigned variable A_4.

—The output event of an AND gate is assigned a Boolean variable equal to the product of the Boolean variables for the input events.

—The output event of an OR gate is assigned a Boolean variable equal to the (Boolean) sum of the input event variables.

—The r/n gate can be regarded as a combination of the two previous cases, noting that it is equivalent to an OR gate assigned $\frac{n!}{(n-r)!\,r!}$ inputs, each of which is the output of an AND gate with r inputs (all the r to r combinations of the n inputs).

—The other logic gates (Table 5) are more complex and do not lend themselves easily, if at all, to conversion to Boolean form. The reader is referred to Ch. 4 to see how some of these cases are treated.

We finally obtain a Boolean expression for the undesirable final event as a function of the Boolean variables associated with each basic event.

Consequently, for the fault tree shown in Fig. 5 we obtain, in turn:

$A_1 + A_2$ for 'transformers not supplied'
$A_3 . A_4$ for 'both transformers failed'
$A_1 + A_2 + A_3 A_4$ for 'simultaneous loss of supply to both computers'.

In the general case (coherent system), the Boolean expression for the undesirable final event assumes the form:

$$F = C_1 + \cdots + C_i + \cdots + C_m$$

where C_i is the intersection of n_i basic events:

$$C_i = B_{i1} \ldots B_{in_i} .$$

In the case of non-coherent systems, C_i will be the intersection of basic events or their complements (e.g. $C_i = B_1 \bar{B}_2 B_3$).

Expression F is termed *reduced* if the absorption laws of Boolean algebra can no longer be applied:

$$A + A = A, \quad A . A = A$$
$$A + B = A \text{ if } B \subset A, \quad A . B = B \text{ if } B \subset A .$$

Therefore, the following fault tree:

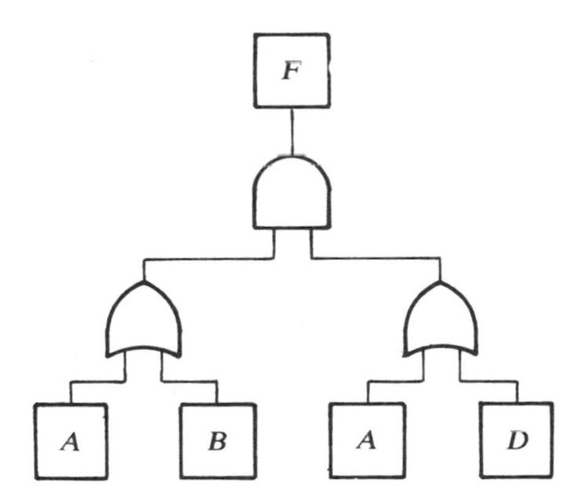

gives $F = (A + B)(A + D) = A^2 + AD + BA + BD = A + BD$.

In the case of coherent systems, *if F is reduced, each polynomial C_i corresponds to a minimal cut set.*

In the previous example, we have two minimal cut sets $\{A\}$ and $\{B, D\}$. Note also in this example that there is a false redundancy, as the failure of A alone results in system failure.

In the example in Fig. 5, there are three minimal cut sets $\{A_1\}$, $\{A_2\}$ and $\{A_3, A_4\}$.

Exercise 3 contains an algorithm for deriving the minimal cut sets from a fault tree comprising only AND and OR gates (cf. [Chatterjee, 1974] MICSUP algorithm: MInimal Cut Set UPward algorithm).

Other algorithms can be defined, such as MOCUS, Fussel's downward algorithm [Fussel, 1973].

4.3. Reliability block diagram approach

In terms of the reliability block diagram, a cut set is ultimately a set of components which intersect all the paths between the input I and the output O of the diagram.

In Exercise 4, the reader will find an algorithm for determining the minimal cut sets from a reliability block diagram.

In the case of the power supply reliability block diagram given in Fig. 2, the following minimal cut sets are found:

11, 12	2nd-order cut set
11, 4, 10	3rd-order cut sets
12, 3, 9	
11, 4, 1, 2	4th-order cut sets
11, 4, 1, 8	
11, 4, 2, 6	
11, 4, 6, 8	
12, 3, 1, 2	
12, 3, 1, 7	
12, 3, 2, 5	
12, 3, 5, 7	

The cut sets are enumerated in order of importance. As the first three cut sets are fundamental, it is the components entering these cut sets which may be the weak points of the system.

5. OTHER REPRESENTATIONS OF OPERATING STATES

Many other ways of representing the operating (or failed) states of a system are described in the literature. We will only mention them briefly, as they are only rarely suitable for system description.

We shall refer each time to the representation of the power supply system of the two computers as described in section 3.3 and whose fault tree is shown in Fig. 5.

5.1. Truth table

This shows the operating (or failed) states of the system.

For the example quoted in section 3.3, the table is as follows:

$\overline{A}_1$	$\overline{A}_2$	$\overline{A}_3$	$\overline{A}_4$
1	1	1	0
1	1	0	1

where the first line, for example, means that the system will operate if the source, cutout and transformer 1 are working.

5.2. Operating function

If each component is assigned a Boolean variable y_i equal to 1 when the component is up and 0 when the component is down, the operating function of the system is a Boolean function $f(y)$ such that $f(y) = 1$ if and only if $y = (y_1, \ldots, y_n)$ represents an operating state of the system.

In the example in section 3.3, we obtain:

$$f(y) = y_1\, y_2\, y_3 + y_1\, y_2\, y_4$$

5.3. Failure function

This is the complementary function of the operating function equal to 0 if and only if y represents an operating state of the system.

The failure function F is therefore equal to f, or

$$F = \overline{f} = \overline{y_1\, y_2\, y_3} \cdot \overline{y_1\, y_2\, y_4} = (\overline{y}_1 + \overline{y}_2 + \overline{y}_3)(\overline{y}_1 + \overline{y}_2 + \overline{y}_4)$$
$$F = \overline{y}_1 + \overline{y}_2 + \overline{y}_3\, \overline{y}_4\,.$$

In practice we define the variable $x_i = \bar{y}_i$ as being equal to 1 if the component i is down, and 0 otherwise.

We then have

$$F = x_1 + x_2 + x_3\, x_4$$

where

$$F = A_1 + A_2 + A_3\, A_4$$

as in section 4.1.

In the case of a coherent system, the monomials of the operating function represent the successful paths and the monomials of the failure function represent the minimal cut sets.

A number of other special representations are possible, such as event or accident trees, Petri networks, etc., and the state transition diagram discussed in the next section.

6. STATE TRANSITION DIAGRAM OR MARKOV DIAGRAM

In order to take account of the dependencies between the various system components, we construct a diagram whose *vertices* correspond to the various system *states* (if each component has two states, operating and failed, and if the system has n components, the maximum number of states is 2^n) and whose *arcs* correspond to the interstate *transitions*. On this diagram, each arc (i,j) is valued by the transition rate from state i to state j. Let us take a very simple example.

We will consider a system comprising two identical components operating in parallel.

When both components are working, each has a failure rate λ. When one or other of the components fails, the other then has a greater failure rate $\lambda' > \lambda$. In addition, there is only one repair facility and the repair rate is μ. If we denote by 2 the system state in which both components are operating, by 1 the system state in which only one component is operating and by 0 the system state in which both components are down, we obtain the following diagram:

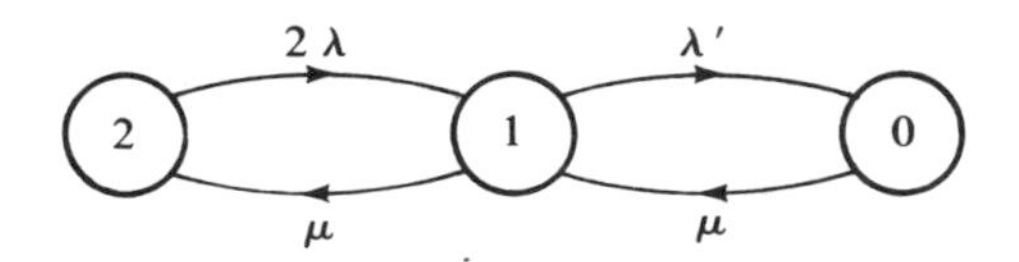

Fig. 8. *State transition diagram.*

This representation enables us to take account of the statistical dependence between these two components.

A few remarks are called for on how to obtain the interstate transition rates. If the probability of transiting from state i to state j between times t and $t + dt$ is $\lambda_{ij} dt + 0(dt)$, then λ_{ij} is the transition rate between states i and j.

When the transition rates are constant, the system is Markovian and the state transition diagram is often called the *Markov diagram*.

The transition rate between 2 and 0 is zero since the probability of going from state 2 to state 0 between t and $t + dt$ is $(2\lambda\, dt)(\lambda'\, dt) = 2\lambda\lambda'(dt)^2$.

In practice, this representation is much more cumbersome than the previous ones. *It will therefore only be used for subsystems of the overall system within which there are strong connections between components.*

Note, however, that this representation will be very important in reliability computations, but as we shall see in Ch. 6, we can often use it implicitly and without representing it explicitly.

Note: A number of authors give a slightly different representation of the transition rates; they represent on each arc (i,j) the probability of going from state i to state j between times t and $t + dt$. In this case, they add a loop to each vertex which corresponds to the probability of remaining in this state between t and $t + dt$. The state transition diagram shown in Fig. 8 then becomes:

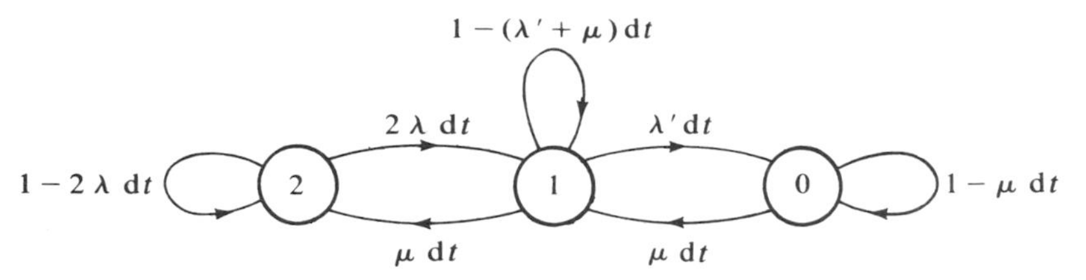

Fig. 9. *State transition diagram.*

7. SUMMARY

Subsystems whose components are strongly dependent are represented by a diagram of its states. In the very general case where the components (or events) are weakly dependent, we have a choice between representing them by the reliability block diagram, fault tree or minimal cut sets.

Direct representation by minimal cut sets is often difficult, as exhaustiveness cannot be assured (cf. [Lievens, 1976]). Therefore, the system is represented *initially* by a reliability block diagram or fault tree, and from one of these representations we then derive the minimal cut sets, as shown in section 4. Furthermore, this representation by minimal cut sets is theoretically basic, because it is unique, unlike the reliability diagram and fault tree representations.

There remains the choice between a reliability block diagram and a fault tree, and there is no clear-cut answer. In the present state of knowledge, *the fault tree method* appears to provide more sophisticated risk analysis. It is therefore preferred in cases where the risks due to system design, manufacture and use are relatively little known. It does, however, have a number of drawbacks as the size of the systems increases: diagram construction time, understanding, possible construction errors, etc.

The *reliability block diagram* has fewer limitations. It is therefore preferred for systems where the risks are relatively well known. In order to remedy the fact that the reliability block diagram appears to provide less detailed risk analysis, it is necessary, before constructing the diagram, to carry out an in-depth study of each system component, its functions, its failure modes, the effects of these failures on the system, etc.

This preliminary '*failure modes and effects analysis*' (cf. Ch. 2, Part II, and Ch. 9), commonly known to reliability engineers as FMEA, is essential: in fact, a given component performs a well determined function in the system. To give a simple example: a valve is normally closed in order to obtain an isolation, but at an emergency signal it will open to allow a through-flow; its operation, i.e. normally closed, can tolerate an external leakage without being affected, but its failure to open when required will have operating consequences, and it is this failure mode which is important.

Detailed FMEA usually makes it possible to construct the reliability block diagram and the fault tree. The formal connection between these approaches is discussed in Ch. 2, Part II.

Note that there is always a reliability block diagram of the type given in section 2 to represent any coherent system (it is sufficient, for example, to arrange the various successful paths in parallel and the various minimal cut sets in series, each cut set being represented by the set of these components in parallel). In order to represent non-coherent systems by means of a reliability block diagram, it is merely necessary to introduce negation blocks $\bar{E}_i$ in addition to the E_i blocks.

At present, we can use the existence of these two representations to make sure that no mistakes have been made in compiling them. In fact it is 'sufficient' to provide both representations and then, in order to check that they correspond to the same system, to find the minimal cut sets for each. Unless identical minimal cut sets are obtained, there is an error in one or both of the representations. There we have a good example of redundancy to think about!

EXERCISES

Exercise 1 — Representing a protection system

Figure 10 shows the diagram of a turbine overspeed protection system.

When an overspeed occurs, it is necessary to cut off the steam supply. For this purpose, it is sufficient that the four stop valves (12/1, 12/2, 12/3 and 12/4) should close or that the four control valves (10/1, 10/2, 10/3 and 10/4) should close.

Overspeed detection is continuously provided by the electrical systems (supercomponent 1) and the mechanical system (supercomponent 2).

When an overspeed is detected by the electrical systems, the instruction to cut off the steam supply is issued by the electrical control unit (component 7), either directly via electrical connections to the eight components (11/1, 11/2, 11/3, 11/4, 9/1, 9/2, 9/3 and 9/4), or indirectly via the electrovalves (3 and 4) and hydraulic relay (5).

When an overspeed is detected by the mechanical system, the instruction to cut off the steam is issued via the hydraulic relay (5).

(1) Construct the reliability block diagram for this protection system.

(2) Combine the components arranged in series or parallel on the diagram to form supercomponents such that there are no more than nine supercomponents.

(3) Determine from this diagram the minimal successful paths and the minimal cut sets.

What can be deduced from this for electrovalves 3 and 4?

Exercise 2 — Constructing a fault tree

Construct a fault tree for the power supply system shown in Fig. 1 (section 2.1).

Exercise 3 — Determining minimal cut sets from the fault tree [Chatterjee, 1974]

Let us consider a fault tree with only AND and OR gates.

(1) We will first take the case where each basic event appears once in the fault tree.

Show how minimal cut sets can be obtained directly without applying the absorption laws of Boolean algebra.

Deduce from this that the minimal cut sets of any subtree satisfying the above condition can be simply computed.

(2) In order to determine the minimal cut sets of any tree, we determine the reduced Boolean expression of the undesirable event starting from the basic events and moving up the tree. At each stage, we determine the reduced Boolean expression of the logic gate output event. Give an algorithm to represent the reduced Boolean expression of the output event for an AND gate and for an OR gate.

(3) Determine the minimal cut sets from the fault tree obtained in Exercise 2. Compare these minimal cut sets with those found in section 4.2.

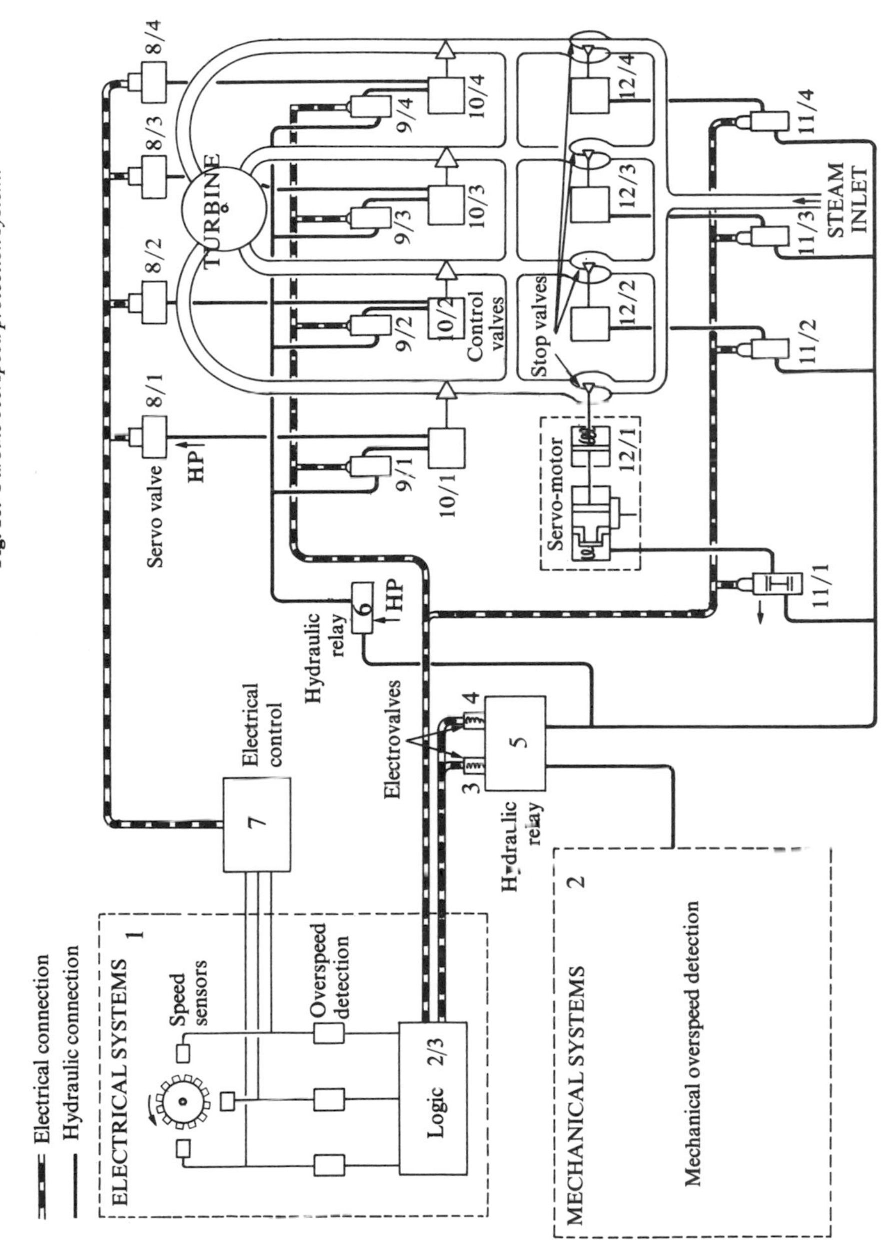

Fig. 10. *Turbine overspeed protection system.*

Exercise 4 — Determining minimal cut sets from the reliability block diagram

Let us consider a reliability block diagram made up of n components.

(1) Noting that the oriented graph corresponding to this reliability block diagram is circuitless, show that it is easy to find the successful paths.

(2) Let $P_1, P_2, \ldots, P_m$ be the various successful paths. For each component i from 1 to n, we shall denote the number of successful paths in which it appears by m_i. Give a method of enumerating the minimal cut sets which use these numbers m_i in a dynamic manner.

(3) Let us consider the $m \times n$ matrix $A = (a_{ij})$ such that

$$a_{ji} = \begin{cases} 1 & \text{if the element } i \in P_j \\ 0 & \text{otherwise} \end{cases}$$

We will consider the set $\mathscr{H}$ of the characteristic vectors of the subsets of n components: $x = (x_i)_{i=1 \text{ to } n}$ with $x_i = 0$ or 1.

Show that the minimal cut sets are identical to the set of the minimal vectors x of $\mathscr{H}$ satisfying the system.

$$Ax \geq \begin{pmatrix} 1 \\ \cdot \\ \cdot \\ \cdot \\ 1 \end{pmatrix}.$$

SOLUTIONS TO EXERCISES

Exercise 1

(1)

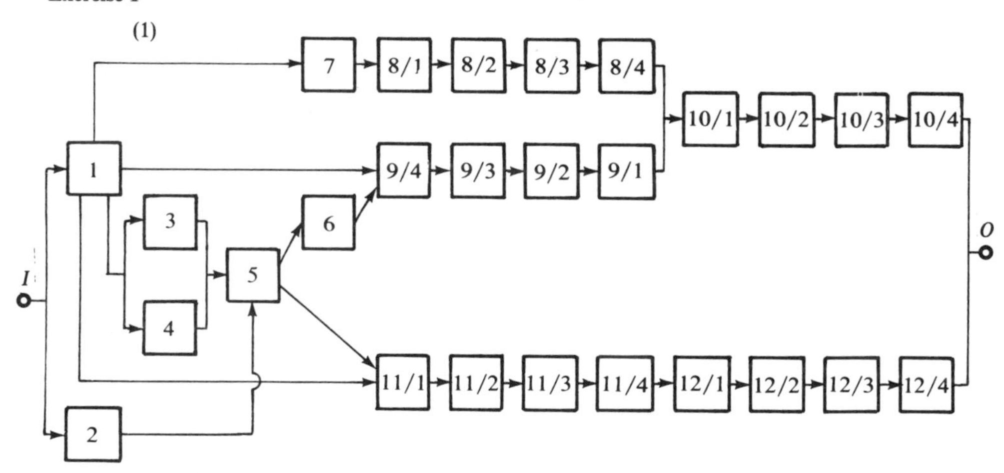

(2) With obvious regroupings, this reduces to:

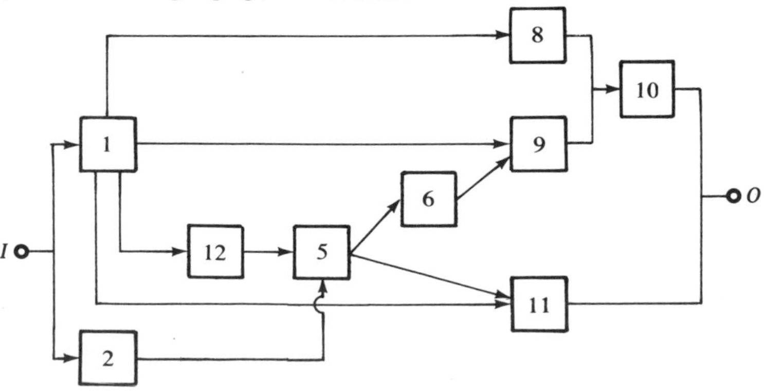

(3) The minimal successful paths are: $\{1, 8, 10\}$, $\{1, 9, 10\}$, $\{1, 11\}$, $\{2, 5, 6, 9, 10\}$, $\{2, 5, 11\}$.
The minimal cut sets are: $\{1, 2\}$, $\{1, 5\}$, $\{10, 11\}$, $\{1, 6, 11\}$, $\{1, 9, 11\}$, $\{1, 10, 11\}$ and $\{8, 9, 11\}$.
The fact that supercomponent 12 (electrovalves 3 and 4) does not figure in the cut sets shows, at first sight, that electrovalves 3 and 4 are perhaps unnecessary for this protection system.

Exercise 3

(2) *OR gate algorithm*

Let us first consider an OR gate with two inputs. The non-reduced Boolean expression F of the output event is then written as $F = E_1 + E_2$, E_1 and E_2 being the reduced Boolean expressions of the two input events.
E_1 is written as $E_1 = B_1 + B_2 + \ldots + B_l$ and
E_2 is written as $E_2 = D_1 + D_2 + \ldots + D_k$, the B_i and D_j being products of basic events.
Let n_i be the number of elements of B_i and m_j the number of elements of D_j.
To reduce F is to eliminate from E_1 the sets B_i contained in a set D_j of E_2 and to eliminate from E_2 the sets D_j contained in a set B_i of E_1.
As this relation is symmetrical, we will study the second case as an example.
The simplest algorithm is to consider each set B_i, i from 1 to l, and for each i to verify whether or not $B_i \subset D_j$ for j from 1 to k (note that the verification will only be achieved if $n_i \leqslant m_j$).
An algorithm which is more complicated to apply but which is faster, is to use inverted files. This involves compiling, for each basic event, the list of the various sets D_j containing this basic element.
In order to verify that B_i is included in one of the D_j, it is therefore sufficient to make the intersection of the lists corresponding to the basic events of B_i. The intersection set of these lists is then the set of the D_j containing B_i.
The advantage is that these lists are compiled once and for all.
When the OR gate has more than two inputs, the previous algorithm is applied successively: for example, if $F = E_1 + E_2 + E_3 + E_4$, we reduce $E_1 + E_2$, $(E_1 + E_2) + E_3$, $((E_1 + E_2) + E_3) + E_4$ in turn.

AND gate algorithm

Consider the case of an AND gate with two inputs; the general case is the same as for the OR gate. We therefore have $F = E_1 . E_2$ with $E_1 = B_1 + B_2 + \ldots + B_1$ and $E_2 = D_1 + D_2 + \ldots + D_k$.
We first form the lk products B_iD_j by making $A^2 = A$ whenever necessary.
Let n_{ij} be the number of elements of B_iD_j reduced in this way. We then order the B_iD_j by increasing n_{ij}.
Then, following this order, we examine each B_iD_j and determine whether it is included in a set classified further on; this set is then eliminated. The resultant expression corresponds to the sum of the non-eliminated sets.

Exercise 4

See [Dubois and Langer, 1977].

REFERENCES

Birnbaum Z. W., Esary J. D. and Sanders S. C. (1961): Multi-components Systems and Structures and their Reliability; *Technometrics*, Vol. 3, no. 1.

Chatterjee P. (1974): *Fault tree analysis: reliability theory and systems safety analysis*; Operations Research Center Report ORC 74–34.

Cohu P. and Pages A. (1976): *Fiabilité de la fonction réfrigération du coeur pendant l'arrêt annuel*; Note EDF-DPT, Service de la Production Thermique D57 7327-02.

Dubois D. and Langer C. (1977): *Rapport de stage: calcul de la fiabilité des systèmes réparables, le code FIABC*; EDF Service IMA.

Fussel J. B. (1973): Fault Tree Analysis—Concepts and Techniques; Proceedings of the NATO Advanced Study Institute of *Generic Techniques in Systems Reliability Assessment*, University of Liverpool, U.K. Published in NATO Advanced Study Institutes Series, Series E, no. 5, Noordhoff-Leyden, 1976.

GONDRAN M. and MINOUX M. (1979): *Graphes et Algorithmes*; Eyrolles, Collection des Etudes et Recherches EDF, Paris.

GREPPO J. F., BOURSIER M., CARNINO A. and BLIN A. (1975): *Détermination par une méthode probabiliste d'une règle d'exploitation relative aux sources d'alimentation électrique* 6.6 kV *des tranches PWR type* 900 MW; Note EDF-DPT, Service de la Production Thermique D57 7381-02.

IEEE (1972): *Trial Use Guide—General Principles for Reliability Analysis of Nuclear Power Generating Station Protection Systems*; ANSI N 41.4, IEEE Std 352.

LIEVENS C. (1976): *Sécurité des systèmes*; Cepadues-Editions, Toulouse.

WASH 1400 (1974): *An Assessment of Accident Risks in U.S. Commerical Nuclear Power Plants*; Reactor Safety Study, WASH 1400, U.S.A.E.C.

CHAPTER 2 PART II

REPRESENTATION OF KNOWLEDGE IN RELIABILITY STUDIES: FCP ANALYSIS

1. INTRODUCTION

In the second part of this chapter, we take the subject of system logic representation a stage further. The aim is to present a new approach to modelling system failures. We regard this approach as essential for making reliability studies the key to complete understanding of a system throughout its entire life cycle. This new method generalizes and formalizes conventional Failure Modes and Effects Analysis (FMEA).

FMEA systematically analyses the failure modes of system components and the effects of these failures on the system. At present, it constitutes an indispensible preliminary to any subsequent analysis; e.g. for the Concorde and Airbus projects [Lievens, 1976], the lunar module LEM [Bussolini, 1971], military systems [IEEE, 1975], the car industry [Yamada, 1977], nuclear power plants [Barbet et al., 1978]. See also the official text of the International Electrotechnical Commission [Llory, 1980; IEC, 1985].

A concrete example of FMEA applied to a travelling crane used in nuclear power plants is given in Ch. 9, section 4.

Although FMEA is extremely useful, it has two drawbacks:

—Its *lack of formal connection* with subsequent reliability analysis, particularly with the fault tree.

—It becomes *superfluous* after reliability analysis has been completed. It is unfortunate that the considerable body of knowledge acquired in the course of FMEA is filed away and forgotten.

In this chapter we shall describe *formal modelling of system failures* as a means of overcoming these two drawbacks. This modelling method generalizes and formalizes conventional FMEA. It is based on the knowledge representation methods developed in the fields of artificial intelligence and expert systems [Gondran, 1983; Laurière, 1983]. It corresponds to a *knowledge base* which can be used in various ways: for *automated fault tree construction*, *failure simulation* or *diagnostic assistance*. This type of modelling is crucial to the development of computer-aided reliability.

In section 2, we describe this new method of representation, known as FCP (functions–components–parameters) analysis. Its use is described in section 3, with particular reference to the automated construction of a fault tree. An application of FCP analysis, including fault tree construction, to an important example is then described in sections 4 and 5. This chapter is based on work by Gondran and Laleuf [Gondran and Laleuf, 1983, 1984] who first developed this method.

2. FCP (FUNCTIONS–COMPONENTS–PARAMETERS) ANALYSIS

The automated construction of fault trees (cf. Ch. 2, section 3) from analysis of the system components has long interested reliability engineers. Unfortunately the various approaches proposed so far, e.g. the CAT code [Apostolakis et al., 1978; Ancelin, 1981] and the GO code [Ancelin et al., 1981], have proved cumbersome to the point of deterring any potential user.

In the CAT code, for example, each system component is described by its input parameters, its output parameters and its internal states, and the logic functions between them.

The initial idea in FCP analysis is to examine the functions and subfunctions of a system in the same way as its components. The introduction of functions in addition to parameters and components provides higher-level concepts which makes analysis much easier.

This procedure is comparable with the failure combination method [Ancelin et al., 1981; Llory and Villemeur, 1980; Villemeur, 1982]. In this method, the 'summarized failures' and 'global failures' are concepts performing a similar role to the function failure modes.

The second idea in FCP analysis is to retain the framework of the FMEA table but giving more precise expression to the relationships between the functions–components–parameters and the failures by means of a thoroughly formalized representation. The interpretation of this new table will therefore be equivalent to a knowledge base of the 'production rules' type used in expert systems [Laurière, 1983; Gondran, 1983].

For the moment we shall consider the four key columns of an FMEA: component identification, failure modes, possible causes of failure and effects on the system (for further discussion of FMEA, the reader is referred to Ch. 9, section 4).

Let us therefore consider the system shown in Fig. 1 (cf. Ch. 2, Part I, section 3).

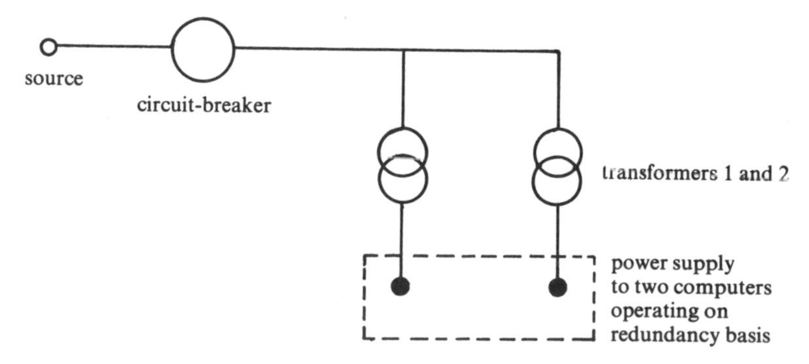

This system constitutes the power supply of two computers in a redundancy configuration. The undesirable event is the simultaneous absence of supply to both computers. For this example, we shall assume that the first four columns of the FMEA are as follows (Table 1):

Table 1.

Components	Failure modes	Possible causes of a failure	Effects on the system
Source	— Loss of source	— Loss of supply system — Breakage of supply line	— No supply to either computer
Circuit-breaker	— Loss of circuit-breaker	— Primary failure — Fire	— No supply to either computer
Transformer 1	— Primary failure		— No supply to computer 1
Transformer 2	— Primary failure		— No supply to computer 2

Table 1 can be represented in FCP analysis as shown in Table 2:

Table 2.

Components	Failure modes	Possible causes of a failure	Effects on the system
Source	— Loss	— Loss (supply system) — Breakage (supply line)	— Failure (supp. computer 1) *and* Failure (supp. computer 2)
Circuit-breaker	— Loss	— Primary failure — Fire (equipment room)	— Failure (supp. computer 1) *and* Failure (supp. computer 2)
Transformer 1	— Primary failure		— Failure (supp. computer 1)
Transformer 2	— Primary failure		— Failure (supp. computer 2)

from which we can automatically derive the following 'production rules':

If loss (supply system) *Then* loss (source)
If breakage (supp. line) *Then* loss (source)
If loss (source) *Then* failure (supp. computer 1) *and* failure (supp. computer 2)
If primary failure (circuit-breaker) *Then* loss (circuit-breaker)
If fire (equipment room) *Then* loss (circuit-breaker)
If loss (circuit-breaker) *Then* failure (supp. computer 1) *and* failure (supp. computer 2)
If primary failure (transformer 1) *Then* failure (supp. computer 1)
If primary failure (transformer 2) *Then* failure (supp. computer 2)

Note that each of the facts appearing in the rules is of the form $R(\alpha)$ where α represents a component or a function and where R represents a property satisfied by that component or that function (in general a failure mode).

In fact, FCP analysis goes further than in Table 2; it not only includes, as a reminder, the analysis of the components 'supply system', 'supp. cable' and 'equipment room', but also analyses the functions 'supp. both computers', 'supp. computer 1' and 'supp. computer 2'.

Table 3.

Components	Failure modes	Possible causes of a failure	Effects on the system
Supply system	— Loss		— Loss (source)
Supp. line	— Breakage		— Loss (source)
Local	— Fire		— Loss (circuit-breaker)

Table 4.

Functions	Failure modes	Possible causes of a failure	Effects on the system
Supp. both computers	Failure	— { Failure (supp. computer 1) Failure (supp. computer 2) }	
Supp. computer 1	Failure	— Primary failure (transfo. 1) — Loss (source) — Loss (circuit-breaker)	*If* Failure (supp. computer 2) *Then* Failure (supp. both computers)
Supp. computer 2	Failure	— Primary failure (transfo. 2) — Loss (source) — Loss (circuit-breaker)	*If* Failure (supp. computer 1) *Then* Failure (supp. both computers)

Table 4 raises a number of points.

First, one should not be deterred by its size. In fact, most of the facts listed are only given for control purposes and can be produced automatically by computer. We shall return to this point in section 3. The brace in the 'causes of failure' column is equivalent to a logical AND; thus the corresponding production rule is, in this case

If failure (supp. computer 1)
and failure (supp. computer 2)
Then failure (supp. both computers)

The *If Then* in the 'effects on the system' column also corresponds to a logical AND. The *If* indicates that there has to be a failure combination with the line failure already taken into account in order to bring about the conclusion indicated by the *Then*. The corresponding production rules are in this case identical to the rule described above.

3. USE OF FCP ANALYSIS

Let us return to the relative unwieldiness of Table 4.

There is in fact only one new rule: '*If* failure (supp. computer 1) and failure (supp. computer 2) *Then* failure (supp. both computers)'. The other rules correspond to the redundancies of Table 1.

This apparent cumbersomeness involving the replication of at least each rule has a considerable advantage in practice, as it ensures data reliability. In order to be validated, each rule of the form '*If* $R(\alpha)$ *Then* $P(\beta)$' must be introduced twice into the FCP analysis, once at function–component–parameter level α with R as failure mode and $P(\beta)$ as the effect on the system, and secondly at function–component–parameter level β with P as failure mode and $R(\alpha)$ as possible cause of failure.

More complex rules such as '*If* $R_1(\alpha_1)$ and $R_2(\alpha_2)$ *Then* $P(\beta)$' must be introduced at least three times.

Computer checking of the data can therefore be performed automatically.

3.1. Fault tree

FCP analysis therefore provides a knowledge base consisting of rules for logically generating propositions [Gondran, 1983]. The fault tree associated with an undesirable event is therefore merely the AND-OR tree having this event as its root [Gondran, 1983; Laurière, 1983].

Automated construction of this fault tree is therefore possible. For the previous example we obtain the fault tree of Fig. 2.

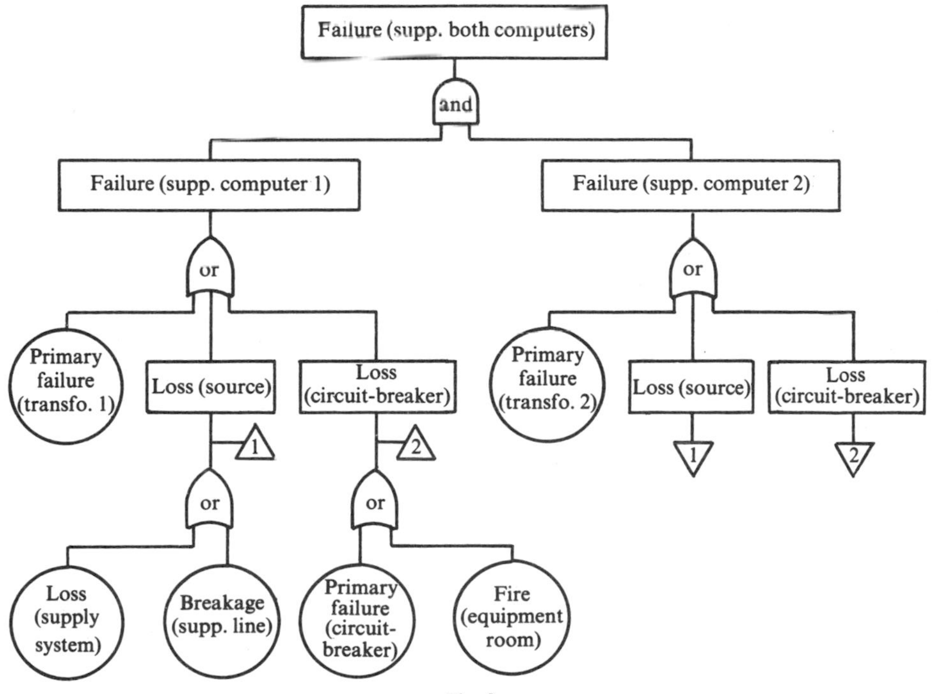

Fig. 2.

3.2. Failure simulation

The resultant knowledge base can be used very simply for simulation purposes. It provides a direct answer to questions of the following type: if a particular fault occurs, what happens to the system?

3.3. Assisted diagnosis and control

The fifth column of an FMEA lists the failure detection modes. We can once again formalize this column which corresponds to the system *data* which is available.

It seems that three types of detection procedure can be identified. These relate to the way the data are acquired: alarms, tests or routine examinations, inspections. Formal modelling of these detection modes is very important as it should provide a means of aiding diagnosis by using the FCP knowledge base.

The sixth column of an FMEA corresponds to the action modes affecting the system. Once again we can attempt to formalize this column which corresponds to possible action taken with respect to the system.

Formal modelling of these data and these actions is currently under development [Gondran and Laleuf, 1985]. Note that an analysis of data and action anomalies should be added to the above analysis.

4. EXAMPLE

In this section we describe the system to which FCP analysis will be applied in section 5.

4.1. Function of the system

This system is used to cool nitric acid before it is fed into a chemical reactor. The output temperature of the nitric acid is monitored by a control valve in the cooling water circuit. A protection system enables the nitric acid supply to the reactor to be cut off if the water flow rate is inadequate. *An excessively large increase in the output temperature T of the nitric acid may cause an explosion in the chemical reactor and therefore constitutes the undesirable event.*

4.2. Description

The system under study can be divided into five subsystems and one 'system' incorporating the environmental effects. This division is shown schematically in Fig. 4.

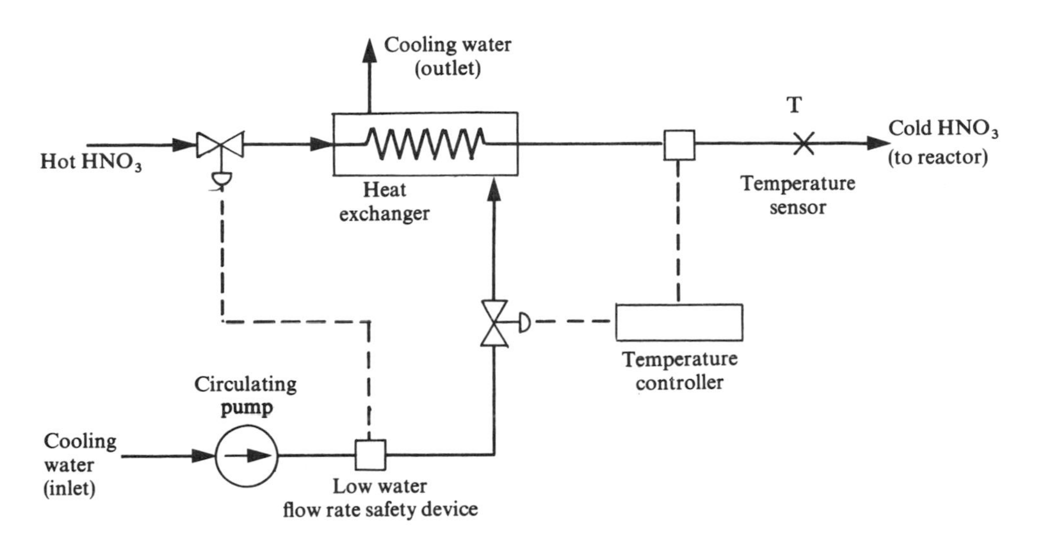

Fig. 3. *System representation.*

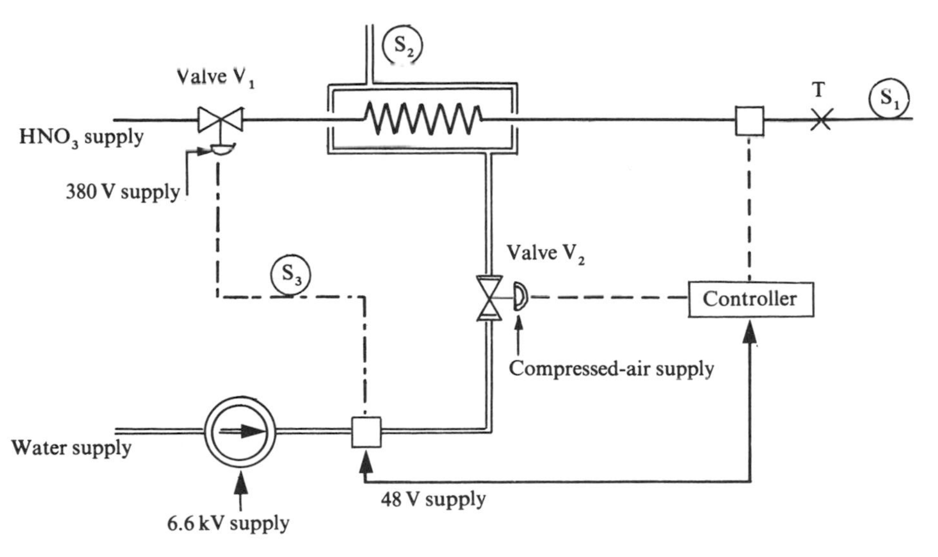

S_1 HNO_3 circuit
S_2 Water circuit
S_3 Low water flow rate safety device

Fig. 4. *System broken down into subsystems.*

4.2.1. *HNO_3 circuit (subsystem 1)*

(a) *Description*

This circuit comprises:

—all-or-nothing valve V_1 (mechanical part and motorization);
—the heat exchanger tubes. The heat exchanger is of the contraflow type with double-wall tubes. The nitric acid flows in the tubes at a lower pressure than the cooling water.

For the sake of simplification, we have excluded the pipework analysis.

(b) *Hypothesis and modelling*

—Valve V_1 closes in response to an emergency signal indicating low water flow rate.
—Valve V_1 remains in the same position on loss of the 380 V supply and closes on loss of the 48 V supply.
—A leakage of water into the nitric acid is possible and would cause a marked increase in T.

4.2.2. *Water circuit (subsystem 2)*

(a) *Description*

This circuit comprises:
—the motor-driven circulating pump supplied at 6.6 kV;
—the pneumatic control valve V_2 (mechanical part and motorization);
—the grille section of the exchanger.

(b) *Hypothesis and modelling*

—Valve V_2 is controlled by the controller so as to maintain a constant output temperature T.
—Valve V_2 closes on lack of compressed air or loss of 48 V supply.
—The tolerances on the output temperature T of the nitric acid are such that slight variations in water flow rate (i.e. if the flow rate remains above the low-flow-rate threshold) have no effect on the reactor located downstream of the system in question. We shall include small leaks (e.g. packing leakage) from valve V_2, the exchanger and the pump or reduced performance of the latter in this category.

4.2.3. *Low water flow rate safety device ('pump safety device') (subsystem 3)*

(a) *Description*

This subsystem comprises:

—the flow-rate sensor and associated instrumentation, supplied at 48 V;
—the transmission system causing valve V_1 to close if the water flow rate falls below a certain threshold; supplied at 48 V.

(b) *Hypothesis and modelling*

—On loss of the 48 V supply, the transmission system closes valve V_1.

—If the water flow rate remains above the threshold and if the nitric acid conditions upstream of the system are 'normal' (for precise definition: cf. section 4.2.5 on subsystem 5).

4.2.4. *Control unit (subsystem 4)*

(a) *Description*

This subsystem comprises:

—the temperature sensor which records the temperature T and transmits an electrical signal to the controller;
—the controller (supplied at 48 V) which controls the opening and closing of valve V_2.

(b) *Hypothesis and modelling*

—On loss of the 48 V supply, the control unit closes valve V_2.
—In the course of monthly maintenance, the controller is switched to manual and the operator then has to select the valve setting himself (precautions are taken to ensure that no compensatable increase in the HNO_3 conditions occurs during maintenance).

4.2.5. *Auxiliary systems (subsystem 5)*

(a) *Description*

There are six auxiliary systems:

—compressed air supply of valve V_2;
—6.6 kV power supply of motor-driven circulating pump;
—380 V power supply of valve V_1;
—48 V supply of low water flow rate transmission system and the control unit;
—water supply to water circuit, ensuring adequate circulating pump suction pressure;
—nitric acid supply destined for the chemical reactor.

(b) *Hypothesis and modelling*

—The effects of losing the electric power or compressed air supplies have been described in connection with each subsystem.
—A significant drop in pump suction pressure causes a marked reduction in the water flow rate.
—The operation of the nitric acid supply can be broken down into three types of situation, namely:

- slight variations in nitric acid temperature and flow rate (normal conditions). In this case, there can be no significant increase in T unless the low water flow rate threshold is reached;
- compensatable variations in nitric acid temperature and flow rate. In this case, proper regulating action is necessary to prevent a large increase in T;
- large (non-compensatable) variations in nitric acid temperature and flow rate. If these occur, even maximum water flow rate cannot prevent a substantial increase in T.

—Loss of the 380 V supply triggers an alarm. We shall disregard failure of the operator to acknowledge this alarm.

4.2.6. *Environmental effects (subsystem 6)*

(a) *Description*

Fire is the only environmental effect taken into account. A fire may occur either in the exchanger room or in the measurement room housing the 48 V busbar and the transmission system of the pump safety device.

(b) *Hypothesis and modelling*

—A fire in the exchanger room results in a significant increase in *T*.
—A fire in the measurement room results in the loss of the pump safety device and 48 V busbar.

5. FCP ANALYSIS OF SYSTEM

Using this analysis, we formally deduce the fault tree of the undesirable event: a large increase in *T*. The results are presented in Tables 5, 6 and 7 and Fig. 5.

Table 5.

Components	Failure modes	Causes of failures	*If*	Effects on system
Valve V_1	Failure to close	Loss (380 V supp.) Jamming		Failure (Protection system)
Exchanger	— Internal leak — External leak			— Large increase (T) — Undetectable failure (Cooling system)
	— Blockage, water side			— Detectable failure (Cooling system)
Motor-driven pump	Unwanted shutdown	Loss (6.6 kV supp.) Primary failure		— Detectable failure (Cooling system)
Valve V_2	— Unwanted closure	Loss (Compressed air)		— Detectable failure (Cooling system)
		Loss (48 V supp.)		
	— Jamming in same position — External leak			— Loss (Control unit) — Undetectable failure (Cooling system)
Flow meter	Failure	Loss (48 V supp.) Primary failure		Failure (protection system)
Transmission system	Failure to operate	Jamming Fire (Measurement room)		Failure (protection system)
T sensor	Incorrect measurement			Loss (control)
Controller	Failure	Control error (Operator) Primary failure		Loss (control)

Table 5 *(cont.)*

Components	Failure modes	Causes of failures	*If*	Effects on system
Compressed air	Loss			Unwanted closure (Valve V_2)
6.6 kV supp.	Loss			Unwanted shutdown (Motor-driven pump)
380 V supp.	Loss			Failure to close (Valve V_1)
48 V supp.	Loss			— Unwanted closure (Valve V_2) — Failure (Flow meter)
HNO_3 supp.	Compensatable variation		— Detectable failure (Cooling system) — Monitoring loss (Cooling system)	— Large increase (T) — Large increase (T)
	Non-compensatable variation			Large increase (T)
Exchanger room	Fire			Large increase (T)
Measuring room	Fire			Failure to operate (Transmission system)
Operator	Control error			Failure (Controller)

Table 6.

Functions	Failure modes	Causes of failures	*If*	Effects on system
Protection system	Failure	— Failure to close (Valve V_1) — Failure to operate (Transmission system)	— Detectable failure (Cooling)	Large increase (T)
		— Failure (Flow meter)	— Monitoring loss (Cooling system)	Monitoring loss (T)
Cooling system	— Undetectable failure	— External leak (Exchanger) — External leak (Valve V_2)		Large increase (T)
	— Detectable failure	— Blockage, water side (Exchanger)	— Failure (Protection system)	Large increase (T)
		— Unwanted shutdown (Motor-driven pump) — Closure (Valve V_2)	— Compensatable variation (HNO_3 supp.)	Large increase (T)
	— Monitoring loss	Loss (Control unit)	— Compensatable variation (HNO_3 supp.)	Large increase (T)
Control unit	Loss	— Incorrect measurement (T sensor) — Failure (Controller) — Jamming in position (Valve V_2)		Monitoring loss (Cooling system)

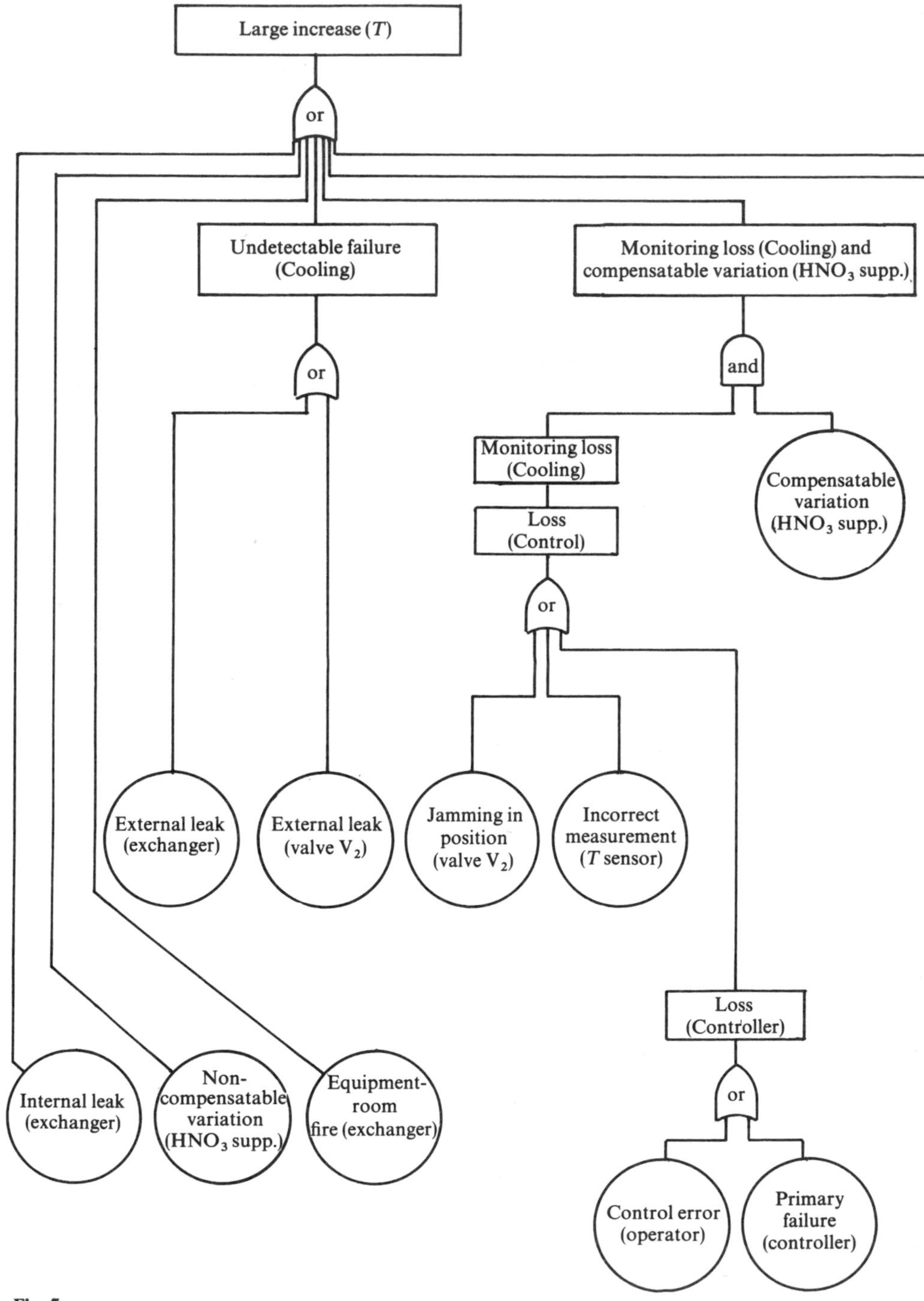

Large increase (T)
or
Undetectable failure (Cooling)
Monitoring loss (Cooling) and compensatable variation (HNO3 supp.)
or
and
Monitoring loss (Cooling)
Compensatable variation (HNO3 supp.)
Loss (Control)
or
External leak (exchanger)
External leak (valve V2)
Jamming in position (valve V2)
Incorrect measurement (T sensor)
Loss (Controller)
or
Internal leak (exchanger)
Non-compensatable variation (HNO3 supp.)
Equipment-room fire (exchanger)
Control error (operator)
Primary failure (controller)

Fig. 5.

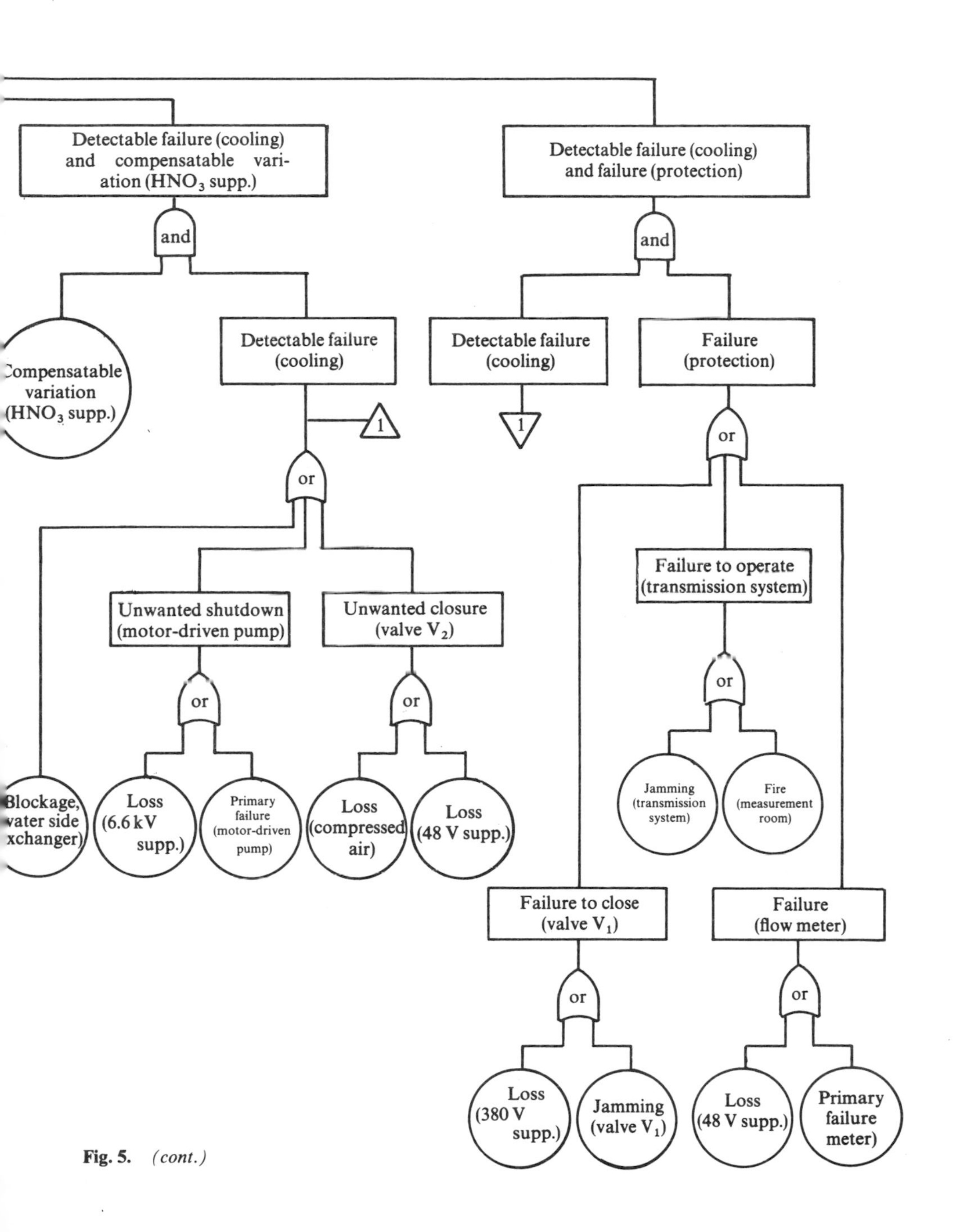

Detectable failure (cooling) and compensatable variation (HNO_3 supp.)
and
Compensatable variation (HNO_3 supp.)
Detectable failure (cooling)
1
or
Blockage, water side exchanger)
Unwanted shutdown (motor-driven pump)
or
Loss (6.6 kV supp.)
Primary failure (motor-driven pump)
Unwanted closure (valve V_2)
or
Loss (compressed air)
Loss (48 V supp.)
Detectable failure (cooling) and failure (protection)
and
Detectable failure (cooling)
1
Failure (protection)
or
Failure to operate (transmission system)
or
Jamming (transmission system)
Fire (measurement room)
Failure to close (valve V_1)
or
Loss (380 V supp.)
Jamming (valve V_1)
Failure (flow meter)
or
Loss (48 V supp.)
Primary failure meter)

Fig. 5. *(cont.)*

Table 7.

Parameters	Failure modes	Causes of failure	*If*	Effects on system
T	— Large increase	— Internal leak (Exchanger) — Compensatable variation (HNO_3 supp.) — Monitoring loss (Cooling system) Compensatable variation (HNO_3 supp.) — Detectable failure (Cooling system) Compensatable variation (HNO_3 supp.) — Fire (Exchanger room) — Failure (Protection system) Detectable failure (Cooling system) — Undetectable failure (Cooling system)		
	— Monitoring failure	— Failure (Protection system) Monitoring loss (Cooling system)		

6. SUMMARY

The feasibility of *FCP analysis* appears established. Larger-scale examples should make it possible to improve and extend the modelling of data and actions.

FCP analysis is based on the methods of knowledge representation developed in the fields of *artificial intelligence* and *expert systems* [Gondran and Laleuf, 1985; Laurière, 1983]. It makes it possible to set up a *computerized knowledge base* whose structure enables consistency to be checked and validated.

This knowledge base can be used for automatically constructing the *fault tree* of any undesirable event defined on the basis of the failure modes of the system functions.

Apart from reliability studies, this knowledge base can be put to various uses:

— *as a diagnostic aid* (data modelling and utilization),
— *as a management aid* (action modelling and utilization),
— *to work out accident sequences* (failure simulation),
— *for computer-aided teaching* (interrogation of knowledge base).

On the other hand, the use of CAD data bases on systems should enable FCP analysis to be set up with computer assistance [Gondran, Héry and Laleuf, 1986].

Finally, it should be possible to incorporate experience-derived feedback to update the data base, thereby simultaneously updating systems reliability studies and other applications.

REFERENCES

ANCELIN J. (1981): *Le Code CAT d'analyse automatique de la fibailité des systèmes – Présentation des problèmes de modélisation — Modification du code*; EDF Note HT 13/31/81.

ANCELIN J., CHATY E. and CARRE M. (1981): *Notice d'utilisation du code GO d'analyse automatique de la fiabilité*; EDF Note HT 13/34/81.

APOSTOLAKIS G. E., SALEM J. L. and WU J. J. (1978): *CAT: A Computer code for the Automated Construction of Fault Trees*; EPRI Report NP-705-Project 297-1.

BARBET J. F., LLORY M., PORTAL R. and VILLEMEUR A. (1978): *Méthodologie utilisée pour l'analyse probabiliste des systèmes de sûreté nucléaire*; International Conference on Reliability and Maintainability, Paris.

BUSSOLINI J. J. (1971): High Reliability Design Techniques Applied to the Lunar Module; Lecture Series no. 47 on *Reliability on Avionics Systems*. A.G.A.R.D., Rome, 16–17 September 1971, London, 20–21 September 1971.

GONDRAN M. (1983): *Introduction aux Systèmes Experts*; Eyrolles, Paris.

GONDRAN M. and LALEUF J. C. (1983): *La représentation des connaissances dans les études de fiabilité*; FCP analysis, EDF Note HI 4600/02.

GONDRAN M. and LALEUF J. C. (1984): La représentation des connaisances dans les études de fiabilité; In: *Operational Research 84*, J. P. Braus (Ed.), North-Holland, pp. 112–123.

GONDRAN M. and LALEUF J. C. (1985): *Le diagnostic dans l'analyse FCP*; EDF Note, in preparation.

GONDRAN M., HÉRY J. F. and LALEUF J. C. (1986): "Système Expert et Fiabilité assistée par ordinateur"; 5th International Conference on Reliability and Maintainability, Biarritz.

IEC (1985): Analysis techniques for system reliability. Procedure for failure mode and effects analysis (FMEA); Publication 812, 1st edition. International Electrotechnical Commission.

IEEE (1975): *IEEE Guide for General Principles of Reliability Analysis of Nuclear Power Generating Station Protection Systems*; American National Standard ANSI N41.4-1976-IEEE Std 352–1975 (revision of IEEE Std 352–1972).

LAURIERE J. L. (1983): Knowledge representation and use. Part I: Expert Systems, Part II: Representations; *Technology and Science of Informatics*, **1**, nos 1 and 2, pp 9–26 and 79–102.

LIEVENS C. (1976): *Sécurité des systèmes*; Cepadues Editions, Toulouse.

LLORY M. (1980): *L'analyse des modes de défaillance et de leurs effets*; Document de travail de la Commission Electrotechnique Internationale, EDF Note HT 13/63/80.

LLORY M. and VILLEMEUR A. (1980): *La méthode des combinaisons de pannes: présentation, application à un exemple simple de systèmes*; EDF Note HT 13/68/80.

VILLEMEUR A. (1982): *Une méthode d'analyse de la fiabilité et de la sécurité des systèmes: la méthode des combinaisons de pannes (M.C.P.)*; EDF Note HT 13/55/81.

YAMADA K. (1977): Reliability Activities at Toyota Motor Company; *Rep. Stat. Appl. Res., JUSE*, **24**, no. 3.

CHAPTER 3

DIRECT ANALYSIS OF SIMPLE SYSTEMS

In this chapter, we will examine small systems which are important either because they will be used in subsequent chapters as the building blocks of complex systems, or because they are frequently encountered in practice and it is comparatively easy to compute their reliability or availability.

We therefore hope to show that, in a number of important practical cases, a few simple calculations will provide the engineer with solutions to his problems.

1. FAILURE AND REPAIR RATE OF A SINGLE-COMPONENT SYSTEM. COMPUTING AVAILABILITY OF THE MARKOVIAN CASE

1.1. Failure rate and uptime distribution

Consider a system S consisting of a single component whose uptime distribution has a cumulative distribution function F. As system failure occurs when the component fails, then:

$$R(t) = 1 - F(t). \tag{1}$$

If the component is not repairable, the availability $A(t)$ is also given by Eqn (1). On the other hand, from Eqn (75) (Ch. 1):

$$\text{MTTF} = \int_0^\infty [1 - F(t)]\,\mathrm{d}t. \tag{2}$$

From Eqn (7) (Ch. 1), the failure rate $\lambda(t)$ of this component is therefore:

$$\lambda(t) = \frac{\frac{\mathrm{d}F}{\mathrm{d}t}(t)}{1 - F(t)}. \tag{3}$$

Statistical analysis of this function $\lambda(t)$ has shown that a number of components have a failure rate of the following form:

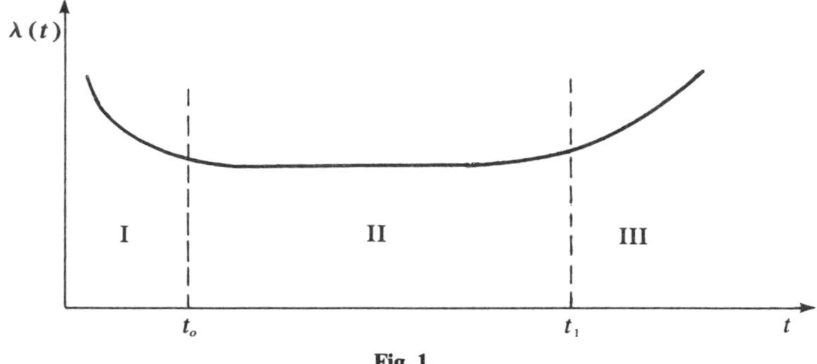

Fig. 1.

This so-called '*bathtub curve*' has three distinct regions:

—region I ($t < t_0$) is the *early failure period* of the component during which the failure rate rapidly decreases;
—region II ($t_0 \leqslant t \leqslant t_1$) is the *useful life period* of the component during which the failure rate is virtually constant. Failures occurring during this period are termed accidental;
—region III ($t > t_1$) is the *wear-out failure period* during which the failure rate increases rapidly.

During the component's useful life period, as the failure rate λ is constant, the distribution function can therefore be written:

$$F(t) = 1 - e^{-\lambda t}.$$

The probability density function of the uptime exists and is given by:

$$f(t) = \frac{dF(t)}{dt} = \lambda\, e^{-\lambda t}$$

this being the density of an exponential distribution with parameter λ.
It follows that:

$$R(t) = e^{-\lambda t} \tag{4}$$

$$\text{MTTF} = \int_0^\infty e^{-\lambda t} = \frac{1}{\lambda}. \tag{5}$$

1.2. Repair rate and downtime distribution

Let us now assume that the system has failed and that repair commences immediately. Let us also assume that the repair time distribution has a distribution function $G(t)$.

The system maintainability $M(t)$ is therefore, by definition:

$$M(t) = G(t) \tag{6}$$

hence, from Eqn (76) (Ch. 1):

$$\text{MTTR} = \int_0^\infty [1 - G(t)]\,\mathrm{d}t. \tag{7}$$

From Eqn (8) (Ch. 1), the repair rate $\mu(t)$, if it exists, of this component is therefore given by:

$$\mu(t) = \frac{\frac{\mathrm{d}G}{\mathrm{d}t}(t)}{1 - G(t)}. \tag{8}$$

In the case where the repair rate $\mu(t)$ is a constant μ, it follows that:

$$M(t) = G(t) = 1 - \mathrm{e}^{\mu t}$$

and the MTTR $= \mu^{-1}$ which will often be denoted by τ (mean repair time of the component).

1.3. Computing the availability of the component when the repair and failure rates are constant

The component is then said to be *Markovian*: the state of the component at a given instant provides sufficient information as to its entire future life. In fact, let us assume that at time t the component is working; if t_0 is the last time the component was restored to an operable condition, the probability of it failing between t and $t + \Delta t$ is equal to $\lambda(t - t_0)\Delta t + 0(\Delta t)$. However, in the case of a Markovian component, this function does not depend on t_0, since $\lambda(t - t_0)\Delta t = \lambda\Delta t$. The same applies if the component is in a failed state at time t. The probability of the end of the repair occurring between t and $t + \Delta t$ is

$$\mu\,\Delta t + 0(\Delta t).$$

Let us try to find a differential equation satisfied by the availability $A(t)$ of a Markovian component.

The total probability theorem (Eqn (11), Ch. 1) enables us to write:

$$\begin{aligned} A(t + \Delta t) = {} & \mathscr{P}\,(\text{component operating at } t \text{ and not failing between } t \text{ and } t + \Delta t) \\ & + \mathscr{P}\,(\text{component out of order at } t \text{ and was repaired between } t \text{ and } t + \Delta t) \end{aligned} \tag{9}$$

As the probability of being in a failed state at time t (i.e. of being unavailable) is equal to $1 - A(t)$, in the case of a Markovian component, expression (9) can therefore be written:

$$A(t + \Delta t) = A(t)(1 - \lambda\,\Delta t - 0(\Delta t)) + (1 - A(t))(\mu\,\Delta t + 0(\Delta t))$$

consequently, dividing by Δt and letting Δt tend to 0:

$$\frac{\mathrm{d}A}{\mathrm{d}t}(t) = \mu - (\lambda + \mu)\,A(t).$$

By applying the Laplace transform to both members of the equation using (Eqn (45), Ch. 1) we obtain:

$$s\bar{A}(s) - A(0) = \frac{\mu}{s} - (\lambda + \mu)\,\bar{A}(s)$$

therefore:

$$\bar{A}(s) = \frac{\mu + A(0)\, s}{s(s + \lambda + \mu)} \tag{10}$$

$$\bar{A}(s) = \frac{\mu}{\lambda + \mu}\,\frac{1}{s} - \frac{\mu(1 - A(0)) - A(0)\,\lambda}{\lambda + \mu}\,\frac{1}{s + \lambda + \mu}.$$

By inversion, we obtain:

$$A(t) = \frac{\mu}{\lambda + \mu} - \frac{\mu(1 - A(0)) - \lambda A(0)}{\lambda + \mu}\,e^{-(\lambda+\mu)t}.$$

When $A(0) = 1$, we have the well-known formula

$$A(t) = \frac{\mu}{\lambda + \mu} + \frac{\lambda}{\lambda + \mu}\,e^{-(\lambda+\mu)t} \qquad \text{(curve 1 in Fig. 2)} \tag{11}$$

When $A(0) = 0$

$$A(t) = \frac{\mu}{\lambda + \mu}(1 - e^{-(\lambda+\mu)t}) \qquad \text{(curve 2 of Fig. 2).} \tag{12}$$

When $A(0) = \frac{\mu}{\lambda + \mu}$

$$A(t) = A(0)$$

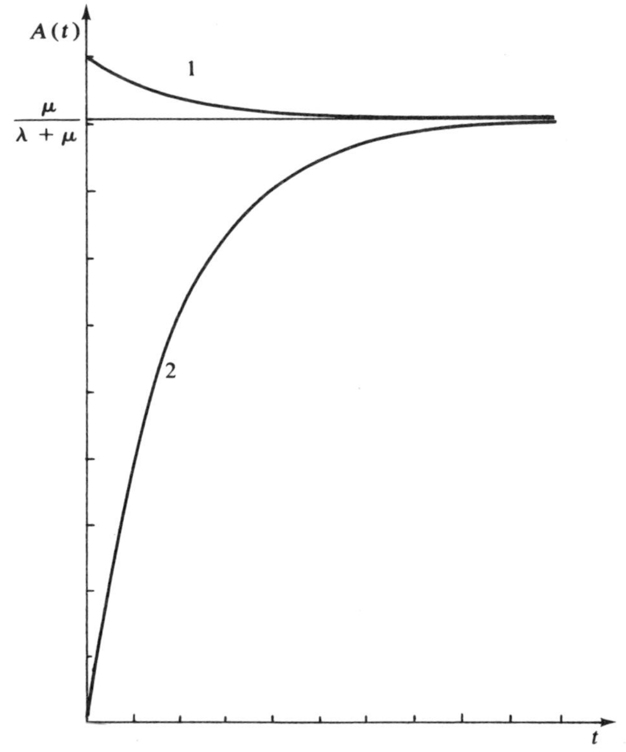

Fig. 2.

On the other hand, irrespective of the value of $A(0)$:

$$\lim_{t\to\infty} A(t) = \frac{\mu}{\lambda + \mu} = \frac{1}{1 + \lambda\tau} \quad \text{with} \quad \tau = \frac{1}{\mu} = \int_0^\infty t\,\mathrm{d}G(t)$$

We will examine the general case of computing the availability of a component in section 3.

2. RELIABILITY OF SOME SIMPLE NON-REPAIRABLE SYSTEMS

2.1. Components in series

An n-component system is termed a *series configuration* system if the failure of any one of the n-components results in system failure.

The reliability block diagram of a system of this type is therefore as shown below:

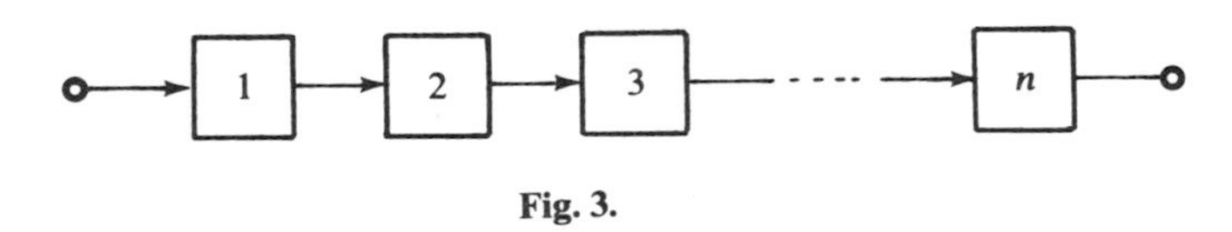

Fig. 3.

If we denote by $M_i(t)$ the event 'component i is operating at time t', the reliability $R(t)$ of the system can be written:

$$R(t) = \mathcal{P}(M_1(t) \cap M_2(t) \cap \ldots \cap M_n(t))$$

therefore, from Eqn (7) (Ch. 1):

$$R(t) = \mathcal{P}(M_1(t)) \,.\, \mathcal{P}(M_2(t)/M_1(t)) \ldots \mathcal{P}(M_n(t)/M_1(t) \cap \ldots \cap M_{n-1}(t))\,.$$

When the events $M_i(t)$ are independent:

$$R(t) = \prod_{i=1}^{n} \mathcal{P}(M_i(t))\,. \tag{13}$$

However, it is also possible to consider the events $P_i(t)$: 'component i is failed at time t'. The reliability $R(t)$ may then be written in the form:

$$R(t) = 1 - \mathcal{P}(P_1(t) \cup P_2(t) \cup \ldots \cup P_n(t))\,,$$

a formula which can be expanded using Poincaré's theorem. It is a matter of finding the most suitable approach. In this case, the first approach is the more useful.

Let us assume that the components are independent. Putting:

$$R_i(t) = \mathcal{P}(M_i(t)) = \exp\left(-\int_0^t \lambda_i(u)\,\mathrm{d}u\right),$$

Eqn (13) can then be written as

$$R(t) = \exp\left(- \int_0^t \left(\sum_{i=1}^{n} \lambda_i(u) \right) du \right).$$

The system failure rate $\Lambda(t)$ is therefore the *sum* of the component failure rates

$$\Lambda(t) = \sum_{i=1}^{n} \lambda_i(t). \tag{14}$$

In the case where the λ_i are constant, we therefore obtain:

$$\text{MTTF} = \frac{1}{\sum_{i=1}^{n} \lambda_i}. \tag{15}$$

2.2. Components in parallel

A system of n components is termed a *parallel configuration* system if all the components have to fail for the entire system to fail. The reliability block diagram of such a system is given below:

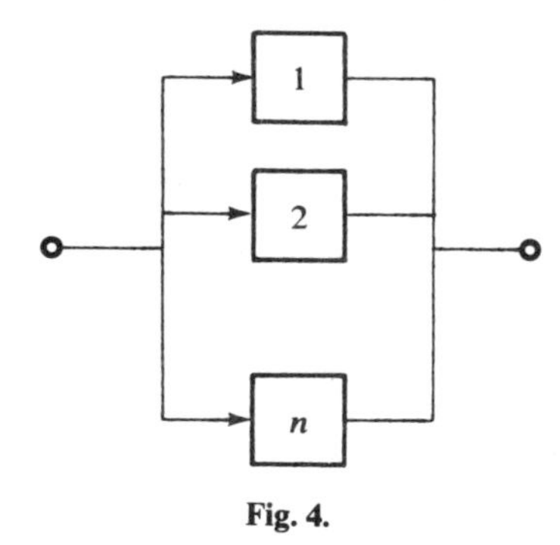

Fig. 4.

Adopting the notations used in the previous section, we therefore have:

$$R(t) = \mathcal{P}(M_1(t) \cup M_2(t) \cup \ldots \cup M_n(t))$$

or

$$R(t) = 1 - \mathcal{P}(P_1(t) \cap P_2(t) \cap \ldots \cap P_n(t)). \tag{16}$$

In this case, the second formulation is the simpler and, when events $P_i(t)$ are independent we obtain:

$$R(t) = 1 - \prod_{i=1}^{n} \mathcal{P}(P_i(t)). \tag{17}$$

If, in addition, all the components are in *continuous operation*, the reliability $R(t)$ of the system is given by:

$$R(t) = 1 - \prod_{i=1}^{n} \left(1 - \exp\left(-\int_0^t \lambda_i(u)\,\mathrm{d}u\right)\right) \tag{18}$$

consequently:

$$R(t) = \sum_{i=1}^{n} \exp\left(-\int_0^t \lambda_i(u)\,\mathrm{d}u\right) - \sum_{i=1}^{n} \sum_{j \neq i} \exp\left(-\int_0^t (\lambda_i(u) + \lambda_j(u))\,\mathrm{d}u\right) + \\ + \sum_{i=1}^{n} \sum_{j \neq i} \sum_{\substack{k \neq j \\ k \neq i}} + \cdots + (-1)^{n+1} \exp\left(-\sum_{i=1}^{n} \int_0^t \lambda_i(u)\,\mathrm{d}u\right). \tag{19}$$

For example, when $n = 2$, we obtain:

$$R(t) = \exp\left(-\int_0^t \lambda_1(u)\,\mathrm{d}u\right) + \exp\left(-\int_0^t \lambda_2(u)\,\mathrm{d}u\right) \\ - \exp\left(-\int_0^t (\lambda_1(u) + \lambda_2(u))\,\mathrm{d}u\right)$$

and when the components forming the system have constant failure rates:

$$R(t) = \mathrm{e}^{-\lambda_1 t} + \mathrm{e}^{-\lambda_2 t} - \mathrm{e}^{-(\lambda_1 + \lambda_2)t}.$$

Let us calculate the system failure rate $\Lambda(t)$ in the case where the component failure rates are constant.

Using Eqn (7) (Ch. 1) and Eqn (18), we obtain:

$$\Lambda(t) = \frac{1}{R(t)} \sum_{i=1}^{n} \lambda_i\, \mathrm{e}^{-\lambda_i t} \prod_{j=1,n}^{j \neq i} (1 - \mathrm{e}^{-\lambda_j t})$$

therefore, when $t \to 0$, for $n \geqslant 2$ we obtain:

$$\Lambda(t) \sim n\left(\prod_{i=1}^{n} \lambda_i\right) t^{n-1}.$$

When $t \to \infty$, Eqn (19) shows that

$$R(t) \sim \exp\left(-\left(\min_i \lambda_i\right) t\right)$$

consequently:

$$\Lambda(\infty) = \min_i \lambda_i.$$

Equation (19) additionally provides a simple means of calculating the system MTTF by integration

$$\mathrm{MTTF} = \sum_{i=1}^{n} \frac{1}{\lambda_i} - \sum_i \sum_{j \neq i} \frac{1}{\lambda_i + \lambda_j} + \sum_i \sum_{j \neq i} \sum_{\substack{k \neq j \\ k \neq i}} \frac{1}{\lambda_i + \lambda_j + \lambda_k} \cdots + \\ + (-1)^{n+1} \frac{1}{\sum_{i=1}^{n} \lambda_i}. \tag{20}$$

In the case where all the components have the same failure rate λ, Eqn (20) reduces to (cf. Exercise 5):

$$\text{MTTF} = \frac{\sum_{i=1}^{n} \frac{1}{i}}{\lambda} . \tag{21}$$

Using $r(t)$ to denote the reliability of this component, the system reliability $R(t)$ can then be written for the following different values of n:

$$n = 1 \quad R(t) = r(t)$$
$$n = 2 \quad R(t) = 2\,r(t) - r^2(t)$$
$$n = 3 \quad R(t) = 3\,r(t) - 3\,r^2(t) + r^3(t)$$
$$n = 4 \quad R(t) = 4\,r(t) - 6\,r^2(t) + 4\,r^3(t) - r^4(t) .$$

This mode of operation, known as *active redundancy*, is not the only possibility in this type of configuration. In fact, as only a single component has to be working for the system to be operating, the other components can be *on standby*. This type of *redundancy* is termed *passive*. In this case, it is necessary to activate at least one sound component when the operating component fails; however, the switching processes may also be the cause of failures.

In the simple case where the components are independent or where the switches

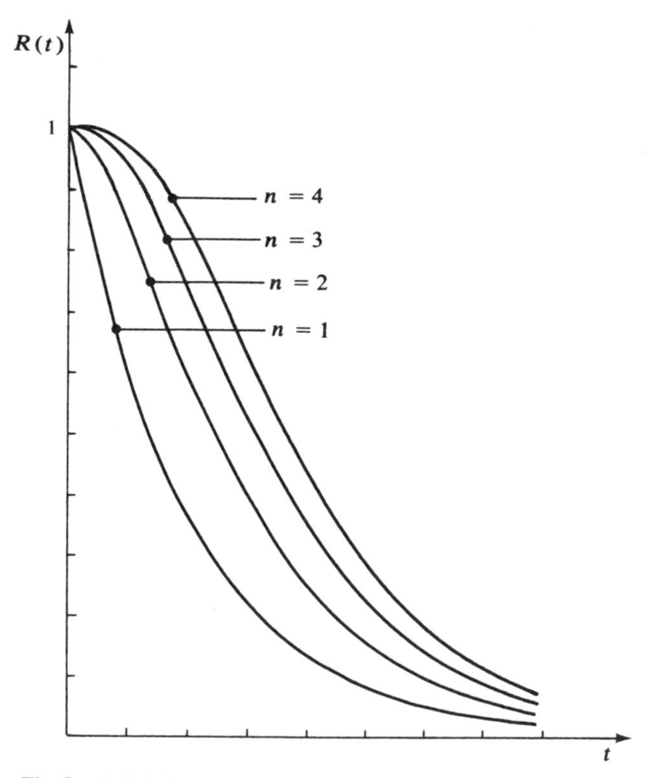

Fig. 5. *Reliability of n components in parallel (active redundancy).*

are perfectly reliable, system reliability can be easily determined. In fact, in this case we have a simple renewal process (cf. Ch. 1, section 6.3).

In addition, when all the components are identical and possess a constant failure rate, we have a Poisson process, hence:

$$R(t) = \sum_{k=0}^{n-1} \frac{(\lambda t)^k e^{-\lambda t}}{k!}. \tag{22}$$

In [Beliav et al., 1972], active and passive redundancies are examined in detail. We will merely note that, in the case of passive redundancies

$$R(s) = \frac{1}{\lambda}\sum_{k=0}^{n-1}\left(\frac{\lambda}{s+\lambda}\right)^{k+1}, \text{hence MTTF} = n/\lambda.$$

2.3. Mixed-configuration systems consisting of independent components

This refers to systems which are a combination of series and parallel configurations.

2.3.1. *Series-parallel configuration*

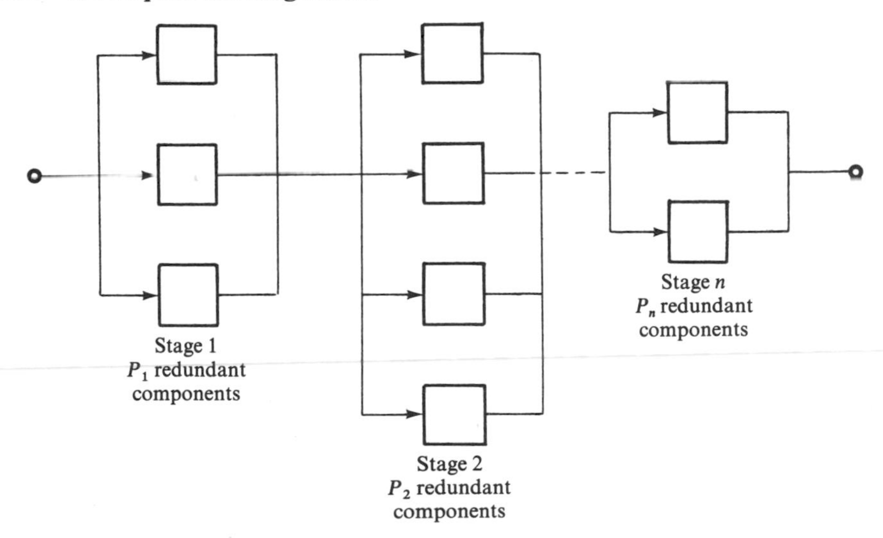

Fig. 6.

Reliability is the product of the reliabilities of each stage. In the case of active redundancy, we therefore obtain, using Eqns (13) and (17):

$$R(t) = \prod_{i=1}^{n}\left[1 - \prod_{j=1}^{p_i}(1 - R_{ij}(t))\right],$$

where $R_{ij}(t)$ represents the reliability of the jth component of stage i.

2.3.2. *Parallel-series configuration*

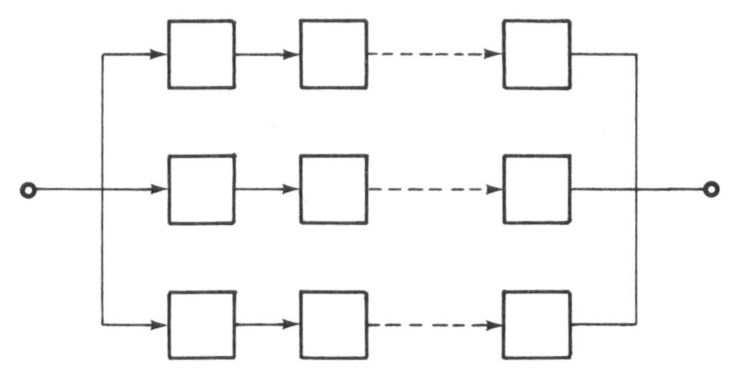

Fig. 7.

Each of the p branches contains n_i components in series. Assuming the redundancy to be active, using Eqns (13) and (17) we obtain:

$$R(t) = 1 - \prod_{i=1}^{p} \left(1 - \prod_{j=1}^{n_i} R_{ij}(t)\right),$$

where $R_{ij}(t)$ represents the reliability of the jth component of branch i.

2.4. Active redundancy system *r/n*

In this active redundancy configuration, the system will operate if at least r out of n components are working. There are two degenerate cases: $r = 1$ (parallel configuration) and $r = n$ (series configuration).

If we assume that the n components are identical, we have:

$$R(t) = \mathscr{P}\,(\text{at least } r \text{ out of } n \text{ components are operating})$$

The number of components operating has a binomial distribution with parameters n and $r(t)$ (reliability of a component). Therefore:

$$R(t) = \sum_{k=r}^{n} C_n^k\, r(t)^k\, (1 - r(t))^{n-k}. \tag{23}$$

2.5. Systems which cannot be reduced to a mixed system

Consider the following classical example (bridge system) whose reliability block diagram is given below:

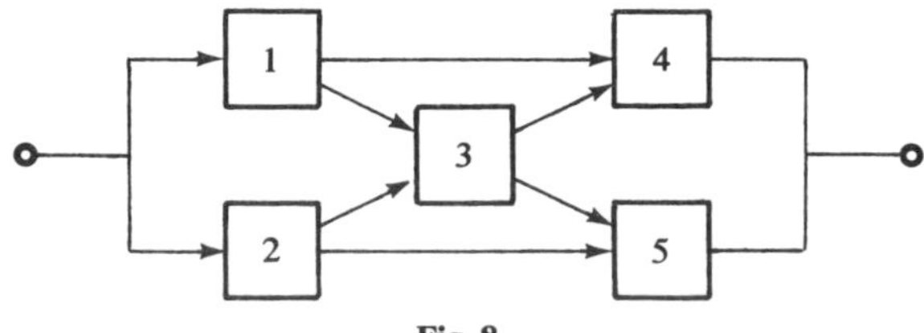

Fig. 8.

We now apply the total probability theorem (Eqn (11), Ch. 1), taking the two events 'component 3 working' and 'component 3 failed' as a set of complete events. These two events have probabilities of $R_3(t)$ and $1 - R_3(t)$ respectively. The system reliability is therefore given by:

$$R(t) = \mathscr{P}\,(S \text{ is operating during } [0, t]/\text{component 3 working at } t)R_3(t) + \\ + \mathscr{P}\,(S \text{ operating during } [0, t]/\text{component 3 failed at } t)(1 - R_3(t)).$$

When the components are independent, using obvious notations we obtain:

$R(t) = R$ [1 4 / 2 5] $R_3(t) + R$ [1 4 / 2 5] $(1 - R_3(t))$

The two remaining subsystems are of the mixed type. For more complicated cases, this theorem is applied as often as necessary.

3. DETERMINING THE AVAILABILITY OF A RENEWED COMPONENT (GENERAL CASE)

Consider a repairable component, the distribution of the uptime having a distribution function $F(t)$ and the repair time distribution having a distribution function $G(t)$.

3.1. Constant failure rate

Let us first examine the case where the failure rate λ is constant. This means that $F(t) = 1 - e^{-\lambda t}$.

We put $A(0) = A_0$ and assume that $G(t)$ has a density function $g(t)$. Equation (9) still holds. Therefore, in accordance with the failure rate definition (Eqn (3), Ch. 1):

$$A(t + \Delta t) = A(t)(1 - \lambda\,\Delta t + 0(\Delta t)) + (1 - A_0)\,g(t)\,\Delta t + \int_0^t g(u)\,\mathscr{P}\,(\text{component}$$

is operating at $t - u$ and fails between $t - u$ and $t - u + \Delta t)\,du$.

When $\Delta t \to 0$, we have the integro-differential equation satisfied by $A(t)$:

$$\frac{dA}{dt}(t) = -\lambda A(t) + (1 - A_0)\,g(t) + \lambda \int_0^t g(u)\,A(t - u)\,du. \qquad (24)$$

When the component is working at the initial time $t = 0\,(A_0 = 1)$, we obtain:

$$\frac{dA}{dt}(t) = -\lambda A(t) + \lambda g(t) * A(t).$$

Equation (24) can be solved using the Laplace transform:

$$s\bar{A}(s) - A_0 = -\lambda \bar{A}(s) + \lambda \bar{A}(s)\,\bar{g}(s) + (1 - A_0)\,\bar{g}(s)$$

giving:

$$\bar{A}(s) = \frac{A_0 + (1 - A_0)\,\bar{g}(s)}{s + \lambda - \lambda \bar{g}(s)}. \tag{25}$$

3.1.1. *System state equation*

In the case where the repair time distribution is exponential with parameter $\mu, g(t) = \mu e^{-\mu t}$ and $\bar{g}(s) = \dfrac{\mu}{\mu + s}$, we obviously have Eqn(10).

However, this is rarely a realistic assumption, and so we shall examine the case of constant repair time. We could also use an Erlangian distribution (cf. Exercise 3).

3.1.2. *Constant repair time*

Let us now assume, therefore, that the duration of the repair time is constant and equal to τ, i.e.:

$$g(t) = \delta(t - \tau)$$

and

$$\bar{g}(s) = e^{-s\tau}.$$

Equation (25) can then be written as:

$$\bar{A}(s) = \frac{A_0 + (1 - A_0)\,e^{-s\tau}}{s + \lambda - \lambda\, e^{-s\tau}}$$

$$\bar{A}(s) = \frac{A_0 + (1 - A_0)\,e^{-s\tau}}{(s + \lambda)\left(1 - \dfrac{\lambda\, e^{-s\tau}}{s + \lambda}\right)} = \frac{A_0 + (1 - A_0)\,e^{-s\tau}}{s + \lambda} \sum_{l=0}^{\infty} \left(\frac{\lambda\, e^{-s\tau}}{s + \lambda}\right)^l$$

$$\bar{A}(s) = A_0 \sum_{l=0}^{\infty} \frac{(\lambda\, e^{-s\tau})^l}{(\lambda + s)^{l+1}} + \left(\frac{1 - A_0}{\lambda}\right) \sum_{l=0}^{\infty} \left(\frac{\lambda\, e^{-s\tau}}{s + \lambda}\right)^{l+1}$$

hence, by inversion (from Eqn (54), Ch. 1):

$$A(t) = \frac{A_0}{\lambda} \sum_{l=0}^{\infty} \frac{\lambda^{l+1} (t - l\tau)^l\, e^{-\lambda(t - l\tau)}}{l\,!}\, Y(t - l\tau) + \\ + \frac{1 - A_0}{\lambda} \sum_{l=0}^{\infty} \frac{\lambda^{l+1} [t - (l+1)\,\tau]^l\, e^{-\lambda[t - (l+1)\tau]}}{l\,!}\, Y(t - (l+1)\,\tau)$$

or if:

$$N\tau \leq t < (N + 1)\,\tau$$

$$A(t) = A_0 \sum_{l=0}^{N} \frac{\lambda^l (t - l\tau)^l\, e^{-\lambda(t - l\tau)}}{l\,!} + \\ + (1 - A_0) \sum_{l=0}^{\max(0, N-1)} \frac{\lambda^l [t - (l+1)\,\tau]^l}{l\,!}\, e^{-\lambda[t - (l+1)\tau]}. \tag{26}$$

Figure 9 shows the curve $A(t)$ for $A_0 = 1$ as well as the line $y = A(\infty) = 1/(1 + \lambda\tau)$.

Note that this curve is continuous. It is differentiable everywhere except at the point $t = \tau$. Note also that the availability oscillates about its limit value $A(\infty)$.

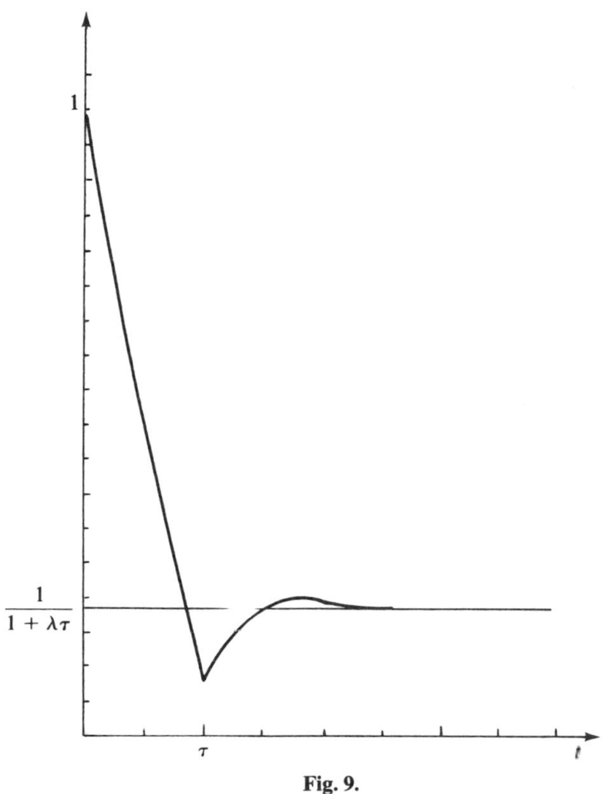

Fig. 9.

3.1.3. *Availability limit*

Observe that whatever the value of A_0, for the two special cases above we obtain:

$$\lim_{t\to\infty} A(t) = \frac{1}{1+\lambda\tau}.$$

Furthermore, the limit is reached after a few τ, if $\lambda\tau \ll 1$.

To show that this result is general, we will assume that the density of the repair time $g(t)$ is such that:

$$\int_0^\infty g(t)\,\mathrm{d}t = 1$$

$$\int_0^\infty tg(t)\,\mathrm{d}t = \tau < +\infty$$

By expanding e^{-st} in series, we therefore have:

$$\bar{g}(s) = \bar{g}(0) + s\frac{\mathrm{d}\bar{g}(0)}{\mathrm{d}s} + \frac{s^2}{2!}\frac{\mathrm{d}^2\bar{g}}{\mathrm{d}s^2}(0) + \cdots$$

Or (from Eqn (52), Ch. 1):

$$\bar{g}(s) = 1 - s\tau + 0(s^2). \tag{27}$$

It follows, using Eqn (55) (Ch. 1) and Eqn (25), that:

$$\lim_{t \to \infty} A(t) = \lim_{s \to 0} \frac{A_0 + (1 - A_0)(1 - s\tau)}{1 + \lambda \frac{s\tau}{s}}$$

hence

$$A(\infty) = \frac{1}{1 + \lambda\tau} \tag{28}$$

giving the unavailability limit:

$$1 - A(\infty) = \frac{\lambda\tau}{1 + \lambda\tau} \simeq \lambda\tau \quad \text{si } \lambda\tau \ll 1 .$$

3.1.4. *Permissible short-duration failures*

Let us assume that an unavailability of this component can be tolerated provided that it does not exceed a certain duration Δt_0. The component therefore has three states:

State 0: the component is up.
State 1: the component is down, the failure less than Δt_0.
State 2: the component is down, the failure lasting more than Δt_0.

Let $Q_0(t)$, $Q_1(t)$ and $Q_2(t)$ be the probabilities of being in states 0, 1 and 2 respectively, at time t.

At the moment of failure, assume that the component passes into states 1 and 2 proportionally to $\int_0^{\Delta t_0} g(t)\,dt$ and $\int_{\Delta t_0}^{\infty} g(t)\,dt$ respectively, $g(t)$ being the repair time distribution density.

Consequently:

$$Q_1(t + \Delta t) = Q_1(t) + \mathscr{P}\begin{pmatrix}\text{up at } t \text{ and failing} \\ \text{between } t \text{ and } t + \Delta t\end{pmatrix} \int_0^{\Delta t_0} g(t)\,dt$$

$$- \mathscr{P}\begin{pmatrix}\text{down at } t \text{ and repaired between } t \text{ and } t + \Delta t, \\ \text{downtime less than } \Delta t_0\end{pmatrix}.$$

Or putting:

$$K = \int_{\Delta t_0}^{\infty} g(u)\,du$$

$$Q_1(t + \Delta t) = (1 - K)\, Q_0(t)\, \lambda\, \Delta t + Q_1(t)$$

$$- \lambda\, \Delta t \int_{t - \Delta t_0}^{t} Y(u)\, Q_0(u)\, g(t - u)\,du .$$

We then put:

$$\tilde{g}(t) = g(t) \quad \text{if} \quad 0 \le t \le \Delta t_0$$
$$\tilde{g}(t) = 0 \quad \text{if} \quad t < 0 \text{ or } t > \Delta t_0 .$$

Then:

$$\frac{\mathrm{d}Q_1(t)}{\mathrm{d}t} = \lambda(1-K)\,Q_0(t) - \lambda Q_0(t) * \tilde{g}(t) .$$

Or, after applying the Laplace transform, assuming $Q_0(0) = 1$.

$$s\bar{Q}_1(s) = \lambda\,.\,(1-K)\,\bar{Q}_0(s) - \lambda\bar{Q}_0(s)\,\bar{\tilde{g}}(s) .$$

As far as state 0 is concerned, the fact that the failed state is divided into two states makes no difference: consequently, $Q_0(t)$ satisfies Eqn (24) and $Q_0(s)$ is given by Eqn (25). Therefore:

$$\bar{Q}_1(s) = \frac{\lambda}{s}\,\frac{1-K-\bar{\tilde{g}}(s)}{s+\lambda-\lambda\bar{g}(s)} .$$

The availability $A(t)$ is equal to $Q_0(t) + Q_1(t)$, hence:

$$\bar{A}(s) = \frac{s+\lambda-\lambda K-\lambda\bar{\tilde{g}}(s)}{s(s+\lambda-\lambda\bar{g}(s))} . \tag{29}$$

Let us assume that the repair time is exponentially distributed. In this case

$$g(t) = \mu\,\mathrm{e}^{-\mu t}, \quad K = \mathrm{e}^{-\mu\,\Delta t_0} .$$

Therefore:

$$\bar{g}(s) = \frac{\mu}{\mu+s}$$

$$\bar{\tilde{g}}(s) = \frac{\mu}{\mu+s} - \frac{\mu}{\mu+s}\,\mathrm{e}^{-(\mu+s)\Delta t_0} .$$

$\bar{A}(s)$ can then be written:

$$\bar{A}(s) = \frac{(s+\lambda-\lambda\,\mathrm{e}^{-\mu\,\Delta t_0})(\mu+s) - \lambda\mu + \lambda\mu\,\mathrm{e}^{-(\mu+s)\Delta t_0}}{s^2(s+\lambda+\mu)} .$$

By inversion, we obtain:

$$A(t) = 1 - \frac{\lambda\,\mathrm{e}^{-\mu\,\Delta t_0}}{\lambda+\mu}\left[1+\mu(t-t_1) - \frac{\lambda\,\mathrm{e}^{-(\lambda+\mu)t}+\mu\,\mathrm{e}^{-(\lambda+\mu)t_1}}{\lambda+\mu}\right] \tag{30}$$

with $t_1 = 0 \quad$ if $\quad t \le \Delta t_0$
$\quad\;\; t_1 = t - \Delta t_0 \quad$ if $\quad t > \Delta t_0$.

hence:

$$A(\infty) = 1 - \frac{\lambda\,\mathrm{e}^{-\mu\,\Delta t_0}}{1+\lambda\tau}(\tau+\Delta t_0) . \tag{31}$$

In the general case, i.e. for any repair time distribution, we obtain:

$$A(\infty) = 1 - \frac{\lambda \int_{\Delta t_0}^{\infty} tg(t)\,dt}{1 + \lambda\tau}. \tag{32}$$

Let us now calculate the failure rate of a component used in these circumstances. For this purpose, it is necessary to calculate the probabilities $P_0(t)$ and $P_1(t)$ of being in state 0 or state 1 respectively without having encountered state 2 during $[0, t]$. Mathematically, one calculates these probabilities by making state 2 *absorbing*, i.e. assuming that there is no subsequent transition from failed state 2 to the operating state. The relationship between $P_0(t)$ and $P_1(t)$ is the same as the relationship between $Q_0(t)$ and $Q_1(t)$ (Eqn (10)). Conversely, as far as $P_0(t)$ is concerned, Eqn (25) no longer holds and, assuming $P_0(0) = 1$, we have:

$$P_0(t + \Delta t) = \mathscr{P}\begin{pmatrix}\text{operating at } t \text{ and not failing}\\ \text{between } t \text{ and } t + \Delta t\end{pmatrix} + \mathscr{P}\begin{pmatrix}\text{down at } t \text{ and repair completed between } t \text{ and } t + \Delta t,\\ \text{downtime less than } \Delta t_0\end{pmatrix}$$

$$= P_0(t)(1 - \lambda\,\Delta t + 0(\Delta t)) + \int_{t-\Delta t_0}^{t} Y(u)\,\lambda\,\Delta t P_0(u)\,g(t-u)\,du$$

hence, when $\Delta t \to 0$:

$$\frac{dP_0(t)}{dt} = -\lambda P_0(t) + \lambda \int_0^t P_0(u)\,\bar{g}(t-u)\,du$$

and

$$\bar{P}_0(s) = \frac{1}{s + \lambda - \lambda\bar{\bar{g}}(s)}.$$

It follows that the Laplace transform of the reliability is:

$$\bar{R}(s) = \frac{s + \lambda(1 - K - \bar{\bar{g}}(s))}{s(s + \lambda - \lambda\bar{\bar{g}}(s))}.$$

Even where g is exponential, it is not easy to invert $\bar{R}(s)$ (cf. [Pagès, 1976] and Exercise 4). However, the MTTF is easily obtained:

$$\text{MTTF} = \bar{R}(0) = \frac{1 + \lambda \int_0^{\Delta t_0} tg(t)\,dt}{\lambda \int_{\Delta t_0}^{\infty} g(t)\,dt}$$

but

$$\int_0^{\Delta t_0} tg(t)\,dt < \int_0^{\infty} tg(t)\,dt = \tau$$

therefore if $\lambda\tau \ll 1$, we have:

$$\text{MTTF} \simeq \frac{1}{\lambda K}.$$

It can also be shown that

$$R(t) \simeq e^{-\lambda K t}.$$

Note that in the case where t_0 was zero (cf. sections 1.2.1–1.2.3) the relevant parameter was the mean of the repair time distribution; the same does not apply here; for example, if $g(t) = \delta(t - \tau)$ (constant repair time):

$$A(t) \text{ unchanged if } \Delta t_0 < \tau$$

$$A(t) = 1 \qquad \text{if } \Delta t_0 > \tau.$$

If the failure can be classified into several modes, each mode having a different repair time distribution, one can ignore the failures such that $g(t) = 0$ over $[\Delta t_0, \infty[$.

Let us now consider the case where the component is only repairable if the repair does not exceed a certain limit.

The availability of such a component is then equal to the reliability of the previous component and it behaves like a non-repairable component, unless $\int_0^{\Delta t_0} g(t)\, dt = 1$.

3.2. Non-constant failure rate

Let $F(t)$ be the distribution function of the uptimes and $G(t)$ the distribution function of the repair times.

We will assume that they possess probability density functions $f(t)$ and $g(t)$ respectively. The times of failure and restoration form an alternating renewal process (cf. Ch. 1, section 6.3).

We will use the following notation:

$A(t)$ Availability of the component.
$N(t)$ Number of operating periods, except for the first if the component is working at time $t = 0$. This is termed the *number of renewals*.
$H(t) = E[N(t)]$ *renewal function.*

$h(t) = \dfrac{dH(t)}{dt}$ *renewal density.*

3.2.1. *Determining the functions H(t), h(t) and A(t)*

In accordance with Eqns (68) and (70) (Ch. 1), when the component is operating at the initial time $t = 0$ ($f_1 = f, f_2 = g, H = H_1, A = \Pi_1$):

$$\overline{H}(s) = \frac{\overline{f}(s)}{s(1 - \overline{f}(s)\,\overline{g}(s))} \tag{33}$$

$$\overline{h}(s) = \frac{\overline{f}(s)}{1 - \overline{f}(s)\,\overline{g}(s)} \tag{34}$$

$$A(s) = \frac{1 - \overline{f}(s)}{s(1 - \overline{f}(s)\,\overline{g}(s))}. \tag{35}$$

And when the component is down at the initial time $t = 0$ ($f_1 = g, f_2 = f, H = H_2$, $A = 1 - \Pi_1$):

$$\overline{H}(s) = \frac{\overline{f}(s)\,\overline{g}(s)}{s(1 - \overline{f}(s)\,\overline{g}(s))} \tag{36}$$

$$\overline{h}(s) = \frac{\overline{f}(s)\,\overline{g}(s)}{1 - \overline{f}(s)\,\overline{g}(s)} \tag{37}$$

$$\bar{A}(s) = \frac{\bar{g}(s)(1-\bar{f}(s))}{s(1-\bar{f}(s)\bar{g}(s))}. \tag{38}$$

Note that in both cases

$$\bar{A}(s) = \frac{1-\bar{f}(s)}{s\bar{f}(s)}\bar{h}(s). \tag{39}$$

In the intermediate cases we obtain the desired result by linear combination of these two extreme cases, which implies that Eqn (39) still holds.

How do we calculate $A(t)$ and $h(t)$ from these equations?

Let us take as an example the case $A(0) = 1$.

We put:

$$K(t) = \int_0^t f(t-u)\,g(u)\,\mathrm{d}u \tag{40}$$

hence

$$\bar{K}(s) = \bar{f}(s)\,\bar{g}(s)$$

Equation (35) can then be written as:

$$A(t) = 1 - F(t) + \int_0^t A(t-u)\,K(u)\,\mathrm{d}u \tag{41}$$

and similarly, Eqn (34) can be written in the form:

$$h(t) = f(t) + \int_0^t h(t-u)\,K(u)\,\mathrm{d}u. \tag{42}$$

When $F(t)$ and $K(t)$ are zero for $t < 0$ and locally summable, the Volterra-type integral equation of the second kind (Eqn (41)) has a unique solution (cf. [Schwartz, 1961]) given by:

$$A(t) = 1 - F(t) + \sum_{l=1}^{\infty} (1-F(t)) * K^{*l}(t) \tag{43}$$

where $K^{*l}(t)$ represents the convolution product of l functions equal to $K(t)$.

Similarly, for $h(t)$ we obtain:

$$h(t) = f(t) + \sum_{l=1}^{\infty} f(t) * K^{*l}(t). \tag{44}$$

Using Eqns (41) and (42) it is possible to calculate $A(t)$ and $h(t)$ numerically. Note that when the component's failure rate is constant, $h(t) = \lambda A(t)$.

3.2.2. *Asymptotic system behaviour*

Assume that the following quantities are finite:

$$\text{MTTF} = \frac{1}{\lambda^*} = \int_0^\infty t f(t)\,\mathrm{d}t \tag{45}$$

$$\sigma_\lambda^2 = \int_0^\infty \left(t - \frac{1}{\lambda^*}\right)^2 f(t)\,\mathrm{d}t \tag{46}$$

$$\text{MTTR} = \tau = \int_0^\infty t g(t)\,\mathrm{d}t \tag{47}$$

$$\sigma_\tau^2 = \int_0^\infty (t-\tau)^2 \, g(t)\,\mathrm{d}t . \tag{48}$$

It can then be demonstrated, using Eqn (52) (Ch. 1), that:

$$\lim_{t\to\infty} h(t) = \frac{\lambda^*}{1+\lambda^* \tau} \tag{49}$$

$$H(t) \underset{t\to\infty}{\simeq} \frac{\lambda^* t}{1+\lambda^* \tau} \tag{50}$$

$$\lim_{t\to\infty} A(t) = \frac{1}{1+\lambda^* \tau} = \frac{\text{MTTF}}{\text{MTTF}+\text{MTTR}} . \tag{51}$$

To interpret Eqn (51), consider a time interval corresponding to n operating periods lasting t_i and n repair periods lasting $\bar{t}_i$.

The ratio $\dfrac{\Sigma t_i}{\Sigma t_i + \Sigma \bar{t}_i}$ is equal to the fraction of time during which the component is working. We can show that this ratio is converging, with a probability of 1, to $\dfrac{\text{MTTF}}{\text{MTTF}+\text{MTTR}}$ which is the limiting availability of the component.

We can also show (cf. [Beliav et al., 1972]) that the distribution of $N(t)$ tends asymptotically to a normal distribution with mean $\dfrac{\lambda^* t}{1+\lambda^* \tau}$ and variance $\dfrac{(\sigma_\lambda^2 + \sigma_\tau^2)t}{\left(\tau + \dfrac{1}{\lambda^*}\right)^3}$ as t tends to infinity. The process becomes stable (for fixed λ and μ) more slowly, the smaller the overall variance $(\sigma_\lambda^2 + \sigma_\tau^2)$.

3.2.3. *Special case of zero repair time*

The probability density of the repair time is a Dirac distribution $\delta(t)$. Therefore, $\bar{g}(s) = 1$ and $A(0) = 1$. This may be regarded as an approximation to the case where the repair time is very short.

The availability is obviously 1. We are interested in the renewal number $N(t)$.

From [Kendall and Stuart, 1963]:

$$\bar{H}(s) = \frac{\bar{f}(s)}{s(1-\bar{f}(s))} .$$

Consequently, $E[N(t)] = H(t)$ is the solution of

$$H(t) = F(t) + \int_0^t H(t-u) f(u)\,\mathrm{d}u \tag{52}$$

and

$$\sigma^2(t) = \sigma^2[N(t)] = 2\int_0^t H(t-u)\,\mathrm{d}H(u) + H(t) - H^2(t) . \tag{53}$$

As the distribution of $N(t)$ is asymptotically normal, Eqns (52) and (53) enable us to determine the value N_0 of the spares which must be stocked for a mission of duration T to ensure that the risk of running out of spares does not exceed a certain probability α.

In fact:

$$\mathcal{P}(N(T) > N_0) = \alpha \Leftrightarrow \int_{N_0}^{\infty} \frac{1}{\sigma(T)\sqrt{2\pi}} \exp\left(-\frac{1}{2}\left(\frac{x - H(T)}{\sigma(T)}\right)^2\right) \mathrm{d}x = \alpha$$

therefore, in accordance with section 4.2.3 of Ch. 1:

$$N_0 = u_\alpha \, \sigma(T) + H(T)$$

with

$$\alpha = \int_{u_\alpha}^{\infty} \frac{1}{\sqrt{2\pi}} \exp\left(-\frac{u^2}{2}\right) \mathrm{d}u \tag{54}$$

u_α is determined using the standard normal distribution tables (Appendix 1).

If $H(t)$ cannot be determined, it is nevertheless possible to determine bounds. It can in fact be shown (cf., for example [Beliav et al., 1972]) that:

$$F(t) \leq H(t) \leq \frac{F(t)}{1 - F(t)} \tag{55}$$

$$\frac{t}{\text{MTTF}} - 1 \leq H(t). \tag{56}$$

In addition, when the failure rate $\lambda(t) = \dfrac{1 - f(t)}{F(t)}$ *is not decreasing*, we get a new upper bound of $H(t)$:

$$H(t) \leq \frac{t}{\text{MTTF}}. \tag{57}$$

We also know the asymptotic behaviour of $H(t)$:

— $\lim_{t \to \infty} \dfrac{H(t)}{t} = \dfrac{1}{\text{MTTF}}$.

— Blackwell's theorem

$$\lim_{t \to \infty} (H(t + \alpha) - H(t)) = \frac{\alpha}{\text{MTTF}} \qquad \forall \alpha > 0.$$

— $N(t)$ is asymptotically distributed according to a normal law $N\left(\dfrac{t}{\text{MTTF}}, \left(\dfrac{\sigma_\lambda^2 \, t}{(\text{MTTF})^3}\right)^{1/2}\right)$.

Consequently, if T is relatively large, one can determine the level of stock N_0 from Eqn (54) by taking

$$H(T) = \frac{T}{\text{MTTF}} \quad \text{and} \quad \sigma(T) = \frac{\sigma_\lambda}{\text{MTTF}} \sqrt{\frac{T}{\text{MTTF}}}.$$

4. ANALYSIS OF SOME SIMPLE REPAIRABLE SYSTEMS

The availability and reliability of these systems are different functions and both have to be computed.

4.1. Series configuration systems

As the failure of a component causes the entire system to fail, the reliability of a series-type repairable system is the same as that of a non-repairable system, i.e. when all the components are independent:

$$R(t) = \exp\left(- \sum_{i=1}^{n} \int_0^t \lambda_i(u)\,du \right).$$

The availability $A(t)$ of the system is the product of the availabilities $A_i(t)$ of the various components:

$$A(t) = \prod_{i=1}^{n} A_i(t) \tag{58}$$

When λ_i and μ_i are constant, assuming $A(0) = 1$, we obtain:

$$A(t) = \prod_{i=1}^{n} \left(\frac{\mu_i}{\lambda_i + \mu_i} + \frac{\lambda_i}{\lambda_i + \mu_i} e^{-(\lambda_i + \mu_i)t} \right).$$

In all cases, using Eqn (28) we obtain:

$$\lim_{t \to \infty} A(t) = A_\infty = \prod_{i=1}^{n} \frac{1}{1 + \lambda_i \tau_i}.$$

When $\lambda_i \tau_i \ll 1$, we have:

$$A_\infty \simeq 1 - \sum_{i=1}^{n} \lambda_i \tau_i. \tag{59}$$

4.2. Parallel configuration systems

We will assume that the components are independent. Note that this assumption implies that we have one repair facility per component and that the redundancy is of the active type. In fact, if the number of repair facilities is less than the number of components, in certain cases some components will be awaiting repair and the waiting time will depend on the distribution of the repair times of the components undergoing repair. In the case of standby redundancy, the operating periods of each of the components depend on the availability of the other components.

The system is available if at least one of the components is available. Using the independence assumption we can therefore write:

$$A(t) = 1 - \prod_{i=1}^{n} (1 - A_i(t))$$

and from Eqn (28):

$$A_\infty = \lim_{t\to\infty} A(t) = 1 - \prod_{i=1}^{n} \frac{\lambda_i \tau_i}{1+\lambda_i \tau_i}.$$

In the case where $\lambda_i \tau_i \ll 1$, we have

$$A_\infty \simeq 1 - \prod_{i=1}^{n} \lambda_i \tau_i. \tag{60}$$

Calculating the availability in the other cases is more difficult and requires methods of the type which will now be used for computing the reliability.

To simplify the notations, we will examine the case of two identical components operating in parallel (active redundancy).

The system therefore has three states:

— State 2: both components are operating.
— State 1: only one component is operating, the other has failed.
— State 0: both components have failed.

Let $P_i(t)$ be the probability of being in state i ($i = 0, 1, 2$) at time t, the system never having encountered state 0:

Therefore $R(t) = P_2(t) + P_1(t)$.

We will assume that $P_2(0) = 1$.

Let us try to find a system of differential equations satisfied by P_0, P_1 and P_2, assuming that λ and μ are constant. So that the system cannot re-enter states 1 and 2 following failure, we shall make state 0 *absorbing*. (No transition from this state to states 1 or 2.)

Using the total probability theorem, we now obtain:

$$P_2(t+\Delta t) = P_2(t)\,.\,\mathscr{P}\left(\begin{matrix}\text{2 components operating} \\ \text{at } t+\Delta t\end{matrix}\Bigg/\begin{matrix}\text{the system is in} \\ \text{state 2 at time } t\end{matrix}\right) + P_1(t)\,.\,\mathscr{P}\left(\begin{matrix}\text{2 components operating} \\ \text{at } t+\Delta t\end{matrix}\Bigg/\begin{matrix}\text{the system is in} \\ \text{state 1 at time } t\end{matrix}\right)$$

$$P_1(t+\Delta t) = P_2(t)\mathscr{P}\left(\begin{matrix}\text{only 1 component operating} \\ \text{at } t+\Delta t\end{matrix}\Bigg/\begin{matrix}\text{the system is in} \\ \text{state 2 at time } t\end{matrix}\right) + P_1(t)\mathscr{P}\left(\begin{matrix}\text{only 1 component operating} \\ \text{at } t+\Delta t\end{matrix}\Bigg/\begin{matrix}\text{the system is in} \\ \text{state 1 at time } t\end{matrix}\right).$$

As the failure rate λ and the repair rate μ of the components are constant, using the definition of λ and μ we obtain:

$$P_2(t+\Delta t) = P_2(t)\,(1 - 2\,\lambda\,\Delta t + 0(\Delta t)) + P_1(t)\,(\mu\,\Delta t + 0(\Delta t))$$
$$P_1(t+\Delta t) = P_1(t)\,(1 - (\lambda+\mu)\,\Delta t + 0(\Delta t)) + P_2(t)\,(2\,\lambda\,\Delta t + 0(\Delta t)).$$

These equations correspond to the following Markov diagram:

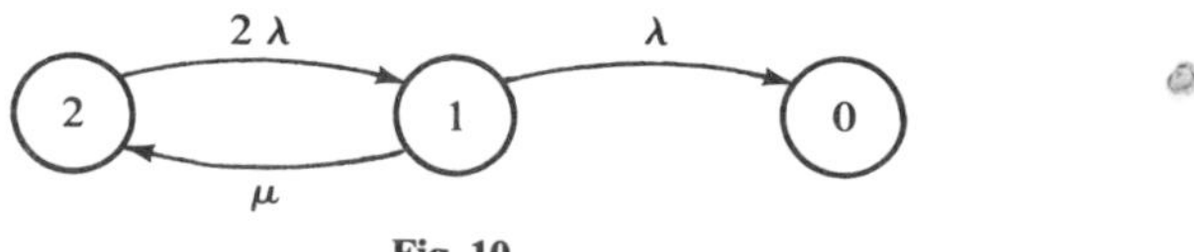

Fig. 10.

Letting Δt tend to zero, we have:

$$\left.\begin{aligned} \frac{dP_2(t)}{dt} &= -2\lambda P_2(t) + \mu P_1(t) \\ \frac{dP_1(t)}{dt} &= 2\lambda P_2(t) - (\mu + \lambda) P_1(t) \end{aligned}\right\} \quad (61)$$

Equation (61) can be solved, for example, by using the Laplace transform. Using Eqn (45) from Ch. 1, then from Eqn (61) we obtain: ($P_2(0) = 1, P_1 = 0$)

$$s\bar{P}_2(s) - 1 = -2\lambda\bar{P}_2(s) + \mu\bar{P}_1(s)$$
$$s\bar{P}_1(s) = 2\lambda\bar{P}_2(s) - (\mu + \lambda)\bar{P}_1(s)$$

therefore:

$$\bar{P}_2(s) = \frac{s + \lambda + \mu}{(s - s_1)(s - s_2)}$$
$$\bar{P}_1(s) = \frac{2\lambda}{(s - s_1)(s - s_2)}$$

where s_1 and s_2 are the roots of:

$$S^2 + (3\lambda + \mu)S + 2\lambda^2 = 0$$

i.e.:

$$s_1 = \frac{-(3\lambda + \mu) + \sqrt{\mu^2 + 6\lambda\mu + \lambda^2}}{2} < 0$$

$$s_2 = \frac{-(3\lambda + \mu) - \sqrt{\mu^2 + 6\lambda\mu + \lambda^2}}{2} < s_1 .$$

It follows that

$$R(t) = P_1(t) + P_2(t) = \frac{s_2}{s_2 - s_1} e^{s_1 t} - \frac{s_1}{s_2 - s_1} e^{s_2 t} \quad (62)$$

$$\text{MTTF} = \bar{P}_1(0) + \bar{P}_2(0) = \frac{\mu + 3\lambda}{2\lambda^2} \quad (63)$$

On the other hand:

$$\Lambda(t) = \frac{-\frac{dR}{dt}(t)}{R(t)} = 2\lambda^2 \frac{e^{s_2 t} - e^{s_1 t}}{s_2 e^{s_1 t} - s_1 e^{s_2 t}}$$

consequently $\quad \Lambda(t) \underset{t \to 0}{\simeq} 2\lambda^2 t$

$$\Lambda_\infty = \lim_{t \to \infty} \Lambda(t) = -s_1 .$$

Now consider the *frequency case* $\lambda \ll \mu$ (mean uptime of a component is large compared with the mean repair time).

Then:

$$s_1 \simeq \frac{-2\lambda^2}{\mu + 3\lambda}, \quad s_2 \simeq -(\mu + 3\lambda)$$

and

$$R(t) \simeq \left(1 + 2\left(\frac{\lambda}{\mu}\right)^2\right) \exp\left(-2\frac{\lambda^2}{\mu + 3\lambda}t\right) - 2\left(\frac{\lambda}{\mu}\right)^2 \exp(-(\mu + 3\lambda)t).$$

We can also take

$$R(t) \simeq \exp\left(-\frac{2\lambda^2}{\mu + 3\lambda}t\right) = \exp\left(-\frac{t}{\text{MTTF}}\right).$$

as an approximation of $R(t)$.

The computation which we have just performed is therefore relatively complicated, much more complicated than that of availability in the independence assumption. In the case of dependence, the availability computation will also be difficult. When there is only one repair facility, one would find, for example:

$$A(t) = \frac{1}{1 + \dfrac{2\lambda^2}{\mu^2 + 2\lambda\mu}} - \frac{2\lambda^2}{s_1 - s_2}e^{s_1 t} + \frac{2\lambda^2}{s_1 - s_2}e^{s_2 t}$$

where s_1 and s_2 are the roots of $S^2 + (2\mu + 3\lambda)S + \mu^2 + 2\lambda\mu + 2\lambda^2 = 0$.

We will see in Ch. 4 that the availability computation can be very easily generalized in the case of independence. As to the other cases (computation of reliability and availability for dependent components), the theory of Markovian processes will enable us to solve small system problems easily. For large systems, Ch. 6 describes the critical operating states method for computing the reliability when the availability can be determined. A few approximation computations by determining bounds can be found in [Shooman, 1968, p. 210].

5. COMPONENTS NOT CONTINUOUSLY MONITORED

5.1. Defining the problem

Until now, we have assumed instantaneous information concerning the state of a component. In practice, in a large system, some components are not continuously monitored and a failure can go unnoticed if redundancy is provided. In order to improve system reliability, it is therefore advisable to inspect these components from time to time so that they can be repaired if necessary.

Let us assume that the interval between two inspections is constant and equal to T_0 and that the repair time is negligible compared with T_0. Therefore

$$A(t) = e^{-\lambda(t - NT_0)} \quad \text{if} \quad NT_0 \leq t \leq (N+1)T_0.$$

This availability is unlimited when $t \to \infty$, but it is the mean value which is of interest:

$$E[A(t)] = \frac{1}{T_0} \int_0^{T_0} e^{-\lambda t}\, dt = \frac{1 - e^{-\lambda T_0}}{\lambda T_0}.$$

In the case where $\lambda T_0 \ll 1$, we obtain:

$$E[A(t)] \simeq 1 - \frac{\lambda T_0}{2}. \tag{64}$$

Intuitively, where the mean time to repair τ is no longer negligible compared with T_0, one is tempted to write, taking account of Eqns (28) and (64):

$$E[A(t)] \simeq 1 - \lambda\left(\tau + \frac{T_0}{2}\right).$$

We will demonstrate this formula in two particular cases:

(a) the repair time distribution is exponential with parameter μ,
(b) the repair time is a constant τ.

5.2. System state equations

The component has three states:

— State 1: the component is operable.
— State 2: the component has failed but this is not known.
— State 3: the component is being repaired.

Calling $P_i(t)$ the probability of being in state i at time t, we have:

$$A(t) = P_1(t).$$

Assume that $P_1(0) = 1$ and the inspections are T_0 apart, the first taking place at T_0. Using a now familiar method, we can write:

$$P_1(t + dt) = \mathscr{P}\begin{pmatrix}\text{component working at } t \text{ and not} \\ \text{failing during } [t, t + dt]\end{pmatrix} + \mathscr{P}\begin{pmatrix}\text{being repaired at } t \text{ and repair ending between} \\ t \text{ and } t + dt\end{pmatrix}$$

$$P_3(t + dt) = \mathscr{P}(\text{being repaired at } t \text{ and not repaired during } [t, t + dt]) + P_2(t) \sum_{n=1}^{\infty} \mathscr{P}(nT_0 \in [t + t + dt]).$$

5.2.1. *Constant repair rate*

Therefore, if the repair time is exponentially distributed with parameter μ:

$$\left.\begin{aligned} \frac{dP_1}{dt}(t) &= -\lambda_1 P_1(t) + \mu P_3(t) \\ \frac{dP_3}{dt}(t) &= -\mu P_3(t) + \sum_{n=1}^{N} P_2(t)\,\delta(t - nT_0) \end{aligned}\right\} \tag{65}$$

with $NT_0 \leq t < (N+1)\,T_0$.

Or, using the Laplace transform

$$s\bar{P}_1(s) - 1 = -\lambda \bar{P}_1(s) + \mu \bar{P}_3(s)$$

$$s\bar{P}_3(s) = -\mu \bar{P}_3(s) + \sum_{n=1}^{\infty} P_2(nT_0)\,e^{-snT_0}.$$

It follows that:

$$P_1(t) = e^{-\lambda t} + \frac{\mu}{\mu - \lambda} \sum_{n=1}^{N} P_2(nT_0)\,[e^{-\lambda(t-nT_0)} - e^{-\mu(t-nT_0)}].$$

The coefficients $P_2(nT_0)$ are calculated by recurrence, noting that

$$\begin{cases} P_2(nT_0) = \displaystyle\int_{(n-1)T_0}^{nT_0} \lambda P_1(t)\,dt \\ P_2(0) = 0 \end{cases}$$

hence:

$$P_2[(N+1)\,T_0] = (1 - e^{-\lambda T_0})\,e^{-\lambda N T_0} + \sum_{n=1}^{N} \frac{P_2(nT_0)}{\mu - \lambda}[\mu\, e^{-\lambda(N-n)T_0}(1 - e^{-\lambda T_0}) - \lambda\, e^{-\mu(N-n)T_0}(1 - e^{-\mu T_0})]. \tag{66}$$

It therefore follows from Eqn (66) (cf. Exercise 3) that:

$$\lim_{N\to\infty} P_2(NT_0) = \frac{(\mu - \lambda)(1 - e^{-\lambda T_0})(1 - e^{-\mu T_0})}{\mu(1 - e^{-\mu T_0}) - \lambda(1 - e^{-\lambda T_0})}$$

hence

$$\lim_{t\to\infty} E[A(t)] = \lim_{N\to\infty} \frac{1}{T_0} \int_{NT_0}^{(N+1)T_0} P_1(t)\,dt = \frac{(\mu - \lambda)(1 - e^{-\lambda T_0})(1 - e^{-\mu T_0})}{\lambda T_0[\mu(1 - e^{-\mu T_0}) - \lambda(1 - e^{-\lambda T_0})]} \tag{67}$$

where $\lambda T_0 \ll 1$, by putting $\tau = \mu^{-1}$ (mean time to repair), we obtain:

$$\lim_{t\to\infty} E[A(t)] \simeq \left(1 - \frac{\lambda T_0}{2}\right)(1 - \lambda\tau)$$

and assuming $\lambda\tau \ll 1$

$$1 - \lim_{t\to\infty} E[A(t)] \simeq \lambda\left(\frac{T_0}{2} + \tau\right). \tag{68}$$

5.2.2. *Constant repair time*

In the case of a constant repair time equal to τ, Eqns (66) and (67) become respectively:

$$P_2[(N+1)\,T_0] = (1-e^{-\lambda T_0})\,e^{-\lambda N T_0} + P_2(NT_0) + e^{-\lambda\tau}(1-e^{-\lambda T_0}) + \\ + \sum_{n=1}^{N-1} P_2(nT_0)\,e^{-\lambda(N-n)T_0-\lambda\tau}(1-e^{-\lambda T_0})$$

with $NT_0 + \tau \leq t < (N+1)\,T_0 + \tau$

$$\lim_{t\to\infty} E[A(t)] = \frac{1-e^{-\lambda T_0}}{\lambda T_0(1+e^{-\lambda(T_0-\tau)}-e^{-\lambda T_0})}\,.$$

The latter expression also leads to Eqn (68).

We will see in Ch. 8 how to determine the monitoring interval T_0 by minimizing certain criteria associated with safety or cost.

6. SUMMARY

In this chapter, we have examined the reliability and availability of some simple systems by using the definitions of these quantities directly and, in some cases, attempting to find a differential equation of which they are solutions.

We shall adopt the following results which are valid for systems consisting of components with constant failure rate λ (and in some cases with a constant repair rate $\mu = \tau^{-1}$), continuously operating and available at time $t = 0$: (see Table 1).

Examination of Table 1 shows that even for very simple repairable systems, it is necessary to use more powerful methods to compute reliability.

As far as components with non-constant failure and repair rates are concerned, there are two points to be made:

(a) The computations are very complicated.

(b) When repair is effected by replacing the component within a negligible time, it is possible to compute the number of spares necessary to provide a fixed level of availability.

EXERCISES

Exercise 1 — Comparison of two non-repairable systems

Consider the two following systems:

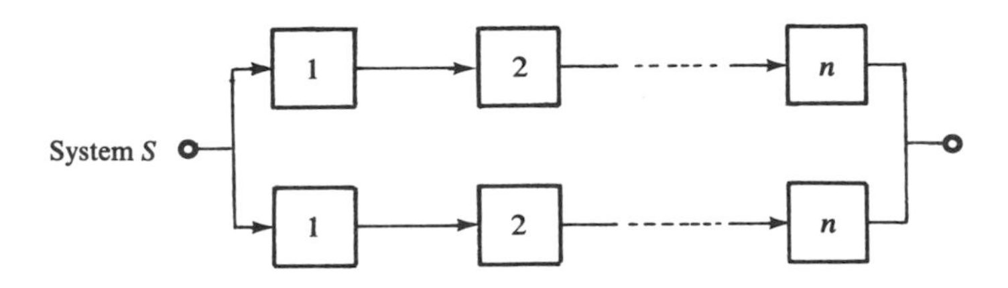

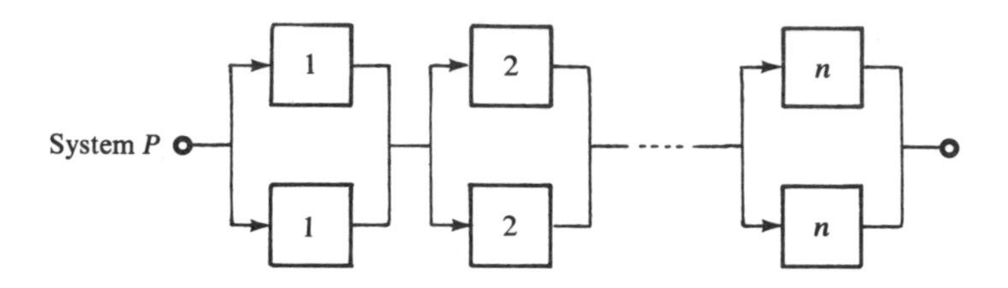

Both systems contain identical components. In the first case, redundancy is achieved by arranging two identical lines in parallel, whereas in the second case the redundancy is at the level of each of the n components.

(1) Compute the reliabilities $R_p(t)$ and $R_s(t)$ of each of these two systems as a function of the reliabilities $R_i(t)$ of each of the n components assumed to be independent.

(2) Putting $y(t) = \dfrac{R_p(t) - R_s(t)}{\prod_{i=1}^{n} R_i(t)}$, show that $R_p(t) \geqslant R_s(t)$.

Deduce that system P is preferable to system S.
What if the components are repairable?

Exercise 2—Non-repairable, n/p active redundancy systems consisting of independent identical components

(1) Compute the system reliability $R_{n,p}$ as a function of the reliability $\Gamma(t)$ of a component.
(2) Show that in the case

- $n = 1$,

$$R_{n,p}(\Gamma) \geq R_{1,1}(\Gamma),$$

equality only being obtained for $\Gamma = 0$ and 1.
- $n > 1$, there exists non-zero Γ_0 so that

$$R_{n,p}(\Gamma_0) = R_{1,1}(\Gamma_0).$$

Exercise 3—Components whose repair time distribution is Erlangian

(1) Compute the Laplace transform of the availability of a component whose uptime distribution is exponential with parameter λ and whose repair time distribution is Erlangian with parameters μ and k:

$$g(t) = \frac{\mu^k t^{k-1} e^{-\mu t}}{(k-1)!}:$$

(2) Compute the availability in the case $k = 2$.

(3) Assume that this component is on standby and is tested at regular intervals T_0. Use the notations in section 5. Calculate $P_1(t)$ and the recurrence equation for calculating $P_2(NT_0)$.

(4) Give a method of calculating $P_2(\infty)$.

Exercise 4—Availability of a quasi-repairable component

Consider a component with constant failure rate λ and whose repair time distribution $g(t)$ is such that:

$g(t) = \mu e^{-\mu t}$ if the time t which elapses from the start of repair is less than or equal to Δt_0.
$g(t) = 0$ otherwise.

This component is not repairable since $\int_0^\infty g(t)\,dt < 1$. Physically, the component is only

Table 1.

TYPE OF SYSTEM	NON-REPAIRABLE SYSTEMS		REPAIRABLE SYSTEMS (constant repair rates)	
	Reliability	MTTF	Reliability	Availability
1 component	$e^{-\lambda t}$	$\frac{1}{\lambda}$	$e^{-\lambda t}$	$a(t) = \frac{\mu}{\lambda+\mu} + \frac{\lambda}{\lambda+\mu} e^{-(\lambda+\mu)t}$ $a(\infty) \simeq 1 - \lambda\tau$ if $\lambda\tau \ll 1$
n independent components in series	$e^{-(\Sigma\lambda_i)t}$	$\frac{1}{\sum_i \lambda_i}$	$e^{-(\Sigma\lambda_i)t}$	$A(t) = \prod_i a_i(t)$ $A(\infty) \simeq 1 - \sum_i \lambda_i\tau_i$ if $\lambda_i\tau_i \ll 1$
n independent components in active redundancy	$1 - \prod_i (1 - e^{-\lambda_i t})$	$\sum_i \frac{1}{\lambda_i} - \sum_i \sum_j \frac{1}{\lambda_i + \lambda_j} + \cdots + (-1)^{n+1} \frac{1}{\Sigma\lambda_i}$	See Ch. 6	$A(t) = 1 - \prod_i (1 - a_i(t))$ $A(\infty) \simeq 1 - \prod_i \lambda_i\tau_i$ if $\lambda_i\tau_i \ll 1$
n identical components in active redundancy	$1 - (1 - e^{-\lambda t})^n$	$\frac{\sum_{i=1}^{n} \frac{1}{i}}{\lambda}$	See Ch. 6 (for $n = 2$, see section 4.2)	$A(t) = 1 - (1 - a(t))^n$ $A(\infty) \simeq 1 - (\lambda\tau)^n$ if $\lambda\tau \ll 1$
1 component, repair time distribution $g(t)$	$A(\infty) = \frac{1}{1+\lambda\tau}$ with $\tau = \int_0^\infty t g(t)\,dt$ $\quad A(\infty) \simeq 1 - \lambda\tau$ if $\lambda\tau \ll 1$			
1 standby component tested at regular intervals T_0	$1 - E[A(t)] = \lambda\left(\tau + \frac{T_0}{2}\right)$ if $\lambda\tau \ll 1$			

repairable if the repair time does not exceed Δt_0. This case may be encountered when failure involves the loss of a refrigeration function and after a certain time Δt_0 the units are so deformed by heat that they cannot be repaired.

Compute the availability of such a component which may be termed quasi-repairable.

Exercise 5

Demonstrate Eqn (21) giving the MTF of a system consisting of n independent components which are identical and operating on an active redundancy basis.

SOLUTIONS TO EXERCISES

Exercise 1

$$(1)\ R_p(t) = \prod_{i=1}^{n} (2R_i(t) - R_i^2(t)) = \left(\prod_{i=1}^{n} R_i(t)\right)\left(\prod_{i=1}^{n}(2 - R_i(t))\right)$$

$$R_s(t) = -\prod_{i=1}^{n} R_i^2(t) + 2\prod_{i=1}^{n} R_i(t).$$

$$(2)\ y(t) = \prod_{i=1}^{n}(2 - R_i(t)) - 2 + \prod_{i=1}^{n} R_i(t).$$

If $R_i = 1\ \forall_i,\quad R_p - R_s = 2^n - 2 > 0.$

If $R_i = 0\ \forall_i,\quad R_p - R_s = 0$

on the other hand

$$\frac{\partial y}{\partial R_j} = -\prod_{i \neq j}(2 - R_i) + \prod_{i \neq j} R_i \leq -1 + \prod_{j \neq i} R_i \leq 0.$$

The same applies to the higher order derivatives. Hence $R_p(t) \geqslant R_s(t)$.

Exercise 2

(1)
$$R_{n,p} = \sum_{k=n}^{p} C_p^k\, \Gamma^k (1-\Gamma)^{p-k}$$
$$R_{1,1} = \Gamma.$$

(2) Putting $x(\Gamma) = R_{n,p} - R_{1,1}$

$$x(1) = 0$$
$$x(0) \underset{\Gamma \to 0}{\simeq} C_p^n\, p^n - p.$$

If $n = 1$ $\quad x(0) \geqslant 0.$

If $n > 1$ $\quad x(0) < 0.$

In the case $n = 1$

$$x(\Gamma) = 1 - (1-\Gamma)^p - \Gamma = \Gamma[(1-\Gamma) + (1-\Gamma)^2 \cdots + (1-\Gamma)^{p-1}] \geq 0.$$

$x(\Gamma)$ only becomes zero for $\Gamma = 0$ and 1.

For $n > 1$

$$x'(\Gamma) = \sum_{k=n}^{p} kC_p^k\, \Gamma^{k-1}(1-\Gamma)^{p-k} - \sum_{k=n}^{p-1}(p-k)\, C_p^k\, \Gamma^k (1-\Gamma)^{p-k-1}$$

$$x'(1) = n - 2 \geq 0.$$

giving the following variation table:

Γ	1	0
x	$0\nearrow$	<0
x'	$+$	

consequently x becomes zero over $]1, 0[$.

Exercise 3

(1) $\bar{g}(s) = \dfrac{\mu^k}{(s+\mu)^k}$

Equation (25) can be written:

$$\bar{A}(s) = \frac{A_0(s+\mu)^k + (1-A_0)\,\mu^k}{(s+\lambda)(s+\mu)^k - \lambda\mu^k}.$$

(2) $k = 2$

$$\bar{A}(s) = \frac{A_0\, s(s+2\mu) + \mu^2}{s(s-s_1)(s-s_2)}$$

with s_1 and s_2 roots of

$$s^2 + s(2\mu + \lambda) + \mu^2 + 2\lambda\mu = 0$$

(note that s_1 and s_2 are generally complex since $\Delta = +\lambda^2 - 4\lambda\mu$)

$$A(t) = \frac{\mu^2}{\mu^2 + 2\lambda\mu} + \frac{A_0(s_1+2\mu)\,s_1 + \mu^2}{s_1(s_1-s_2)}\,e^{s_1 t} + \frac{A_0(s_2+2\mu)\,s_2+\mu^2}{s_2(s_2-s_1)}\,e^{s_2 t}$$

(3) $\dfrac{dP_1}{dt} = -\lambda P_1(t) + \sum_{k=1}^{N} P_2(kT_0)\, g(t-kT_0)$ for $NT_0 \le t < (N+1)\,T_0$

$$\bar{P}_1(s) = \frac{1}{s+\lambda} + \sum_{j=1}^{\infty} P_2(jT_0)\,\frac{\bar{g}(s)}{s+\lambda}\,e^{-s_j T_0}$$

therefore, after using the inversion formula given in Ch. 1, section 5.2.4:

$$P_1(t) = e^{-\lambda t} + \sum_{j=1}^{N} P_2(jT_0)\left(\frac{\mu}{\mu-\lambda}\right)^k \left[e^{-\lambda(t-jT_0)} - e^{-\mu(t-jT_0)}\,\frac{(1+(\mu-\lambda)(t-jT_0))^{k-1}}{(k-1)\,!}\right].$$

When $k = 1$, we again obtain the solution given in section 5.
When $k = 2$, we obtain:

$$P_1(t) = e^{-\lambda t} + \sum_{j=1}^{N} P_2(jT_0)\left(\frac{\mu}{\mu-\lambda}\right)^2 [e^{-\lambda(t-jT_0)} - e^{-\mu(t-jT_0)}\,(1+(\mu-\lambda)(t-jT_0))].$$

The recurrence formula for determining the $P_2(jT_0)$ is therefore as follows:

$$P_2((N+1)\,T_0) = e^{-\lambda NT_0}(1-e^{-\lambda T_0}) + \sum_{j=1}^{N} P_2(jT_0)\left(\frac{\mu}{\mu-\lambda}\right)^k e^{\lambda_j T_0}\,e^{-\lambda NT_0}(1-e^{-\lambda T_0}) -$$

$$- \sum_{j=1}^{N} P_2(jT_0)\,\frac{\mu^{k-1}\lambda}{(\mu-\lambda)^k}\,e^{\mu_j T_0}\int_{NT_0}^{(N+1)T_0} \frac{\mu[1+(\mu-\lambda)(t-jT_0)]^{k-1}}{(k-1)\,!}\,e^{-\mu t}\,dt.$$

(4) One can obtain $P_2(\infty)$ by placing oneself in the interval $[0, T_0]$ and assuming that the

phenomenon has been observed since time $t = -\infty$ (see Ch. 1, section 6.3). The differential equation giving $P_1(t)$ can therefore be written:

$$\frac{dP_1}{dt}(t) = -\lambda P_1(t) + P_2(\infty) \sum_{n=0}^{\infty} g(t + nT_0) \text{ avec } P_1(0) = 1 - P_2(\infty) - y$$

where y represents the probability of the component being under repair (from nT_0 ($n = 1, \dots \infty$) to time $t = 0$).

Putting:

$$\Phi_k(t) = \sum_{n=0}^{\infty} g(t + nT_0) = \sum_{n=0}^{\infty} \frac{\mu^k (t + nT_0)^{k-1} e^{-\mu(t+nT_0)}}{(k-1)!}.$$

it follows that:

$$\bar{P}_1(s) = \frac{1 - x - y}{s + \lambda} + P_2(\infty) \frac{\bar{\Phi}_k(s)}{s + \lambda}.$$

By definition:

$$y = \sum_{n=0}^{\infty} P_2(\infty) \int_{T_0}^{\infty} g(t + nT_0)\,dt = P_2(\infty) \sum_{n=1}^{\infty} \int_{nT_0}^{\infty} g(t)\,dt$$

hence $y = P_2(\infty) \sum_{n=1}^{\infty} e^{-\mu n T_0} \sum_{j=1}^{k} \frac{(\mu n T_0)^{j-1}}{(j-1)!}$.

Inversion of $\bar{P}_1(s)$ gives $P_1(t)$. It therefore follows that:

$$P_2(0) = \int_0^{T_0} \lambda P_1(t)\,dt.$$

Example:

- $k = 1$

$$\Phi_1(t) = \frac{\mu e^{-\mu t}}{1 - e^{-\mu t_0}}$$

$$\bar{\Phi}_1(s) = \frac{1}{1 - e^{-\mu T_0}} \frac{\mu}{s + \mu}$$

$$P_1(t) = \frac{\mu - \lambda + \lambda(x + y)}{\mu - \lambda} e^{-\lambda t} - \frac{\mu(x + y)}{\mu - \lambda} e^{-\mu t} \quad \text{with} \quad x = P_2(\infty)$$

hence

$$P_2(\infty) = \frac{\mu - \lambda + \lambda(x + y)}{\mu - \lambda}(1 - e^{-\lambda T_0}) - \frac{\lambda(x + y)}{\mu - \lambda}(1 - e^{-\mu T_0})$$

on the other hand

$$y = P_2(\infty) \frac{e^{-\mu T_0}}{1 - e^{-\mu T_0}}$$

giving the formula in section 5.2.

- $k = 2$

$$\Phi_2(t) = \frac{\mu^2 e^{-\mu t}}{(1 - e^{-\mu T_0})^2}(t(1 - e^{-\mu T_0}) + T_0)$$

$$\bar{\Phi}_2(s) = \frac{1}{1 - e^{-\mu T_0}} \left(\frac{\mu}{s + \mu}\right)^2 + \frac{\mu T_0}{(1 - e^{-\mu T_0})^2} \frac{\mu}{\mu + s}$$

and

$$y = \frac{P_2(\infty)\,e^{-\mu T_0}}{(1-e^{-\mu T_0})^2}(1+\mu T_0 - e^{-\mu T_0}) .$$

Exercise 4

Let $P_1(t)$ be the probability of being in the operating state.

Let $P_2(t)$ be the probability of being under repair, the repair having commenced less than Δt_0 previously.

$$P_1(t+dt) = P_1(t)(1-\lambda\,dt + 0(dt)) + \int_{\max(t-\Delta t_0, 0)}^{t} (\lambda\,dt + 0(dt))\,P_1(u)\,g(t-u)\,du$$

hence

$$\frac{dP_1}{dt}(t) = -\lambda P_1(t) + \lambda\int_0^t P_1(t-u)\,\tilde{g}(u)\,du$$

with

$$\tilde{g}(t) = g(t) \quad \text{if} \quad 0 \leq t \leq \Delta t_0$$
$$\tilde{g}(t) = 0 \quad \text{otherwise,}$$

hence

$$\overline{P}_1(s) = \frac{1}{s+\lambda-\lambda\bar{\tilde{g}}(s)} .$$

Application to the case where $g(t)$ is exponential

$$\bar{\tilde{g}}(s) = \frac{\mu}{\mu+s} - \frac{\mu}{\mu+s}\,e^{-(\mu+s)\Delta t_0}$$

and

$$\overline{P}_1(s) = \frac{\mu+s}{s(s+\lambda+\mu)}\sum_{n=0}^{\infty}\left(\frac{-\lambda\mu\,e^{-\mu\,\Delta t_0}\,e^{-s\,\Delta t_0}}{s(s+\lambda+\mu)}\right)^n$$

then, by inversion

$$P_1(t) = \frac{\mu}{\lambda+\mu} + \frac{\lambda}{\lambda+\mu}e^{-(\lambda+\mu)t} + \sum_{n=1}^{N} K_0^n \sum_{l=1}^{n+1} k_{n,l(t)}\frac{[(\lambda+\mu)(t-n\,\Delta t_0)]^{n+1-l}}{(n+1-l)\,!}$$

with

$$N\,\Delta t_0 \leq t < (N+1)\,\Delta t_0$$

$$K_0 = \frac{\lambda\mu\,e^{-\mu\,\Delta t_0}}{(\lambda+\mu)^2}$$

$$k_{n,l}(t) = (-1)^{n+l-1}\left(C_{n+l-1}^{n}\frac{\mu}{\lambda+\mu} - \frac{l-1}{n}C_{n+l-2}^{n-1}\right) + \left(C_{n+l-2}^{n-1} - C_{n+l-1}^{n}\frac{\mu}{\lambda+\mu}\right)e^{-(\lambda+\mu)(t-n\,\Delta t_0)}$$

Exercise 5

$$\text{MTTF} = \int_0^{\infty} R(t)\,dt = \int_0^{\infty}\left[1-\prod_{i=1}^{n}(1-e^{-\lambda u})\right]du$$

Putting $x = 1 - e^{-\lambda\mu}$

$$\text{MTTF} = \frac{1}{\lambda}\int_0^1 \frac{1 - x^n}{1 - x}\,dx = \frac{1}{\lambda}\sum_{i=0}^{n-1}\int_0^1 x^i\,dx = \frac{\sum_{i=1}^{n}\frac{1}{i}}{\lambda}.$$

REFERENCES

Beliaev Y., Gnedenko B. and Soloviev A. (1972): *Méthodes mathématiques en théorie de la fiabilité*; MIR, Moscow.

Pages A. (1976): *Calcul de la fiabilité et de la disponibilité de systèmes réparables dont les pannes de faible durée sont admissibles*; Internal EDF report, HI 2203/02.

Schwartz L. (1961): *Méthodes mathématiques pour les sciences physiques*; Hermann, Paris.

Shooman M. L. (1968): *Probabilistic Reliability: An Engineering Approach*; McGraw-Hill, New York.

CHAPTER 4

EVALUATING SYSTEM AVAILABILITY

1. INTRODUCTION

In Ch. 2 we divided systems into three types:

— *coherent systems* with statistically independent components,
— *non-coherent systems* with statistically independent components,
— systems with *statistically dependent* components (and which can therefore be represented by a state transition diagram).

Availability is calculated differently for each type of system.

In sections 2 and 3 we shall examine the first type which corresponds to the systems most frequently encountered in applications; section 2 will deal with simple systems represented either by a reliability block diagram, fault tree or successful paths. Section 3 will consider the general case of coherent systems represented by minimal cut sets. Lastly, in sections 4, 5 and 6 we shall examine systems of the second and third type.

In the case of systems whose components are statistically independent, system availability depends only on the availability of each component and on the system logic. In the last four sections, we shall therefore consider a system S consisting of n independent components.

The following notation will be used:

$$A(t) = \text{availability of system } S \text{ at time } t;$$
$$\bar{A}(t) = 1 - A(t), \text{ its unavailability;}$$
$$q_i(t) = \text{availability of component } i \text{ at time } t;$$
$$\bar{q}_i(t) = 1 - q_i(t), \text{ the unavailability of } i \text{ at } t.$$

Calculation of the availabilities $q_i(t)$ was examined in detail in Ch. 3. It will therefore be assumed in sections 1–4 that the $q_i(t)$ are given.

2. AVAILABILITY IN SOME SIMPLE CASES

Here, we shall examine systems whose availability can be calculated directly from the reliability block diagram, fault tree or successful paths.

2.1. Reliability block diagram

Consider a reliability block diagram in which the components are *not replicated* and are arranged either *in series* or *in parallel*. In this diagram we can then replace each set of components $E_1\ E_2$ in series by

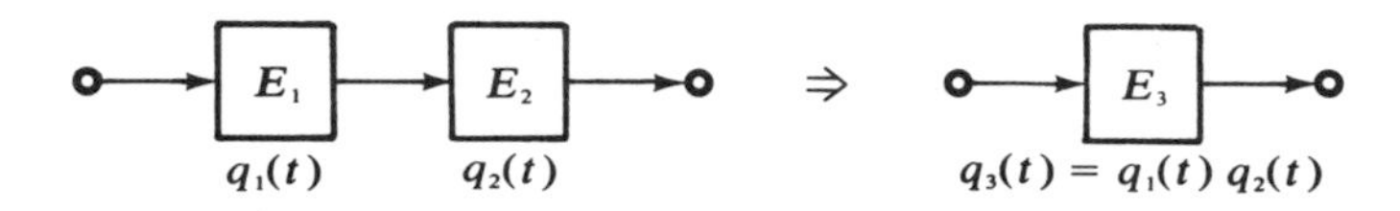

a single component E_3 with availability $q_3(t) = q_1(t)q_2(t)$, and each set of components $E_1\ E_2$ in parallel by

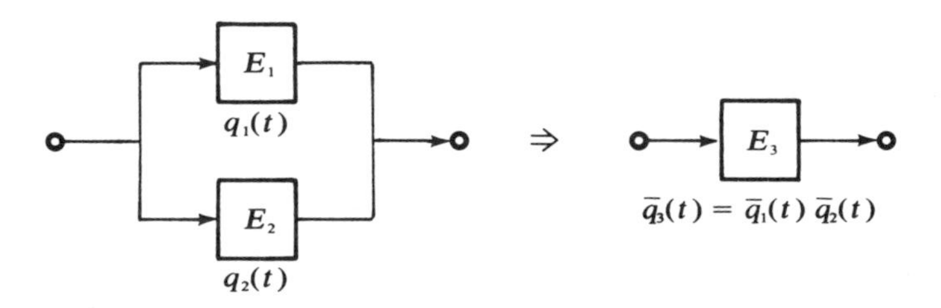

a single component E_3 with availability $q_3(t) = 1 - (1 - q_1(t))(1 - q_2(t))$.

By *successively* applying these reductions, we obtain the system availability directly. This is the method most frequently employed for analysing simple systems (cf. Ch. 3, section 2).

Note that, since $q_1(t)$ generally approximates to unity ($\bar{q}_i(t) \ll 1$), the calculations are greatly simplified by using the unavailabilities: in fact, in the case of components in series, we have

$$\bar{q}_3(t) = 1 - q_1(t)\,q_2(t) = 1 - (1 - \bar{q}_1(t))\,(1 - \bar{q}_2(t)) \simeq \bar{q}_1(t) + \bar{q}_2(t) \tag{1}$$

and in the case of components in parallel, we have

$$\bar{q}_3(t) = \bar{q}_1(t)\,\bar{q}_2(t)\,. \tag{2}$$

Example:

Consider the following reliability block diagram:

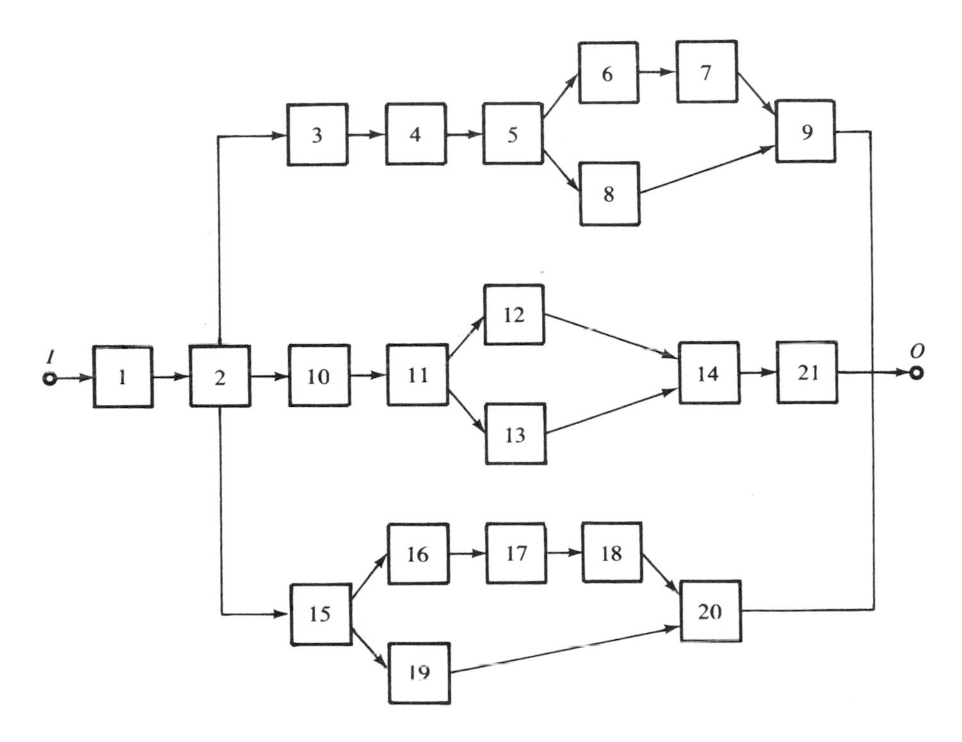

Fig. 1.

with $\bar{q}_3(\infty) = \bar{q}_4(\infty) = \bar{q}_6(\infty) = \bar{q}_8(\infty) = \bar{q}_{11}(\infty) = \bar{q}_{14}(\infty) = \bar{q}_{21}(\infty) = \bar{q}_{15}(\infty) = \bar{q}_{20}(\infty) = 10^{-3}$, $\bar{q}_5(\infty) = \bar{q}_7(\infty) = \bar{q}_{16}(\infty) = \bar{q}_{17}(\infty) = \bar{q}_{19}(\infty) = 3 \,.\, 10^{-3}$, $\bar{q}_9(\infty) = \bar{q}_{10}(\infty) = \bar{q}_{12}(\infty) = \bar{q}_{18}(\infty) = 10^{-2}$, $\bar{q}_{13}(\infty) = 10^{-1}$, $\bar{q}_1(\infty) = \bar{q}_2(\infty) = 10^{-7}$.

The set $\{6, 7\}$ has an asymptotic unavailability $\bar{q}_6 + \bar{q}_7 = 4 \times 10^{-3}$.
The set $\{6, 7, 8\}$ therefore has an unavailability $4 \times 10^{-3}\bar{q}_8 = 4 \times 10^{-6}$.
The set $\{3, 4, 5, 6, 7, 8, 9\}$ therefore has an unavailability

$$\bar{q}_3 + \bar{q}_4 + \bar{q}_5 + 4 \times 10^{-6} + \bar{q}_9 = 1.5 \times 10^{-2}.$$

The set $\{10, 11, 12, 13, 14, 21\}$ has an unavailability

$$\bar{q}_{10} + \bar{q}_{11} + \bar{q}_{12}\bar{q}_{13} + \bar{q}_{14} + \bar{q}_{21} = 1.4 \times 10^{-2}.$$

The set $\{15, 16, 17, 18, 19, 20\}$ has an unavailability

$$\bar{q}_{15} + (\bar{q}_{16} + \bar{q}_{17} + \bar{q}_{18})\bar{q}_{19} + \bar{q}_{20} = 2 \times 10^{-3}.$$

The set $\{1, 2\}$ has an unavailability

$$\bar{q}_1 + \bar{q}_2 = 2 \times 10^{-7}.$$

Finally, the system shown in Fig. 1 has an asymptotic unavailability:

$$\bar{A}_\infty = 2 \times 10^{-7} + (1.5 \times 10^{-2})(1.4 \times 10^{-2})(2 \times 10^{-3}) = 6.2 \times 10^{-7}.$$

2.2. Fault tree

Consider a fault tree in which the *basic events are not replicated.*

It can be shown that this corresponds to the reliability block diagram given in section 2.1 above (if the fault tree is properly constructed). In this case, we can therefore compute the probability of the undesirable event from the basic event probabilities $p_i(t)$. We move up the fault tree, starting from the basic events, by assigning probability $p_3(t) = p_1(t)p_2(t)$ to the 'output' event E_3 of an AND gate having events E_1 and E_2 as 'inputs',

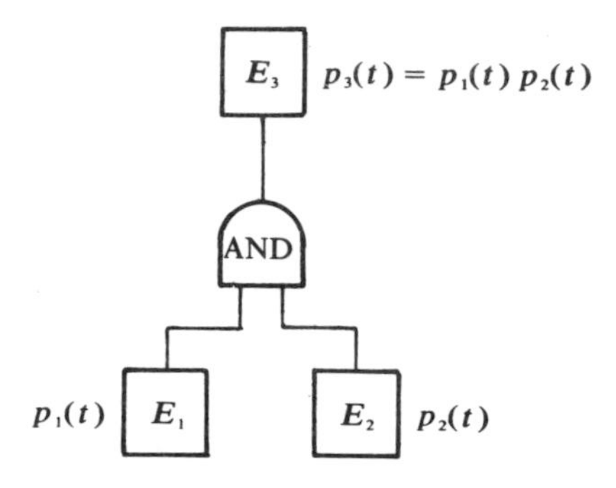

and assigning to 'output' event E_3 of an OR gate having events E_1 and E_2 as 'inputs', the probability

$$p_3(t) = 1 - (1 - p_1(t))\,(1 - p_2(t))\,.$$

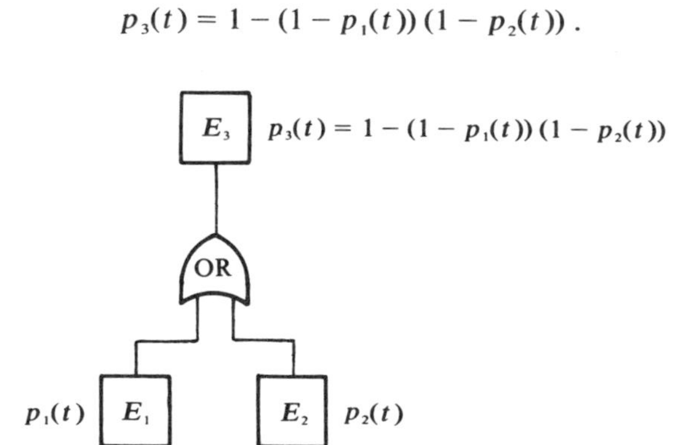

By *successively* applying these two rules, we obtain the probability of the undesirable event.

Note that event i generally represents the failure of a component and $p_i(t)$ therefore corresponds to the unavailability of that component; consequently we often have $p_i(t) = \bar{q}_i(t)$. The comment made in section 2.1 therefore applies, namely that since the $p_i(t)$ are very small ($p_i(t) \ll 1$), the computations in the fault tree are very simple. In fact, in the case of an AND gate, we have:

$$p_3(t) = p_1(t)\,p_2(t) \tag{3}$$

and, in the case of an OR gate, we have:

$$p_3(t) = 1 - (1 - p_1(t))(1 - p_2(t)) \simeq p_1(t) + p_2(t). \tag{4}$$

Example:
Consider the following fault tree:

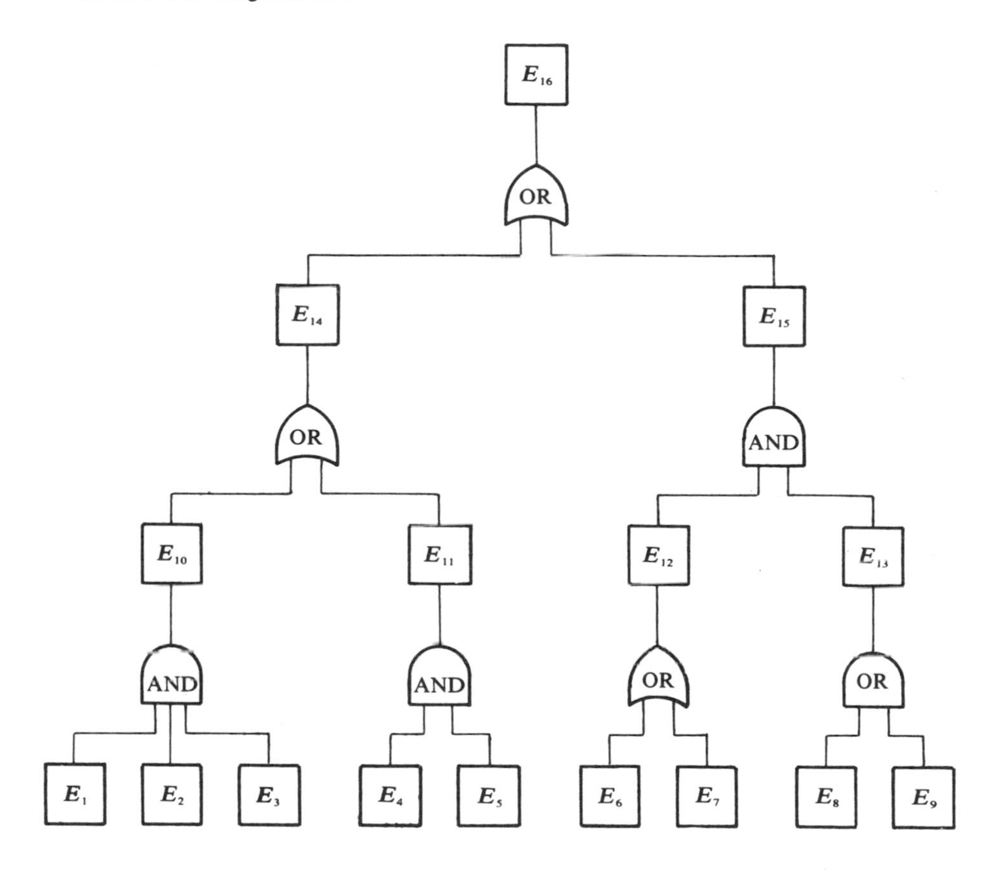

Fig. 2.

with $p_1(\infty) = p_2(\infty) = p_6(\infty) = p_8(\infty) = p_9(\infty) = 10^{-2}$, $p_3(\infty) = p_7(\infty) = 2 \,.\, 10^{-2}$,
$p_4(\infty) = p_5(\infty) = 10^{-3}$.

From Eqns (3) and (4) we extract, in turn:

$p_{10} = p_1 \, p_2 \, p_3 = 2 \,.\, 10^{-6}$, $p_{11} = p_4 \, p_5 = 10^{-6}$, $p_{12} = p_6 + p_7 = 3 \,.\, 10^{-2}$, $p_{13} = p_8 \, p_9 = 10^{-4}$,
$p_{14} = p_{10} + p_{11} = 3 \,.\, 10^{-6}$, $p_{15} = p_{12} \, p_{13} = 3 \,.\, 10^{-6}$, $p_{16} = p_{14} + p_{15} = 6 \,.\, 10^{-6}$.

2.3. Successful paths

We now consider a more general case than those examined in sections 2.1 and 2.2: the system logic is represented by the set of minimal successful paths P_j, j from 1 to m.

The availability $A(t)$ is then the probability that at least one of the paths is working, i.e.

$$A(t) = \text{Prob}\,(P_1 \cup P_2 \cup \cdots \cup P_m)\,. \tag{5}$$

It is therefore given by Poincaré's formula:

$$A(t) = \sum_{j=1,m} \mathcal{P}(P_j) - \sum_{j=2}^{m} \sum_{i=1}^{j-1} \mathcal{P}(P_i\,P_j) + \sum_{j=3}^{m} \sum_{i=2}^{j-1} \sum_{k=1}^{i-1} \mathcal{P}(P_k\,P_i\,P_j) + \cdots + (-1)^m\,\mathcal{P}(P_1\,P_2 \ldots P_m) \tag{6}$$

where $\mathcal{P}(P_j)$ represents the probability that at time t the components i of set P_j are working; therefore

$$\mathcal{P}(P_j) = \prod_{i \in P_j} q_i(t)\,.$$

Let us apply these formulae to the case of a system whose successful paths are: $E_1\,E_3$, $E_1\,E_4$, $E_2\,E_4$, $E_2\,E_5$ (cf. the example in Ch. 2, section 2). We therefore have:

$$\begin{aligned} A(t) = {} & (q_1\,q_3 + q_1\,q_4 + q_2\,q_4 + q_2\,q_5) - \\ & - (q_1\,q_3\,q_4 + q_1\,q_2\,q_3\,q_4 + q_1\,q_2\,q_3\,q_5 + q_1\,q_2\,q_4 + q_1\,q_2\,q_4\,q_5 + q_2\,q_4\,q_5) + \\ & + q_1\,q_2\,q_3\,q_4 + q_1\,q_2\,q_3\,q_4\,q_5 + q_1\,q_2\,q_3\,q_4\,q_5 + q_1\,q_2\,q_4\,q_5) - (q_1\,q_2\,q_3\,q_4\,q_5) \end{aligned}$$

which represents a somewhat lengthy computation.

Note that if the availability $q_i(t)$ of each component is good ($q_i \sim 1$), each term in formula (6) is approximately 1 and no simplification is possible. On the other hand, if the *availabilities $q_i(t)$ are very bad* ($\bar{q}_i(t) \sim 1$), then in formula (6) we can stop at the initial terms:

$$A(t) \simeq \sum_{j=1,m} \mathcal{P}(P_j)\,.$$

Even then, we have the two-sided estimation:

$$\sum_{j=1,m} \mathcal{P}(P_j) - \sum_{j=2}^{m} \sum_{i=1}^{j-1} \mathcal{P}(P_i\,P_j) \leq A(t) \leq \sum_{j=1,m} \mathcal{P}(P_j)\,.$$

As the component availabilities are generally good ($q_i \sim 1$), we never compute the availability of a system directly from the successful paths; to do this we use the minimal cut sets (section 3).

3. AVAILABILITY AND MINIMAL CUT SETS

We shall now consider a coherent system with independent components. The system logic can therefore be represented by the minimal cut set C_j, j from 1 to m (cf. Ch. 2, section 4).

3.1. General case

The unavailability $\bar{A}(t) = 1 - A(t)$ is therefore the probability of at least one of the cut sets failing, i.e.:

$$\bar{A}(t) = \mathcal{P}(\bar{C}_1 \cup \bar{C}_2 \cup \ldots \cup \bar{C}_m)\,. \tag{7}$$

The unavailability is therefore given by

$$\overline{A}(t) = \sum_{j=1}^{m} \mathcal{P}(\overline{C}_j) - \sum_{j=2}^{m} \sum_{i=1}^{j-1} \mathcal{P}(\overline{C}_i \, \overline{C}_j) + \sum_{j=3}^{m} \sum_{i=2}^{j-1} \sum_{k=1}^{i-1} \mathcal{P}(\overline{C}_k \, \overline{C}_i \, \overline{C}_j) + \cdots + (-1)^m \, \mathcal{P}(\overline{C}_1 \, \overline{C}_2 \ldots \overline{C}_m) \tag{8}$$

where $\mathcal{P}(\overline{C}_j)$ represents the probability that at time t the elements i of set C_j have failed; therefore

$$\mathcal{P}(\overline{C}_j) = \prod_{i \in C_j} \overline{q}_i(t) \,. \tag{9}$$

Let us apply these formulae to the case of the system analysed in section 2.3, i.e. a system whose minimal cut sets are $E_1 E_2$, $E_1 E_4 E_5$, $E_2 E_3 E_4$, $E_3 E_4 E_5$ (cf. Ch. 2, sections 2 and 4). We therefore have

$$\begin{aligned} \overline{A}(t) = {} & (\overline{q}_1 \overline{q}_2 + \overline{q}_1 \overline{q}_4 \overline{q}_5 + \overline{q}_2 \overline{q}_3 \overline{q}_4 + \overline{q}_3 \overline{q}_4 \overline{q}_5) \\ & - (\overline{q}_1 \overline{q}_2 \overline{q}_4 \overline{q}_5 + \overline{q}_1 \overline{q}_2 \overline{q}_3 \overline{q}_4 + \overline{q}_1 \overline{q}_2 \cdot \overline{q}_3 \overline{q}_4 + \overline{q}_1 \overline{q}_2 \overline{q}_3 \overline{q}_4 \overline{q}_5 + \overline{q}_1 \overline{q}_2 \overline{q}_3 \overline{q}_4 \overline{q}_5 + \overline{q}_1 \overline{q}_3 \overline{q}_4 \overline{q}_5 + \overline{q}_2 \overline{q}_3 \overline{q}_4 \overline{q}_5) \\ & + (\overline{q}_1 \overline{q}_2 \overline{q}_3 \overline{q}_4 \overline{q}_5 + \overline{q}_1 \overline{q}_2 \overline{q}_3 \overline{q}_4 \overline{q}_5 + \overline{q}_1 \overline{q}_2 \overline{q}_3 \overline{q}_4 \overline{q}_5 + \overline{q}_1 \overline{q}_2 \overline{q}_3 \overline{q}_4 \overline{q}_5) - (\overline{q}_1 \overline{q}_2 \overline{q}_3 \overline{q}_4 \overline{q}_5) \end{aligned} \tag{10}$$

which represents a somewhat lengthy computation.

However, in most cases (e.g. for repairable systems) the availability $q_i(t)$ of each component i is very good ($q_i(t) \sim 1$) and the initial terms of formula (8) suffice:

$$\overline{A}(t) \simeq \sum_{j=1}^{m} \mathcal{P}(\overline{C}_j) = \sum_{j=1}^{m} \prod_{i \in C_j} \overline{q}_i(t) \,. \tag{11}$$

If we wish to check the error made, we can use the two-sided estimation:

$$\sum_{j=1,\, m} \mathcal{P}(\overline{C}_j) - \sum_{j=2}^{m} \sum_{i=1}^{j-1} \mathcal{P}(\overline{C}_i \, \overline{C}_j) \leq \overline{A}(t) \leq \sum_{j=1}^{m} \mathcal{P}(\overline{C}_j) \,. \tag{12}$$

This estimation may sometimes involve a long computation if the number of minimal cut sets is large. It can then be shown (cf. Exercise 3) that the following two-sided estimation is also available:

$$\sum_{j=1}^{m} \prod_{i \in C_j} \overline{q}_i(t) \prod_{i \notin C_j} q_i(t) \leq \overline{A}(t) \leq \sum_{j=1}^{m} \prod_{i \in C_j} \overline{q}_i(t) \,. \tag{13}$$

3.2. Network case

The above formulae therefore enable availability to be computed provided the minimal cut sets are known. In Ch. 2, section 4, we showed how the latter are determined.

However, when the number of components in a system is large, determining the minimal cut sets may become tedious. Now, the approximation formula (11) shows that it is not absolutely necessary to determine them all, but only those which are crucial to the unavailability. We shall show that this is a feasible approach if we use a special reliability block diagram.

Consider, for example, an electric power system represented by a graph (cf.

[Gondran and Minoux, 1979]) whose vertices correspond to the junctions of the lines and whose edges correspond to the power lines themselves:

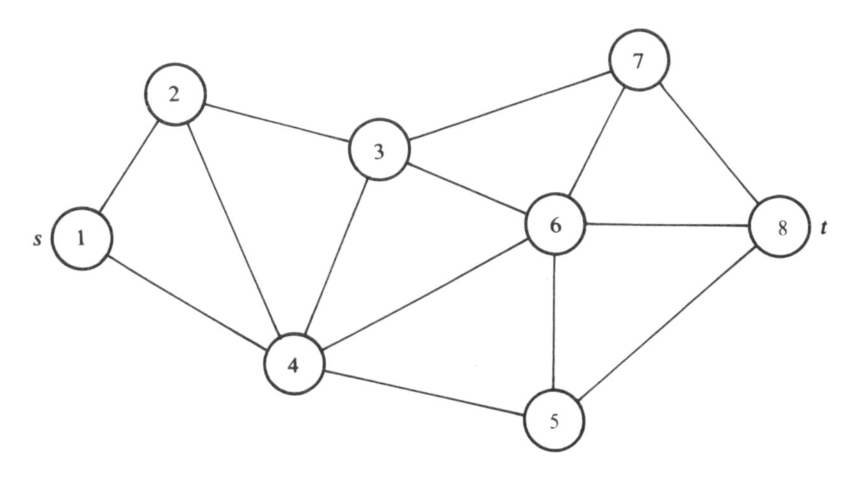

Fig. 3. *Electric power system.*

Assume that, for each edge (ij) of the graph $G = (X, U)$, the probability p_{ij} of the line being down is known and that these probabilities are independent. Consider the stationary case, the p_{ij} are not therefore time-dependent and consequently correspond to the mean (or asymptotic) unavailability of the edge (ij).

The problem is to find the probability of the vertex $t \in X$ being supplied by one of the lines from the generating source located at vertex $s \in X$.

The conventional method described in Ch. 2 is to enumerate all the successful paths and then all the minimal cut sets. The unavailability of the link between s and t is then given by formula (8) or (11).

Let us find a way of directly defining the minimal cut set which has the greatest effect on $\bar{A}(t)$ given by Eqn (11). To do this, we assign to each edge (ij) a capacity $C_{ij} = -\log p_{ij} \geqslant 0$. Note that a minimal cut set in the reliability sense corresponds to a cut set between s and t of the theory of flows (cf. [Gondran and Minoux, 1979]).

Each set A of vertices of X containing s and not containing t can be assigned the cut set:

$$w(A) = \{(i, j) \in U, i \in A, j \notin A\}. \tag{14}$$

The capacity of the cut set $w(A)$ is then equal to:

$$C(A) = \sum_{(i,j) \in w(A)} C_{ij} = -\log\left(\prod_{(i,j) \in w(A)} p_{ij}\right). \tag{15}$$

Moreover, $\prod_{(i,j) \in w(A)} p_{ij}$ represents the unavailability of the minimal cut set $w(A)$.

It follows that the minimal cut set with the greatest effect on unavailability is the cut set $w(A)$ which has the smallest capacity.

Now, in accordance with the Ford Fulkerson theorem (cf. [Gondran and Minoux,

1979]) this minimum capacity cut set is obtained by finding the maximum flow between the vertices s and t. The subsequent cut sets correspond to the 2nd, 3rd, ... solution of the dual linear program of the maximum flow problem. The reader is referred to [Gondran and Minoux, 1979] for algorithms for determining these cut sets. A more complex problem, evaluating the mathematical expectation of failure of an electric power system, may be approached in a similar manner (cf. Exercise 4).

3.3. Importance factors

It is useful to be able to take account of the importance of a particular system component for system availability. A large number of methods for measuring this importance have been proposed, all of which consist of assigning to each component a function known as the *importance factor*.

In this section, we shall only give the importance factors associated with system availability; others associated with the reliability of repairable systems will be given in Ch. 6, section 2.3.

Equations (7), (8), (9) and (11) show that system unavailability $\bar{A}(t)$ is a function of the unavailability $\bar{q}_i(t)$ of the system components.

$$\bar{A}(t) = g[\bar{q}_i(t)]. \tag{16}$$

For coherent systems, this equation can be written as:

$$\bar{A}(t) = (1 - \bar{q}_i(t))\, g[\bar{q}_j(t)/\bar{q}_i(t) \equiv 0] + \bar{q}_i(t)\, g[\bar{q}_j(t)/\bar{q}_i(t) \equiv 1]. \tag{17}$$

It follows that:

$$\frac{\partial \bar{A}}{\partial \bar{q}_i}(t) - g[\bar{q}_j(t)/\bar{q}_i(t) \equiv 1] - g[\bar{q}_j(t)/\bar{q}_i(t) \equiv 0]. \tag{18}$$

From these equations, we can define the following importance factors.

3.3.1. *Marginal importance factor* (cf. [Birnbaum, 1969])

$$B_i(t) = \frac{\partial \bar{A}}{\partial \bar{q}_i}(t). \tag{19}$$

This enables us to measure the system unavailability variation as a function of the unavailability variation of a component.

3.3.2. *Critical importance factor* (cf. [Lambert, 1975])

$$C_i(t) = \frac{\bar{q}_i(t)}{\bar{A}(t)} \frac{\partial \bar{A}}{\partial \bar{q}_i}(t). \tag{20}$$

This represents the probability of component i having caused system failure (it is not necessarily the only failed component, but in this case it is the last to have failed), given that the system has failed.

3.3.3. *'Diagnostic' importance factor* (cf. [Vesely, 1974; Fussell, 1975])

$$VF_i(t) = \frac{\bar{q}_i(t)\, g[\bar{q}_j(t)/\bar{q}_i(t) \equiv 1]}{\bar{A}(t)}. \tag{21}$$

This represents the probability of component i failing, given that the system has failed. This may be useful for diagnosing the causes of system failure.

Readers are referred to Ch. 6, section 2.3 for a more detailed treatment and examples concerning importance factors (cf. also [Lambert 1975]).

4. AVAILABILITY OF NON-COHERENT SYSTEMS

In this section, the components are still independent, but the system is no longer coherent. It is therefore assumed that the system logic can be represented by a Boolean equation (cf. Ch. 2, section 5).

Consider the case where the system has failed when one of the events E_j, j from 1 to m, has occurred. Each event E_j is the conjunction of the failure of a certain number of components with the *operation of a certain number of others*. For a four-component system, the Boolean system failure function F may be:

$$F = \underbrace{\bar{X}_1\, \bar{X}_2\, X_3}_{E_1} + \underbrace{\bar{X}_1\, \bar{X}_4}_{E_2} + \underbrace{X_2\, \bar{X}_3\, \bar{X}_4}_{E_3} \tag{22}$$

where X_i corresponds to the operation of component i and $\bar{X}_i$ corresponds to its failure.

In general, the system will approximate to a coherent system and each of the E_j will approximate to a cut set. In our example, E_2 is a cut set and E_1 and E_3 approximate to cut sets $\bar{X}_1\, \bar{X}_2$ and $\bar{X}_3\, \bar{X}_4$.

In order to compute the system unavailability:

$$\bar{A}(t) = \text{Prob}\,(E_1 \cup E_2 \cup \ldots \cup E_m) \tag{23}$$

we again use Poincaré's formula

$$\bar{A}(t) = \sum_{j=1}^{m} \mathcal{P}(E_j) - \sum_{j=2}^{m} \sum_{i=1}^{j-1} \mathcal{P}(E_i \,.\, E_j) + \cdots + (-1)^m\, \mathcal{P}(E_1\, E_2 \ldots E_m)\,. \tag{24}$$

For the four-component example described above, we therefore have:

$$\bar{A}(t) = (\bar{q}_1\, \bar{q}_2\, q_3 + \bar{q}_1\, \bar{q}_4 + q_2\, \bar{q}_3\, \bar{q}_4) - (\bar{q}_1\, \bar{q}_2\, q_3\, \bar{q}_4 + 0 + \bar{q}_1\, q_2\, \bar{q}_3\, \bar{q}_4) + 0\,. \tag{25}$$

Note, therefore, that when the E_j are virtually minimal cut sets and the availability of each component is good, we obtain a good approximation for the unavailability by:

$$\bar{A}(t) \simeq \sum_{j=1}^{m} \mathcal{P}(E_j)\,. \tag{26}$$

5. AVAILABILITY AND NON-INDEPENDENT COMPONENTS

In all the availability calculations in the previous sections it has been assumed that the components are *s*-independent. Now, this is rarely completely the case in practical problems. (In general, a functional dependence between the components creates a probability dependence):

—a component is only activated if another has failed (standby redundancy),
—there are a limited number of repair facilities, which means that not all the components can be repaired at the same time,
—the failure rate of a component operating in parallel with another (active redundancy) may increase when the second fails (cf. Ch. 2, section 6),
—etc....

We have seen that all these cases can be represented by a state transition diagram: cf. Ch. 2, section 6. The general theory will be examined in Ch. 5, section 2.4.1.

We will show, in the following sections, that the availability of these systems can be computed relatively simply. It will then be possible to use these results in the context of the previous sections.

5.1. Rapid computation of asymptotic unavailability

In a large number of cases the asymptotic availability can be easily computed using the results given below.

5.1.1. *'Flow conservation' method*

Consider the state transition diagram of a system (cf. Ch. 2, section 6). Between two states i and j, we assume that the transition rate λ_{ij} is constant (independent of t). Such a system corresponds to a time-homogeneous Markov process with discrete state space (cf. Ch. 5, sections 1 and 2).

Let $Q_i(\infty)$ be the probability of being in state i under steady-state conditions. Under these conditions, the transition frequency from state i to state j is $Q_i(\infty)\,\lambda_{ij}$ (cf. Ch. 5, section 2.4.4).

Let $\mathscr{F}$ be any set of system states.

A 'cut set' in the graph theory sense (cf. [Gondran and Minoux, 1979] and section 3.2 of this chapter for the non-oriented case) associated with $\mathscr{F}$ is the set of edges having one and only one extremity in $\mathscr{F}$.

Theorem 1 [Krakowski, 1973; Lemaire, 1977; Singh, 1972]—*In the steady state domain, the frequency of the transitions out of any cut set is equal to the frequency of the transitions into it. Or*

$$\sum_{i\in\mathscr{F}}\sum_{j\notin\mathscr{F}} Q_i(\infty)\,\lambda_{ij} = \sum_{j\notin\mathscr{F}}\sum_{i\in\mathscr{F}} Q_j(\infty)\,\lambda_{ji}\,. \tag{27}$$

This can be simply demonstrated by recurrence on the cardinal of $\mathscr{F}$ (cf. [Lemaire, 1977]).

For each cut set, there is therefore conservation of the transition flow.

Let us give an example in which the application of Eqn (27) very quickly gives the asymptotic availability. Consider the following state transition diagram:

$$n \underset{\mu_{n-1}}{\overset{\lambda_n}{\rightleftarrows}} n-1 \underset{\mu_{n-2}}{\overset{\lambda_{n-1}}{\rightleftarrows}} n-2 \cdots i+1 \underset{\mu_i}{\overset{\lambda_{i+1}}{\rightleftarrows}} i \cdots 2 \underset{\mu_1}{\overset{\lambda_2}{\rightleftarrows}} 1 \underset{\mu_0}{\overset{\lambda_1}{\rightleftarrows}} 0$$

A diagram of this kind is often encountered in availability problems (cf. sections 5.1 and 5.3 and Ch. 8, section 3).

Taking $\mathscr{F} = \{n, n-1, n-2, \ldots, i+1\}$, Eqn (27) can be written:

$$Q_{i+1}(\infty)\,\lambda_{i+1} = \mu_i\,Q_i(\infty)\,. \tag{28}$$

A recursive equation providing the classical result:

$$Q_i(\infty) = \frac{\lambda_{i+1}\cdot\lambda_{i+2}\cdots\lambda_n}{\mu_i\cdot\mu_{i+1}\cdots\mu_{n-1}}\,Q_n(\infty)$$

We therefore obtain $Q_n(\infty)$, noting that

$$\sum_{i=0,n} Q_i(\infty) = 1 \tag{29}$$

or

$$Q_n(\infty) = \frac{1}{1 + \displaystyle\sum_{i=0,n-1} \frac{\lambda_{i+1}\cdot\lambda_{i+2}\cdots\lambda_n}{\mu_i\cdot\mu_{i+1}\cdots\mu_{n-1}}}\,. \tag{30}$$

Numerically when $\left[\prod_{j=i}^{n-1} \frac{\lambda_{j+1}}{\mu_j} \ll 1\right]$ then $Q_n(\infty) \simeq 1$.

If the system is unavailable when it is in state 0, then:

$$\bar{A}_\infty = Q_0(\infty) = \frac{\prod_{i=1,n} \lambda_i}{\prod_{i=0,n-1} \mu_i}\,Q_n(\infty)\,. \tag{31}$$

If the system is unavailable when it is in one of states $\{0, 1, \ldots, r-1\}$ (cf. r/n redundancy, for example), then:

$$\bar{A}_\infty = \sum_{i=0,r-1} Q_i(\infty) = Q_n(\infty) \sum_{i=0,r-1} \frac{\lambda_{i+1}\cdot\lambda_{i+2}\cdot\cdots\lambda_n}{\mu_i\cdot\mu_{i+1}\cdot\cdots\cdot\mu_{n-1}}\,. \tag{32}$$

A system of this type will be discussed in Ch. 8, section 3, in connection with stock control of spares.

In the case of more complex systems, Eqns (27) and (32) are no longer adequate and we use the method described in section 5.1.2 below.

5.1.2. *Direct 'reading' of a Markov diagram*

In the case where the λ_i/μ_j are small compared with 1, the results from Ch. 5, section 5.4 and Ch. 6, section 2.2 provide an excellent approximation for the asymptotic availability. A brief outline of this method follows.

In the state transition diagram, we denote by α the vertex for which all the components are available and by B the set of vertices β for which the system is regarded as failed.

For a vertex i of the state transition graph, we denote by $|a_{ii}|$ the sum of the transition rates resulting from this vertex (cf. Ch. 6).

We therefore have:

Theorem 2—*If the state transition graph is strongly connected, then, under certain conditions (cf. theorem 2, Ch. 6), an excellent approximation for system unavailability is given by:*

$$\bar{A}_\infty = \sum_{\text{paths from } \alpha \text{ to } \beta} \frac{\prod (\text{path transition rate})}{\prod\limits_{\substack{i \in \text{path} \\ i \notin \alpha}} |a_{ii}|}. \tag{33}$$

The demonstration is identical to that of theorem 2, Ch. 6 (cf. also Exercise 2 and section 6 of Ch. 4).

This theorem is very important, as it will provide a very simple means of computing the unavailability of complex systems (cf., for example, Exercise 7).

We easily obtain Eqn (31), for example. In fact, there is only one path from n to 0 (from α to β) and we therefore have:

$$\bar{A}_\infty \simeq \frac{\prod\limits_{i=1,n} \lambda_i}{\prod\limits_{i=0,n-1} (\mu_i + \lambda_i)} \simeq \frac{\prod\limits_{i=1,n} \lambda_i}{\prod\limits_{i=0,n-1} \mu_i}.$$

For non-Markovian systems (λ_{ij} non-constant), see Ch. 4, section 6, Ch. 5, sections 3 and 4 and Ch. 6, section 3.

5.2. Standby (passive) redundancy

Consider two identical components with failure rate λ, repair rate $\mu = \tau^{-1}$ in a standby redundancy configuration. When the first fails, the second takes over and repair of the first commences immediately. When both are down, there are two repair facilities.

The system has three states:

—state 2, in which the first component is operating and the second is standing by in an operable condition,

—state 1, in which one of the components is operating and the other is down,
—state 0, where both components are down.

The state transition diagram is therefore:

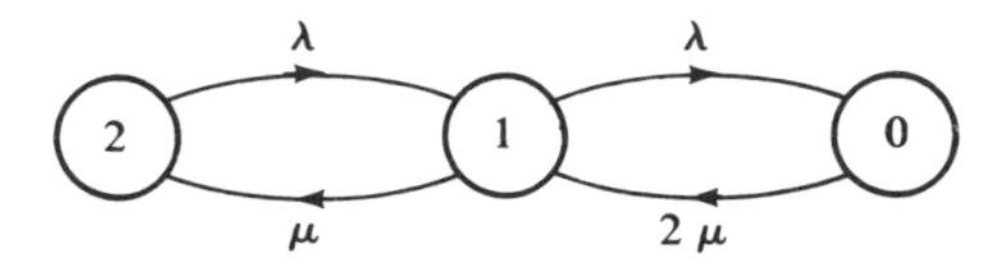

If $Q_i(t)$ is the probability of being in state i, the system availability $A(t)$ is equal to $Q_2(t) + Q_1(t)$ and the $Q_i(t)$ satisfy the differential equation (cf. Ch. 3):

$$\begin{cases} \dfrac{dQ_2}{dt} = -\lambda Q_2(t) + \mu Q_1(t) \\ \dfrac{dQ_1}{dt} = \lambda Q_2(t) - (\lambda + \mu)\, Q_1(t) + 2\,\mu Q_0(t) \\ \dfrac{dQ_0}{dt} = \lambda Q_1(t) - 2\,\mu Q_0(t) \end{cases} \tag{34}$$

with $Q_2(0) = 1$ and $Q_1(0) = Q_0(0) = 0$.

This system can be solved manually using the Laplace transform (cf. Ch. 3) or on a computer by discretization (cf. Ch. 5).

We then have $\bar{A}(t) = Q_0(t)$.

In practice (as generally $(\lambda/\mu) \ll 1$, cf. Ch. 3, section 3), $\bar{A}(t)$ will rapidly tend to a limit $\bar{A}_\infty$ (asymptotic unavailability). This limit is calculated by solving a linear system. In our example, for $t = +\infty$, $dQ/dt = 0$; Eqn (34) therefore gives:

$$\begin{cases} Q_1(\infty) = \dfrac{2\,\mu}{\lambda} Q_0(\infty) \\ Q_2(\infty) = \dfrac{\mu}{\lambda} Q_1(\infty) = \dfrac{2\,\mu^2}{\lambda^2} Q_0(\infty) \end{cases} \tag{35}$$

which, taking account of $Q_0(\infty) + Q_1(\infty) + Q_2(\infty) = 1$, ultimately gives:

$$\bar{A}_\infty = Q_0(\infty) = \frac{\lambda^2}{2\,\mu^2 + 2\,\mu\lambda + \lambda^2}. \tag{36}$$

This result is obtained directly from Eqns (30) and (31).

Two more complex examples of standby redundancy (where there is a probability of non-start-up γ) are examined in Exercises 5 and 7.

5.3. Parallel (active) redundancy

This is the same case as in section 5.2, but here both components are normally operating. In addition, when one of the components is down, the other has a failure rate $\lambda' \geqslant \lambda$ (as it is supporting a greater load). There is only one repair facility (cf. Ch. 2, section 6).

The state transition diagram is then:

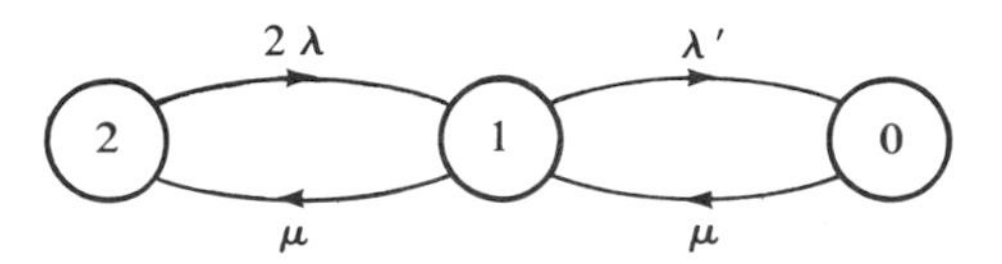

Equation (28) gives:

$$Q_1(\infty) = \frac{\mu}{\lambda'} Q_0(\infty)$$

$$Q_2(\infty) = \frac{\mu}{2\lambda} Q_1(\infty) = \frac{\mu^2}{2\lambda\lambda'} Q_0(\infty).$$

Writing $Q_0(\infty) + Q_1(\infty) + Q_2(\infty) = 1$, we derive

$$\bar{A}_\infty = Q_0(\infty) = \frac{2\lambda\lambda'}{\mu^2 + 2\lambda\mu + 2\lambda\lambda'}. \tag{37}$$

Let us now assume that $\lambda' = \lambda$ and compare the unavailability $\bar{A}_\infty$ obtained in Eqn (37) with the unavailability when there are two repair facilities.

In this case, both components are independent and each has an asymptotic unavailability $\bar{q}(\infty) = \lambda/(\lambda + \mu)$. Thc asymptotic unavailability of the system is therefore:

$$\bar{A}'_\infty = (\bar{q}(\infty))^2 = \frac{\lambda^2}{\mu^2 + 2\lambda\mu + \lambda^2}. \tag{38}$$

In the very general case where $(\lambda/\mu) \ll 1$, we therefore have:

$$\bar{A}_\infty \simeq 2\bar{A}'_\infty. \tag{39}$$

The dependence due to restricted repair facilities has doubled the system unavailability.

5.4. A maintenance policy

The following example is intended as a teaching exercise. A more realistic study of maintenance policy is given in section 6.3 of this chapter.

Let us now consider the case of two different components with failure rates λ_1 and λ_2, repair rates $\mu_1 = \tau_1^{-1}$ and $\mu_2 = \tau_2^{-1}$ in a parallel redundancy configuration. There is only one repair facility and the maintenance policy is to give priority to repairing the first failed component.

The chronological order of the occurrence of failures must therefore be adhered to: this involves duplicating the failed state of the system. The various states are as follows:

{1 1} Both components are working;

$\{1\ 0\}$ Component 1 is working, component 2 has failed;

$\{0\ 1\}$ Component 1 has failed, component 2 is working;

$\{\underline{0}\ 0\}$ Both components have failed, component 1 having failed first;

$\{0\ \underline{0}\}$ Both components have failed, component 2 having failed first.

The state transition diagram is therefore:

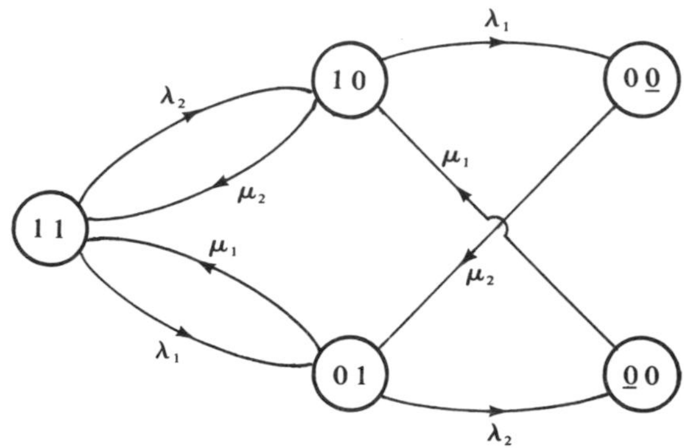

Applying Eqn (27) with

$$\mathscr{F} = \{0\,\underline{0}\}, \mathscr{F} = \{\underline{0}\,0\}, \mathscr{F} = \{1\,0\} \quad \text{et} \quad \mathscr{F} = \{0\,1\}.$$

we obtain the equations

$$\left.\begin{aligned} Q_{0\underline{0}}\,\mu_2 &= Q_{10}\,\lambda_1 \\ Q_{\underline{0}0}\,\mu_1 &= Q_{01}\,\lambda_2 \\ Q_{10}(\lambda_1 + \mu_2) &= Q_{\underline{0}0}\,\mu_1 + Q_{11}\,\lambda_2 \\ Q_{01}(\lambda_2 + \mu_1) &= Q_{0\underline{0}}\,\mu_2 + Q_{11}\,\lambda_1. \end{aligned}\right\} \tag{40}$$

By substituting the two first equations in the last two, we obtain:

$$\begin{aligned} (\lambda_1 + \mu_2)\,Q_{10} - \lambda_2\,Q_{01} &= \lambda_2\,Q_{11} \\ -\lambda_1\,Q_{10} + (\lambda_2 + \mu_1)\,Q_{01} &= \lambda_1\,Q_{11} \end{aligned}$$

giving:

$$\left.\begin{aligned} Q_{10} &= \frac{\lambda_2(\mu_1 + \lambda_2 + \lambda_1)}{(\lambda_1 + \mu_2)(\lambda_2 + \mu_1) + \lambda_1\,\lambda_2}\,Q_{11} \\ Q_{01} &= \frac{\lambda_1(\mu_2 + \lambda_2 + \lambda_1)}{(\lambda_1 + \mu_2)(\lambda_2 + \mu_1) + \lambda_1\,\lambda_2}\,Q_{11} \\ Q_{0\underline{0}} &= \frac{\dfrac{\lambda_1}{\mu_2}\,\lambda_2(\mu_1 + \lambda_2 + \lambda_1)}{(\lambda_1 + \mu_2)(\lambda_2 + \mu_1) + \lambda_1\,\lambda_2}\,Q_{11} \\ Q_{\underline{0}0} &= \frac{\dfrac{\lambda_2}{\mu_1}\,\lambda_1(\mu_2 + \lambda_2 + \lambda_1)}{(\lambda_1 + \mu_2)(\lambda_2 + \mu_1) + \lambda_1\,\lambda_2}\,Q_{11} \end{aligned}\right\} \tag{41}$$

Q_{11} being obtained from Eqn (41) and from $Q_{11} + Q_{10} + Q_{01} + Q_{\underline{0}0} + Q_{0\underline{0}} = 1$.

The unavailability $\bar{A}_\infty$ is equal to $Q_{0\underline{0}} + Q_{\underline{0}0}$.

Assume that the $\lambda_i (i = 1, 2)$ are small compared with the μ_j $(j = 1, 2)$. Equations (41) then give $Q_{11} \simeq 1$ and

$$\bar{A}_\infty \simeq \lambda_1 \lambda_2 \left(\frac{1}{\mu_1^2} + \frac{1}{\mu_2^2} \right). \tag{42}$$

Note that this equation can be obtained directly from Eqn (33) of theorem 2 with $\alpha = \{11\}$ and $B = \{0\underline{0}, \underline{0}0\}$.

Other maintenance policies can be selected (cf. for example [Marguin and Marguinaud, 1973]).

5.5. Availability of a minimal cut set

In sections 2 and 3 the components were assumed to be independent. This assumption enabled us to give simple expressions for system unavailability, e.g. Eqns (7), (8), (9), (11), (12) and (13). Now, Eqns (7) and (8) are exact even if the system components are not independent. These equations can be used to compute system availability if we can compute $\mathscr{P}(\bar{C}_j)$, i.e. the probability of the components i of set C_j being down at time t.

Here, since the components i are not independent, we no longer have Eqn (9):

$$\mathscr{P}(\bar{C}_j) = \prod_{i \in C_j} \bar{q}_i(t)$$

but we can calculate $\mathscr{P}(\bar{C}_j)$ directly as follows.

We restrict the system to the set of components i of cut set C_j.

We examine the unavailability of this restricted system as we did in sections 5.1, 5.2, 5.3 and 5.4 using the state transition diagram.

This value $\mathscr{P}(\bar{C}_j)$ is then substituted in Eqn (8) giving the system unavailability.

As, in practice, an excellent approximation for the system unavailability $\bar{A}(t)$ is given by Eqn (11):

$$\bar{A}(t) \simeq \sum_{j=1}^{m} \mathscr{P}(\bar{C}_j)$$

the system unavailability computation reduces to computing the unavailability of the m minimal cut sets of the system.

Consequently, if two identical components in a standby redundancy configuration are in the same minimal cut set, the asymptotic unavailability of both components will be given by Eqn (36). In the case of parallel redundancy and of maintenance on a first-failed-first-repaired basis, it will be given by Eqns (37) and (42). For standby redundancy with two different components and a probability of non-start-up γ, it will be given by:

$$\bar{A}_\infty = \frac{\lambda_1}{\mu_1 + \mu_2} \gamma + \frac{\lambda_1 \lambda_2}{\mu_1(\mu_1 + \mu_2)} \tag{43}$$

(cf. Exercise 5).

6. AVAILABILITY WITH CONSTANT REPAIR TIMES

In section 5, we assumed that the system under study was Markovian, i.e. that the failure and repair rates were constant. Although, in reality, the failure rates can often be taken as constant (they actually vary slowly and may therefore be assumed to be constant over a comparatively long period), this is by no means a valid assumption for the repair rates. In fact, the most frequent assumption is that *the repair time is constant* (the repair time, therefore, being a Dirac distribution).

Consequently, if λ_i is the failure rate of component i and τ_i its repair time, in most problems the products $\lambda_i\tau_i$ are small compared with 1. In that case, it will be recalled (Ch. 3, section 3.1) that the availability of component i converges very rapidly (in 2 to 3 τ_i) towards its asymptotic availability $1/(1 + \lambda_i\tau_i) \simeq 1 - \lambda_i\tau_i$. On the other hand, the method of 'interpreting' the Markov diagram described in theorem 2 of this chapter (section 5.1.2) can be generalized to cover non-Markovian cases, particularly those in which the components have constant repair times, cf. [Gondran and Laleuf, 1981].

It is these cases which we shall examine in this section.

6.1. Accidental sequences method [Gondran and Laleuf, 1981]

In the system transition-state diagram, the changes of state depend on the residence time in each state. We denote by α the vertex for which all the components are available and by B the set of vertices β for which the system is regarded as failed.

Let us consider all the *elementary paths* Γ from α to a vertex β, i.e. all the paths from α to β not passing through the same vertex twice. A path of this type corresponds to an *accidental sequence*, i.e. a sequence of successive failures of several components resulting in a failed state of the system.

We denote by $\bar{A}(\Gamma)$ the probability of being in the failed state β having followed the accidental sequence Γ. Consequently, for a very reliable system (i.e. such that $\lambda_i\tau_i \ll 1$), we can show ([Laleuf, 1983]) that we have:

Theorem 3—*If all the system components are repairable with $\lambda_i\tau_i \ll 1$, an excellent approximation for the asymptotic unavailability is given by*:

$$\bar{A}_\infty \simeq \sum_{\substack{\text{paths } \Gamma \\ \text{from } \alpha \text{ to } \beta}} \bar{A}(\Gamma) \tag{44}$$

In fact, the theorem is valid for any very reliable non-Markovian system, i.e. whenever the mean times to repair of the components are small compared with the mean uptimes. There remains the problem of determining the unavailability $\bar{A}(\Gamma)$ due to an accidental sequence Γ.

6.2. Calculating the unavailability due to an accidental sequence [Gondran and Laleuf, 1981]

Here, we shall confine ourselves to the case where the failure rates and repair times are constant, but the methodology may be easily extended to more complex cases [Laleuf, 1983].

Let us assume initially a sequence Γ corresponding to successive failures of components 1 and 2 with failure rates λ_1 and λ_2 and constant repair times τ_1 and τ_2 respectively; they are in an active redundancy setup and system failure corresponds to simultaneous failure of both components. The accidental sequence Γ can be represented by the following diagram:

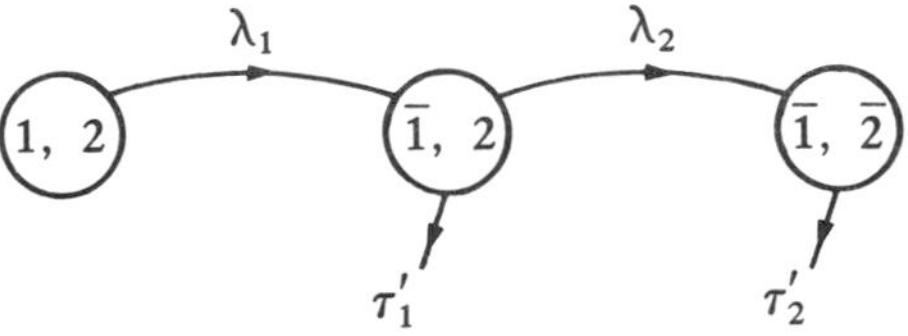

Where τ_1' and τ_2' *represent the smallest remaining repair times (on average) in states* $\{\bar{1}, 2\}$ *and* $\{\bar{1}, \bar{2}\}$, *when they have just been encountered.*

For state $\{\bar{1}, 2\}$, we obviously have $\tau_1' = \tau_1$. The calculation of τ_2' is more complex.

Let S_1 be the *residence time* in state $\{\bar{1}, 2\}$ during the accidental sequence. The distribution of S_1, given that $S_1 < \tau_1$, has a distribution function

$$P(S_1 < t/S_1 < \tau_1) = \frac{1 - e^{-\lambda_2 t}}{1 - e^{-\lambda_2 \tau_1}} \simeq \frac{\lambda_2 t}{\lambda_2 \tau_1} = \frac{t}{\tau_1}$$

from which it follows that the expectation of S_1, given that $S_1 < \tau_1$, is

$$E(S_1/S_1 < \tau_1) = \int_0^{\tau_1} t \frac{dP}{dt}(S_1 < t/S_1 < \tau_1)\, dt \simeq \frac{\tau_1}{2}.$$

During the accidental sequence Γ, the *mean residence time* in state $\{\bar{1}, 2\}$ is therefore $\tau_1/2 = \tau_1'/2$ when the transition $\{\bar{1}, 2\} \to \{\bar{1}, \bar{2}\}$ takes place, and the *mean repair time remaining* on component 1, is $\tau_1 - \tau_1/2 = \tau_1/2$.

Component 1 is, on average, 'half-repaired'.

When state $\{\bar{1}, \bar{2}\}$ is encountered, we are therefore faced with a component 1 which is (on average) half-repaired and a component 2 which is still completely unrepaired. If we continue to repair both components simultaneously, then τ_2', the smallest remaining repair time (on average), is given by

$$\tau_2' = \min\left(\frac{\tau_1}{2}, \tau_2\right) \tag{45}$$

The *system unavailability* due to the accidental sequence Γ is the *mean residence time* in state $\{\bar{1}, \bar{2}\}$ with respect to the mean residence time in the other system states, i.e. $1/\lambda_1$ for state $\{1, 2\}$ and $\tau_1/2$ for state $\{\bar{1}, 2\}$ (note that $1/\lambda_1 \gg \tau_1/2$).

We therefore have

$$\bar{A}(\Gamma) \simeq \frac{P_\Gamma \tau_2'}{1/\lambda_1 + \tau_1/2}$$

where P_Γ is the probability of the accidental sequence Γ occurring, i.e.

$$P_\Gamma = P(S_1 < \tau_1') = \lambda_2 \tau_1'$$

giving ultimately

$$\bar{A}(\Gamma) \simeq \frac{\lambda_2 \tau_1' \tau_2'}{1/\lambda_1} = (\lambda_1 \tau_1')(\lambda_2 \tau_2') = \lambda_1 \lambda_2 \tau_1 \min\left(\frac{\tau_1}{2}, \tau_2\right) \tag{46}$$

The above results can be generalized to apply to any accidental sequence [Gondran and Laleuf, 1981; Laleuf, 1983].

Consider a sequence Γ corresponding to the successive failures of components $1, 2, \ldots, m$. Component i has a failure rate λ_i and a repair time τ_i. The sequence can be represented by the following diagram:

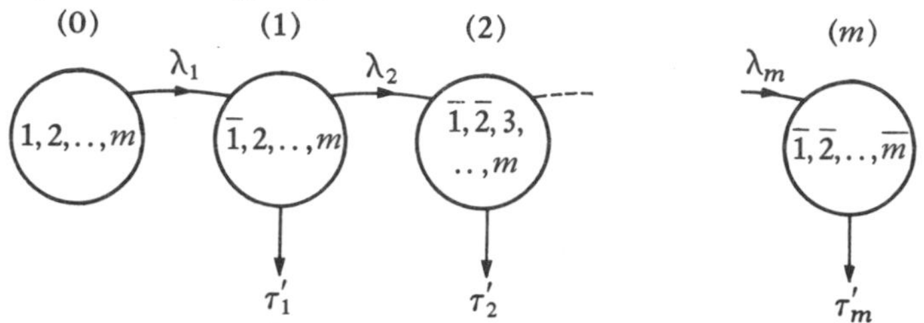

Let (i) be the state in which components $1, 2, \ldots, i$ are failed and components $(i + 1), \ldots, m$ are operating properly. Transition from state (i) to state $(i + 1)$ is caused by the failure of component $i + 1$ with failure rate λ_{i+1}.

We denote by τ'_i *the smallest remaining repair time (on average) in state i for the sequence in question, just after entering state i.*

Using a similar demonstration to the above, we obtain by recursion:

Theorem 4 [Gondran and Laleuf, 1981]: *Given the assumptions of theorem 3, the unavailability due to the accidental sequence $\bar{A}(\Gamma)$ is given by:*

$$\tau'_{i+1} = \min\left(\tau_{i+1}, \frac{\tau'_i}{2}\right) \tag{47}$$

$$\bar{A}(\Gamma) = \prod_{i=1}^{m} \lambda_i \tau'_i \tag{48}$$

Example: accidental sequence involving four components as follows:

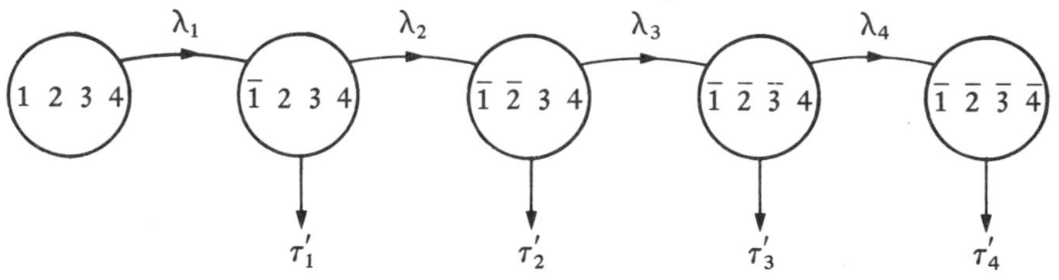

$$\tau'_1 = \tau_1$$

$$\tau'_2 = \min\left(\tau_2, \frac{\tau_1}{2}\right)$$

$$\tau'_3 = \min\left(\tau_3, \frac{\tau_2}{2}, \frac{\tau_1}{4}\right) = \min\left(\tau_3, \frac{\tau'_2}{2}\right)$$

$$\tau'_4 = \min\left(\tau_4, \frac{\tau_3}{2}, \frac{\tau_2}{4}, \frac{\tau_1}{8}\right) = \min\left(\tau_4, \frac{\tau'_3}{2}\right)$$

$$\bar{A}(\Gamma) \simeq \lambda_1 \tau'_1 \lambda_2 \tau'_2 \lambda_3 \tau'_3 \lambda_4 \tau'_4$$

Equations (47) and (48) are essential and enable us to conduct a more realistic examination of a large number of actual problems.

6.3. Maintenance policy with constant repair times

We return to the subject of maintenance policy planning discussed in section 5.4 of this chapter, but in this case with constant repair times.

We have two different components with failure rates λ_1 and λ_2 and repair times τ_1 and τ_2, respectively, in an active (parallel) redundancy configuration. There is only one repair facility adopting the following maintenance policy: priority given to repairing the first failed component.

The system therefore has two accidental sequences:

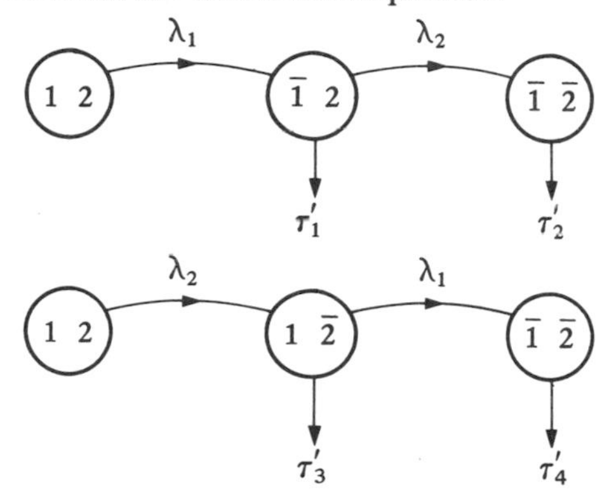

with $\tau'_1 = \tau_1$, $\tau'_2 = \tau_1/2$ as there is only one repair facility and component 1 is being repaired;

$\tau'_3 = \tau_2$, $\tau'_4 = \tau_2/2$ as there is only one repair facility and component 2 is being repaired.

Theorems 3 and 4 therefore give:

$$\bar{A}_\infty \simeq \lambda_1\tau_1\lambda_2\frac{\tau_1}{2}+\lambda_2\tau_2\lambda_1\frac{\tau_2}{2} = \frac{1}{2}\lambda_1\lambda_2(\tau_1^2 + \tau_2^2). \tag{49}$$

This value is half that found in the case of constant repair rates, cf. Eqn (42), section 5.4.

This very large difference shows the type of error which can be made in maintenance policies if the constant repair rate approximation is made. This is *an extremely important point.* By way of illustration, consider another maintenance policy: priority given to repairing the last failed component. For constant repair rates, the transition-state diagram becomes:

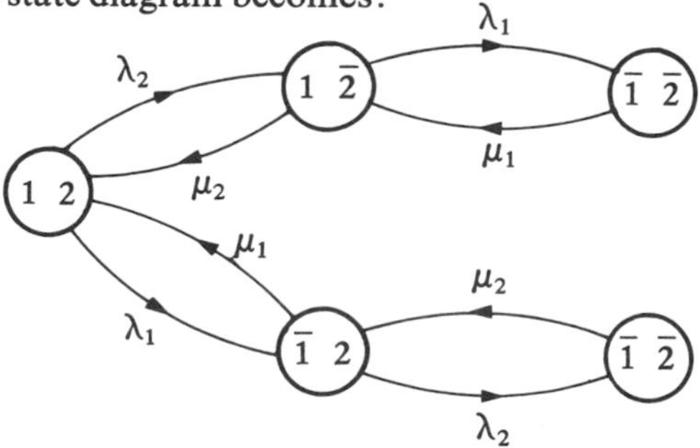

and theorem 2 gives

$$\bar{A}_\infty \simeq 2\frac{\lambda_1\lambda_2}{\mu_1\mu_2} = 2\lambda_1\lambda_2\tau_1\tau_2. \tag{50}$$

For constant repair times, the accidental sequences remain the same as before, but with $\tau_1' = \tau_1, \tau_2' = \tau_2, \tau_3' = \tau_2$ and $\tau_4' = \tau_1$, hence

$$\bar{A}_\infty = 2\lambda_1\lambda_2\tau_1\tau_2. \tag{51}$$

In the case of two identical components, we come to the following conclusions:

—If the *repair times are constant*, the maintenance policy 'priority given to repairing the first failed component' is considerably better than the policy of 'priority given to repairing the last failed component', as the unavailability of the former is half that of the latter $\bar{A}_\infty = \lambda^2\tau^2$ given by Eqn (49) $< \bar{A}_\infty = 2\lambda^2\tau^2$ given by Eqn (51).

—If the *repair rates are constant*, the two maintenance policies are identical from the point of view of unavailability $\bar{A}_\infty = 2\lambda^2\tau^2$ given by Eqn (42) $= \bar{A}_\infty = 2\lambda^2\tau^2$ given by Eqn (50).

Now, it is well known in practice that the first policy is better than the second. This example clearly shows the danger of the Markovian approach (constant repair rates) in designing maintenance policies. We shall therefore use the results of theorems 3 and 4 when considering maintenance policies.

6.4. Case of two components in a passive (standby) redundancy configuration

Let the system comprise two components in a standby redundancy set-up. When component 1 fails, component 2 takes over with a probability $\gamma(\ll 1)$ of failing to start. They have constant failure rates λ_1 and λ_2 and constant repair times τ_1 and τ_2 respectively.

If we consider non-repairable components, the transition-state diagram is as follows:

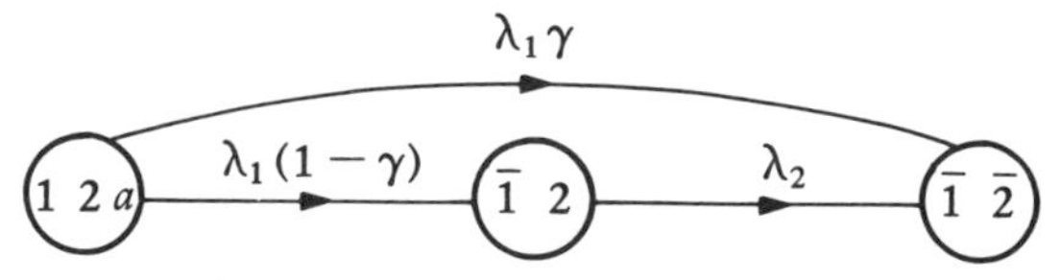

There are two accidental sequences:

$$\Gamma_1: \{1 \ \ 2a\} \to \{\bar{1} \ \ 2\} \to \{\bar{1} \ \ \bar{2}\}$$
$$\Gamma_2: \{1 \ \ 2a\} \to \{\bar{1} \ \ \bar{2}\}$$

For sequence Γ_1, we obtain:

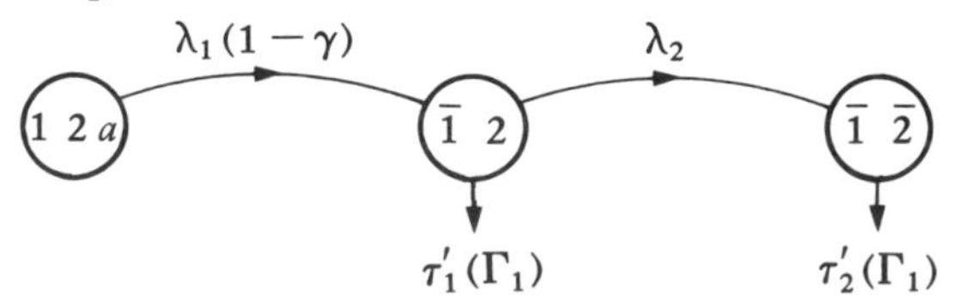

with $\tau'_1(\Gamma_1) = \tau_1, \tau'_2(\Gamma_1) = \min\left(\tau_2, \frac{\tau_1}{2}\right)$

$$\bar{A}(\Gamma_1) \simeq \lambda_1 \lambda_2 \tau_1 \min\left(\tau_2, \frac{\tau_1}{2}\right).$$

For sequence Γ_2, we obtain

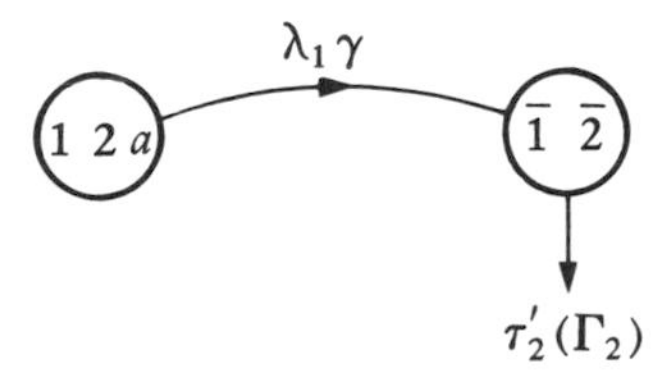

with $\tau'_2(\Gamma_2) = \min(\tau_1, \tau_2)$

$$\bar{A}(\Gamma_2) = \lambda_1 \gamma \min(\tau_1, \tau_2).$$

Note that $\tau'_2(\Gamma_1) \neq \tau'_2(\Gamma_2)$, which is normal as the problem is non-Markovian, and that the τ'_i therefore depend on the path followed to arrive at the corresponding state.

Finally, the asymptotic unavailability is given by

$$\bar{A}_\infty \simeq \lambda_1 \gamma \min(\tau_1, \tau_2) + \lambda_1 \lambda_2 \tau_1 \min\left(\tau_2, \frac{\tau_1}{2}\right) \tag{52}$$

which can be compared with Eqn (43), section 5.5, in the case of constant repair rates.

7. SUMMARY

Apart from very simple cases with components in series and parallel (cf. section 2 and Ch. 3, section 2), the availability of a system is generally determined from the minimal cut sets (cf. section 3) using Eqns (7), (8) and (11):

$$\bar{A}(t) \simeq \sum_{j=1}^{m} \mathcal{P}(\bar{C}_j).$$

When the components of cut set C_j are independent, then $\mathcal{P}(\bar{C}_j)$ is given by Eqn (9):

$$\mathcal{P}(\bar{C}_j) = \prod_{i \in C_j} \bar{q}_i(t).$$

When the components of C_j are not independent, $\mathcal{P}(\bar{C}_j)$ is computed using the state transition diagram as described in section 5 (cf. *the crucial role of Eqn (33) of theorem 2*) and in section 6 (cf. Eqns (44), (47) and (48) of theorems 3 and 4).

Table 1 summarizes the asymptotic availability computation for a number of systems comprising non-independent components (for independent components, cf. summary table in Ch. 3, section 6).

Table 1. *Unavailability for non-independent components.*

System (Components with a failure rate λ and a mean repair time τ, $\lambda\tau \ll 1$)	Asymptotic unavailability $\bar{A}_\infty$ (approximate value)
— two identical components in standby redundancy configuration — two identical components in standby redundancy configuration with only one repair facility (Eqn 36)	$\frac{1}{2}(\lambda\tau)^2$ $(\lambda\tau)^2$
— two different components in standby redundancy configuration with startup failure rate γ — two identical components in standby redundancy configuration with only one repair facility (Exercise 5)	$\frac{\lambda_1}{\mu_1+\mu_2}\gamma + \frac{\lambda_1\lambda_2}{\mu_1(\mu_1+\mu_2)}$ $(\lambda\tau)^2 + (\lambda\tau)\gamma$
— two components in parallel redundancy configuration with only one repair facility (Eqn 37)	$2\lambda\lambda'\tau^2$
— two different components in parallel redundancy configuration with only one repair facility operating on first-failed-first-repaired basis. (Eqn 42)	$\lambda_1\lambda_2(\tau_1^2+\tau_2^2)$
— general case (Markovian) (theorem 2)	$\sum_{\substack{\text{paths from} \\ \alpha \text{ to } B}} \frac{\Pi(\text{path transition rate})}{\prod_{\substack{i \in \text{path} \\ \text{except } \alpha}} \lvert a_{ii} \rvert}$
— general case (non-Markovian) (theorem 3) (theorem 4)	$\sum_{\substack{\text{paths } \Gamma \\ \text{from } \alpha \text{ to } \beta}} \bar{A}(\Gamma)$ $\bar{A}(\Gamma) = \prod_{i=1}^{n} \lambda_i \tau_i'$

EXERCISES

Exercise 1 — Availability of an electricity supply

Consider the example of the electricity supply described in Ch. 2, sections 2.1 and 2.2.

We shall take the case described in Ch. 2, section 2.2: dependence between the 380 kV and 220 kV systems and functional connection through supervisory control.

Supercomponents 1, 2, 5, 6, 7, 8, 9, 10, 11 and 12 are continuously operating, supercomponents 3 and 4, representing diesel engines, are normally on standby and are started up if the supply to their respective busbars 11 and 12 fails.

The system reliability data are given in the following table:

Supercomponent no.	$\lambda_i\,(\text{h}^{-1})$	$\tau_i = \frac{1}{\mu_i}$ (h)	$\lambda_i\tau_i$	γ_i
1	$5\,.\,10^{-5}$	0.2	10^{-5}	
2	10^{-6}	9	$9\,.\,10^{-6}$	
3, 4	$7\,.\,10^{-4}$	100	$7\,.\,10^{-2}$	0.03
5, 6, 9, 10, 11, 12	$5\,.\,10^{-6}$	90	$4.5\,.\,10^{-4}$	
7, 8	$3\,.\,10^{-6}$	8	$2.4\,.\,10^{-5}$	
13	10^{-7}	0.2	$2\,.\,10^{-8}$	

Determine the asymptotic availability of the electricity supply (using the results from Exercises 5 and 6).

Exercise 2 — Demonstration of theorem 2

Consider the state transition diagram of a Markovian system.
Consider a critical failed state β (cf. Ch. 6).
Let M_β be the sum of the transition rates from this state, $M_\beta = |a_{\beta\beta}|$.
Let i be a critical operating state, $Q_i(\infty)$ its availability and Λ_i the sum of the transition rates to β (cf. Ch. 6).

(1) Using theorem 1, show that

$$Q_\beta(\infty) \simeq \frac{\sum_{i \in M_{c\beta}} Q_i(\infty)\,\Lambda_i}{M_\beta}$$

is a good approximation for the probability of being in state β, where $M_{c\beta}$ is the set of critical operating states i having a transition to β.

(2) Using theorem 2 from Ch. 6, deduce Eqn (33) of theorem 2.

Exercise 3 — A two-sided estimation of availability

Prove the formula

$$\sum_{j=1,m} \prod_{i \in C_j} \bar{q}_i(t) \prod_{i \notin C_j} q_i(t) \leq \bar{A}(t) \leq \sum_{j=1}^{m} \prod_{i \in C_j} \bar{q}_i(t)\,. \tag{13}$$

Exercise 4 — Failure expectation of an electric power system

Consider the following method of determining the mean failure of an electric power system (grid), in the approximation of Kirchoff's first law, as a function of the unavailability of the generating sources and power lines:

1. To the electric power systems we add edges representing the generating sources.
In the resulting diagram, we determine all the cut sets of α-minimal capacity (cf. [Gondran and Minoux, 1979]).

2. For each of these cut sets, we define a set of '*critical cut sets*' corresponding to a set of edges whose simultaneous failure results in a demand failure.

3. These 'critical cut sets' are then grouped and sorted. The mean failure is then obtained from the contributions of each of the 'critical cut sets' adopted.

Apply this method in order to determine the unavailability and mean failure of the following system.

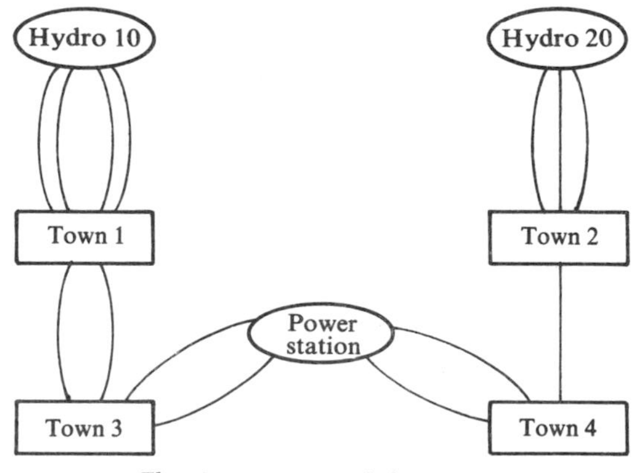

Electric power transmission system

Generation characteristics

	No. of generating sets	Output per set (MW)	Set unavailability
Hydro 10	10	300	0.01
Hydro 20	5	300	0.01
Power station	3	350	0.05

Demand characteristics

Town 1 = 2000, Town 2 = 1700, Town 3 = 500, Town 4 = 800.

Transmission characteristics

	No. of transmission lines	Line capacity (MW)	Line unavailability
Hydro 10 Town 1	4	1 100	0.015
Town 1 Town 3	2	325	0.000 4
Hydro 20 Town 2	3	750	0.009
Town 2 Town 4	1	250	0.000 2
Town 3 Power station	2	250	0.003
Town 4 Power station	2	500	0.005

Exercise 5 — Standby redundancy with start-up failure

1. Consider two identical components with failure rate λ, repair rate $\mu = \tau^{-1}$ in a standby redundancy configuration. When the first fails, the second starts up. The non-start probability is a constant γ. Determine the asymptotic availability of this system.

2. What does it become when there is only one repair facility?

3. Now consider that the two components are different (λ_1, μ_1) and (λ_2, μ_2). When number 1 fails, the second starts up with a non-start probability of γ. Determine the asymptotic availability of this system.

Exercise 6 — Simultaneous parallel and standby redundancy

Consider two different components (λ_1, μ_1) and (λ_2, μ_2) operating in parallel and a third component (λ_3, μ_3) on standby. Let γ be its start-up probability.

In practice, such a system corresponds to a minimal cut set formed from components 1, 2 and 3, where component 3 only starts up if both components 1 and 2 have failed.

Determine the asymptotic availability of this system.

Exercise 7 — Standby redundancy with two spares

Consider a system consisting of three redundant components. Component 1 is normally operating and the other two (identical) are normally standing by.

When component 1 fails (failure rate λ_1), the other two start up simultaneously (the start-up probability of each being γ).

Determine the asymptotic availability of this system, given that the failure rate of components 2 and 3 is λ, that the repair rate of component 1 is μ_1 and that of components 2 and 3 is μ (assuming that there are always adequate repair facilities).

SOLUTIONS TO EXERCISES

Exercise 1

When there is no dependence between the 380 kV and 220 kV networks and no functional connection due to supervisory control, the minimal cut sets are as determined in Ch. 2, section 4.2. We obtained:

11, 12	2nd-order cut set
11, 4, 10	3rd-order cut sets
12, 3, 9	
11, 4, 1, 2	4th-order cut sets
11, 4, 1, 8	
11, 4, 2, 6	
11, 4, 6, 8	
12, 3, 1, 2	
12, 3, 1, 7	
12, 3, 2, 5	
12, 3, 5, 7	

We take account of the dependence between the networks (supercomponent 13) by adding the cut sets $\{11, 4, 13\}$, $\{12, 3, 13\}$ obtained from cut sets $\{11, 4, 1, 2\}$ and $\{12, 3, 1, 2\}$.

We take account of the functional connection due to supervisory control ('supercomponent 1 is down if supercomponent 11 has failed', cf. Ch. 2, section 2.2) by eliminating supercomponent 1 from the minimal cut sets containing component 11.

The minimal cut sets of the system are therefore:

11, 12	2nd-order cut set
11, 4, 10	3rd-order cut sets
12, 3, 9	
11, 4, 13	
12, 3, 13	
11, 4, 2	
11, 4, 8	
11, 4, 2, 6	4th-order cut sets
11, 4, 6, 8	
12, 3, 1, 2	
12, 3, 1, 7	
12, 3, 2, 5	
12, 3, 5, 7	

The unavailabilities of each cut set

$$\mathscr{P}(\{\overline{11, 12}\}) = (\lambda_{11}\,\tau_{11})\,(\lambda_{12}\,\tau_{12}) = 2.025.\ 10^{-7}.$$

still have to be computed.

—The six third-order cut sets all contain a diesel engine (supercomponents 3 and 4).

—Compute $\mathscr{P}(\{\overline{11, 4, 10}\})$. Diesel 4 is independent of supercomponent 11. For supercomponent 10, diesel 4 is a standby component having a start-up probability of $\gamma = 0.03$. The unavailability of $\{4, 10\}$ is therefore given by Eqn (43)

$$\bar{A}_\infty = \frac{\lambda_{10}}{\mu_{10} + \mu_4}\gamma + \frac{\lambda_4\,\lambda_{10}}{\mu_{10}(\mu_4 + \mu_{10})} = 2.2\,10^{-5}$$

hence $\mathscr{P}(\{\overline{11, 4, 10}\}) = (4.5\,10^{-4})(2.2\,10^{-5}) = 10^{-8}$.

Similarly, we have $\mathscr{P}(\{\overline{12, 3, 9}\}) = 10^{-8}$.

—Compute $\mathscr{P}(\{\overline{11, 4, 13}\})$. Diesel 4 is independent of supercomponent 11. For supercomponent 13, diesel 4 is a standby component with $\gamma = 0.03$. The unavailability of $\{4, 13\}$ is therefore given by Eqn (43)

$$\bar{A}_\infty = \frac{\lambda_{13}}{\mu_4 + \mu_{13}}\gamma + \frac{\lambda_4\,\lambda_{13}}{\mu_{13}(\mu_4 + \mu_{13})} \simeq 6\,.\,10^{-10}\,.$$

hence $\mathscr{P}(\overline{\{11, 4, 13\}}) = (6\,10^{-10})(4.5\,10^{-4}) = 2.7\,10^{-13}$, which is negligible.

Similarly $\mathscr{P}(\overline{\{12, 3, 13\}}) = 2.7\,10^{-13}$ (negligible).

— Compute $\mathscr{P}(\overline{\{11, 4, 2\}})$. Supercomponents 2 and 11 are independent. Diesel 4 is on standby to these two supercomponents with a start-up probability $\gamma = 0.03$. The unavailability of $\{11, 2, 4\}$ is therefore given by the equation in Exercise 6.

$$\mathscr{P}(\{11, 2, 4\}) = \frac{\lambda_{11}\,\lambda_2\,\gamma}{\mu_{11} + \mu_2 + \mu_4}\left(\frac{1}{\mu_2} + \frac{1}{\mu_{11}}\right) + 2\frac{\lambda_{11}\,\lambda_2\,\lambda_4}{\mu_{11}\,\mu_2(\mu_{11} + \mu_2 + \mu_4)} = 1.1\,10^{-10} \text{ (negligible).}$$

Similarly we have $\mathscr{P}(\overline{\{11, 4, 8\}}) = 1.1\,10^{-10}$ (negligible).

— The fourth-order minimal cut sets are also negligible.

The system unavailability is therefore equal to:

$$\underline{\bar{A}_\infty = 2.2\,10^{-7}}.$$

Exercise 2

(1) We use theorem 1 with $\mathscr{F} = \{\beta\}$ and disregard the non-critical failed states.

(2) We write $\bar{A}_\infty \simeq \sum_{\beta \in P_c} Q_\beta$, where P_c is the set of critical failed states, and apply theorem 2 from Ch. 6.

Exercise 4

To find the α-minimal cut sets, consider the following diagram (and its construction):

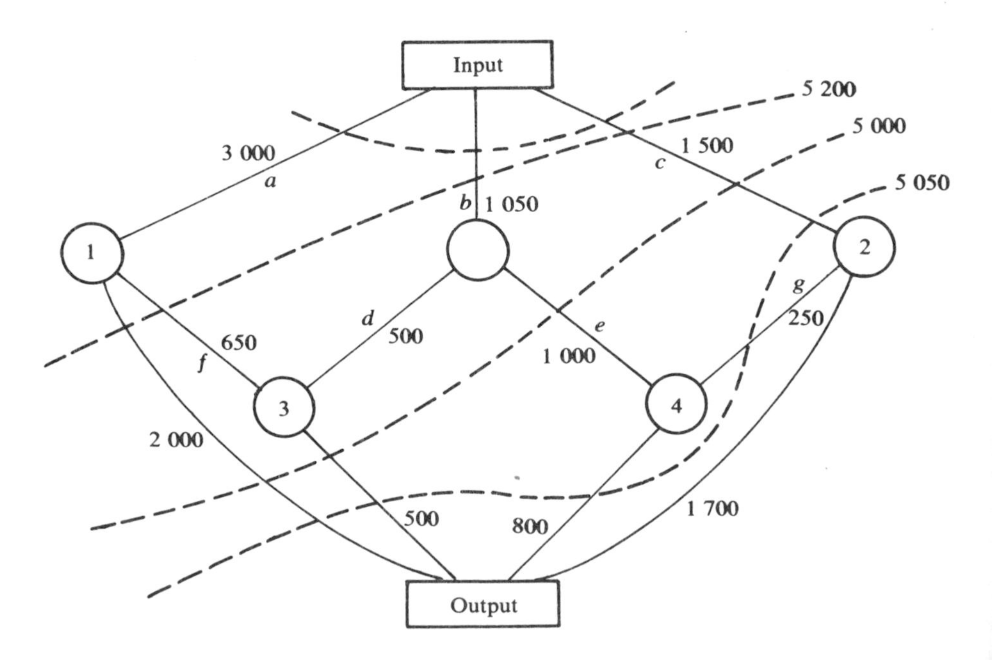

We determine from this diagram the minimal cut sets, or $\{c, e\}$, $\{c, g\}$, $\{c, b, f\}$, $\{a, b, c\}$, etc. ... with respective capacities of 5000, 5050, 5200, 5500, etc. ...

Analysis of cut set $\{c, e\}$

As demand amounts to 5000, the slightest failure on $\{c, e\}$ results in total failure.
Now, edge $\{c\}$ corresponds to the following diagram:

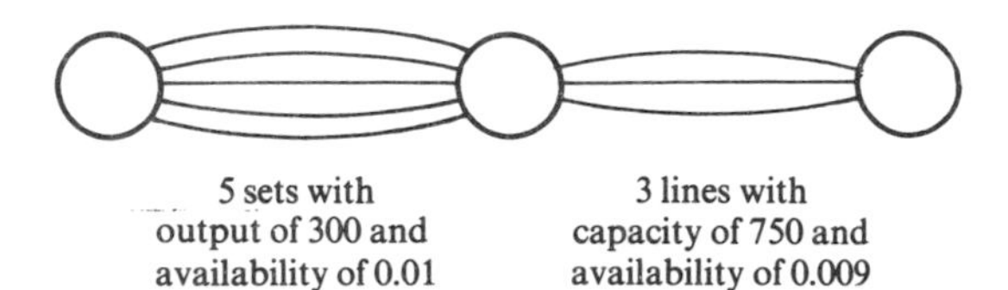

5 sets with output of 300 and availability of 0.01

3 lines with capacity of 750 and availability of 0.009

The contribution of edge $\{c\}$ in cut set $\{c, e\}$ to overall unavailability is therefore $5 \times 0.01 \simeq 0.05$; its contribution to mean failure is $300 \times 0.05 = 15$ MW (the other terms being negligible).
The contribution of edge $\{e\}$ in cut set $\{c, e\}$ to overall unavailability is $2 \times 0.005 = 0.01$; its contribution to mean failure is $500 \times 0.01 = 5$ MW (the other terms being negligible).

Analysis of cut set $\{c, g\}$

The contribution of edge $\{c\}$ in cut set $\{c, g\}$ must be eliminated as it was taken into account in cut set $\{c, e\}$.
The contribution of edge $\{g\}$ in cut set $\{c, g\}$ to overall unavailability is 0.0002; its contribution to mean failure is $200 \times 0.0002 = 0.04$ MW.

Analysis of cut set $\{c, b, f\}$

Contribution of $\{b\}$ to unavailability $\simeq 3 \times 0.05 = 0.15$
to mean failure $= (350-200) \times 0.15 \simeq 22.5$ MW
Contribution of $\{f\}$ to unavailability $\simeq 2 \times 0.0004 = 0.0008$
to mean failure $\simeq (325-200) \times 0.0008 \simeq 0.1$ MW.

Analysis of cut set $\{a, b, c\}$

The contributions of $\{b\}$ and $\{c\}$ have already been taken into account.
Representation of $\{a\}$

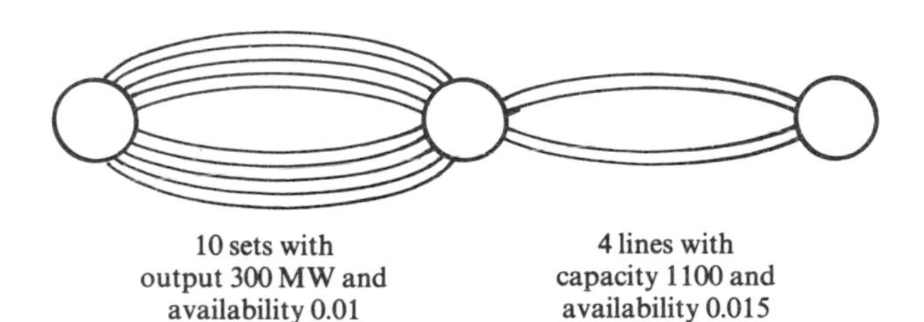

10 sets with output 300 MW and availability 0.01

4 lines with capacity 1100 and availability 0.015

In order to obtain total failure, there has to be a failure of $\{a\}$ of at least $5500 - 5000 = 500$ MW, which requires the simultaneous failure of at least two components. Contribution of $\{a\}$ to:

— unavailability $\simeq C_{10}^2(0.01)^2 + C_4^2(0.015)^2 = 0.006$
— mean failure $\simeq C_{10}^2(0.01)^2 \times 100 + C_4^2(0.015)^3 \times 300 = 2.2$ MW.

Finally, system unavailability is equal to $0.05 + 0.0002 + 0.15 + 0.006 \simeq 0.2$ and the mean failure amounts to $15 + 5 + 0.04 + 22.5 + 0.1 + 2.2 \simeq 45$ MW.

Exercise 5

1. The state transition diagram is

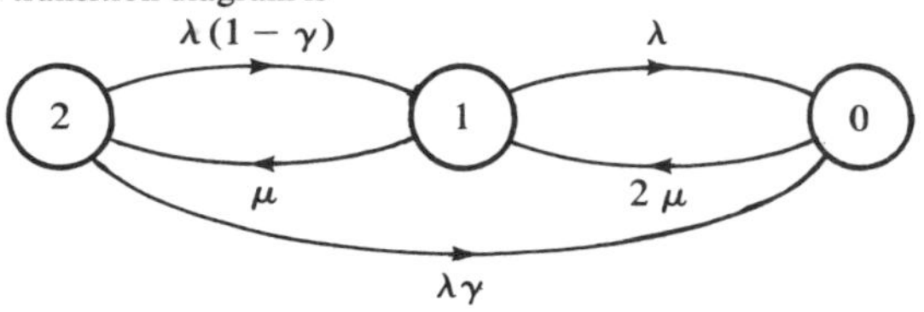

Applying Eqn (27) with $\mathscr{F} = \{2\}$ and $\mathscr{F} = \{1\}$, we find

$$\overline{A}_\infty = \frac{\dfrac{\lambda}{2\mu}\left(\dfrac{\lambda}{\mu} + \gamma\right)}{1 + \dfrac{\lambda}{\mu} + \dfrac{\lambda}{2\mu}\left(\dfrac{\lambda}{\mu} + \gamma\right)} \simeq \frac{1}{2}(\lambda\tau)^2 + \frac{1}{2}(\lambda\tau)\,\gamma\,.$$

2. If there is only one repair facility

$$\overline{A}'_\infty \simeq (\lambda\tau)^2 + (\lambda\tau)\,\gamma\,.$$

3. The state transition diagram is

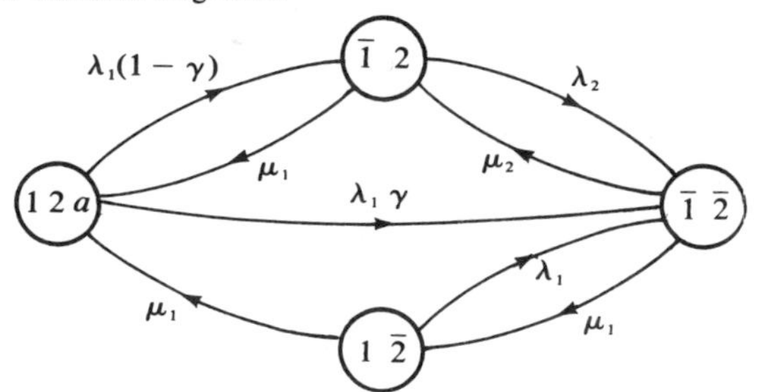

Theorem 2 gives

$$\overline{A}_\infty \simeq \frac{\lambda_1}{\mu_1 + \mu_2}\,\gamma + \frac{\lambda_1\lambda_2}{\mu_1(\mu_1 + \mu_2)}\,.$$

Exercise 6

The state transition diagram is shown below:

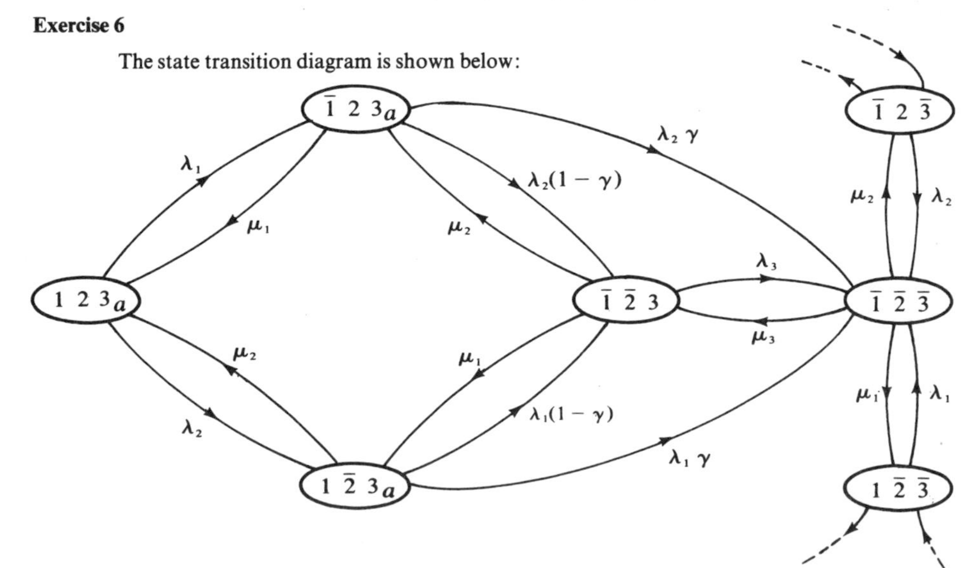

the edges and vertices not drawn correspond to negligible probability vertices.

Theorem 2 gives:

$$\bar{A}_\infty = \frac{\lambda_1\lambda_2\gamma}{\mu_2(\mu_1 + \mu_2 + \mu_3)} + \frac{\lambda_1\lambda_2\lambda_3}{\mu_1\mu_2(\mu_1 + \mu_2 + \mu_3)} + \frac{\lambda_1\lambda_2\gamma}{\mu_1(\mu_1 + \mu_2 + \mu_3)}.$$

Exercise 7

The state transition diagram is shown below:

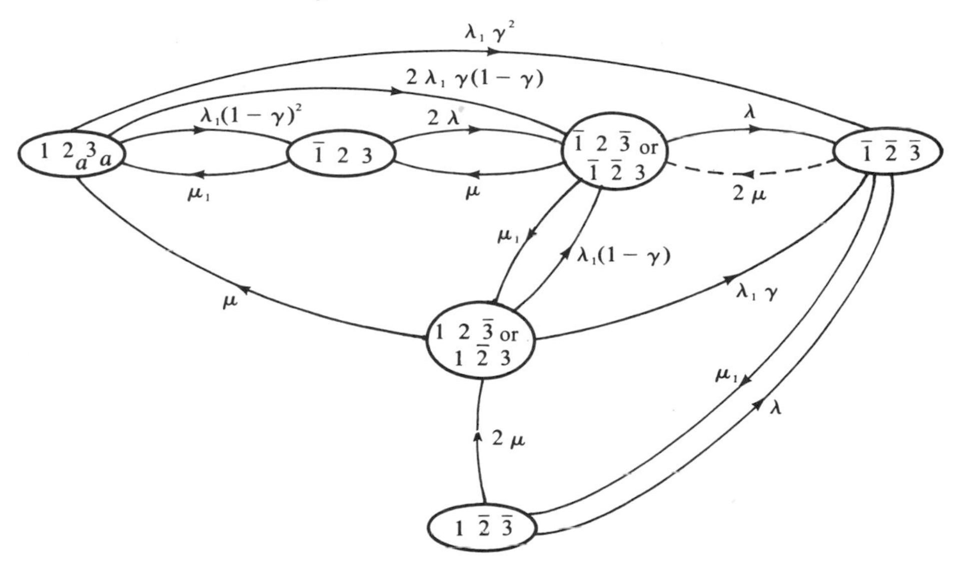

Theorem 2, Equ (33) gives:

$$\bar{A}_\infty = \frac{\lambda_1\gamma_2}{\mu_1 + 2\mu} + \frac{2\lambda_1\gamma(1-\gamma)\lambda}{(\mu_1 + 2\mu)(\mu + \mu_1 + \lambda)} + \frac{\lambda_1(1-\gamma)^2\, 2\lambda^2}{(\mu_1 + 2\lambda)(\mu + \mu_1 + \lambda)(\mu_1 + 2\mu)}$$

the fourth path $(12_a3_a, \bar{1}\,2\,3, \bar{1}\,2\,\bar{3}, 1\,2\,\bar{3}, \bar{1}\,\bar{2}\,3)$ making a negligible contribution.

REFERENCES

BIRNBAUM Z. W. (1969): On the importance of different components and a multicomponent system, In: *Multivariable Analysis II*, P. R. Korishnaiah (Ed.); Academic Press, New York.

FUSSELL, J. B. (1975): How to hand-calculate system reliability characteristics; *IEEE Trans. on Reliability*, **R24**, no. 3.

GONDRAN M. and LALEUF J. C. (1981): Indisponibilité des systèmes à taux de dèfaillance et à temps de réparation constants; *Bulletin de la Direction des Etudes et Recherches d'Electricité de France*, série C, **1**, pp. 19–26.

GONDRAN M. and MINOUX M. (1979): *Graphes et algorithmes*; Eyrolles, Collection des Etudes et Recherches EDF, Paris.

GONDRAN M. and PAGES A. (1976): *Calcul de la fiabilité des systèmes réparables*; FIABC, EDF Note, HI 2199/02.

KRAKOWSKI M. (1973): Conservation Methods in Queuing Theory; *R.A.I.R.O. série verte*, **1**, pp. 63–84.

LALEUF J. C. (1983): Calcul de la fiabilité des systèmes; cours de *l'Ecole Nationale de la Statistique et de l'Administration Economique*, Paris.

LAMBERT H. E. (1975): *Fault Trees for Decision Making in Systems Analysis*, Lawrence Livermore Laboratory, Livermore, UCRL — 51829.

LEMAIRE B. (1977): Méthode de conservation et blocage dans les files d'attente; *R.A.I.R.O. recherche opérationnelle*, **11**, no. 4, pp. 363–377.

MARGUIN J. and MARGUINAUD A. (1973): Calcul exact de la disponibilité opérationelle d'un système complexe; *R.A.I.R.O. série verte*, **1**, pp. 3–20.

SINGH C. (1972). *Reliability Modelling and Evaluation in Electric Power Systems*; University of Saskatchewan, Canada.

VESELY W. E. (1974): Div. of Reactor Safety Research, Nuclear Regulatory Commission, Washington, D.C. (private communication).

CHAPTER 5

CLASSICAL METHODS FOR EVALUATING THE RELIABILITY OF REPAIRABLE SYSTEMS

Evaluating the reliability of repairable systems is made difficult by the inter-component dependence introduced by the conjunction of the possibility of repairing individual components and the impossibility, by definition of reliability, of restoring the system following failure. Of course, the majority of systems are restored after failure and in this case we may also be interested in the failure density in the system renewal process as a function of time. We shall show the relationship which exists between this failure density and the system failure rate in the next chapter.

We shall demonstrate that, under certain virtually non-restrictive *conditions*, these functions are numerically equivalent; this property is fundamental to the critical operating states method which provides a good approximation for the reliability of large repairable systems. In fact, we shall see that the conventional methods described in this chapter are inadequate, either because they cannot be applied to large systems (analytical methods) or because they are very time-consuming and inaccurate (Monte Carlo simulation).

1. PRINCIPLE OF MODELLING BY STOCHASTIC PROCESSES

Analytical methods call for modelling by stochastic processes. The system whose reliability we wish to evaluate consists of p repairable components having a certain number of states. All the combinations of these various states determine the set E of the system states.

At this point, we shall examine only the most frequent case, where each component has a finite number of states (this occurs particularly when the failures are catastrophic). Each component generally possesses two states: 'operating' and 'failed'. The set E of the system states is therefore *finite*; it is therefore possible to enumerate the states. If we observe the system in the course of time, it will occupy different states of E in turn. Each of these observations is termed a *trajectory* or realization and we shall denote the set of these trajectories by Ω. The changes occurring in our system can be modelled using a *stochastic process* if we can probabilize the pair $(\Omega, \mathscr{A})$, $\mathscr{A}$ representing the set of subsets of Ω. In other words, we can define the probabilities:

$$\mathcal{P}(E_{t_1} \in e_1, \ldots, E_{t_i} \in e_i, \ldots, E_{t_n} \in e_n) \quad \begin{array}{l} \forall n \\ \forall t_i \\ \forall e_i \subseteq E \end{array}$$

where E_{t_i} is the state of the system at time t_i. As the states are numbered, the state of the system at a given time t is therefore a real random variable defined on $(\Omega, \mathscr{A}, \mathscr{P})$.

$$\omega \in (\Omega, \mathscr{A}, \mathscr{P}) \xrightarrow{v.a} E_t(\omega) \in E .$$

If we now consider the transformation which produces correspondence of the random variable $E_t(\omega)$ at time t, we thus define a random process (cf. Ch. 1, section 6) whose set of indices T is generally constituted by the real half-line $[0, +\infty[$, but which may also be a discrete set.

We obtain a particularly simple case when the process is *Markovian* (cf. Ch. 1, section 6). In this case:

$$\begin{aligned} \mathcal{P}(E_{t_n} = e_n \mid E_{t_1} = e_1, E_{t_2} = e_2, \ldots, E_{t_{n-1}} = e_{n-1}) &= \\ = \mathcal{P}(E_{t_n} = e_n \mid E_{t_{n-1}} = e_{n-1}) \, \forall t_i \in T \qquad e_i \in E \qquad (1) \end{aligned}$$

The future course of the system depends only on the state which it is occupying and not on the trajectory which it has described to get there. Note that the Markovian nature of a process depends on the definition of the system states.

Let us take a simple example from mechanics. If the state of a moving point is defined by its coordinates, the process describing its evolution is not Markovian. However, if its state is defined by its coordinates and its velocity, the process becomes Markovian. We shall use this observation in section 3. Stretching the terminology, we shall state that a system is Markovian with discrete state space when the stochastic process describing its behaviour is Markovian, the system states being simply defined as the combination of the states of the different components, without introducing other parameters.

Markov processes are very useful, as they enable us to model with accuracy a large number of systems, yielding simple computations when the number of system states is not excessively large.

2. MARKOVIAN SYSTEMS WITH DISCRETE STATE SPACE

2.1. Process generation

The associated process is called a discrete or a continuous parameter *Markov chain*, depending on the nature of the set T.

Definition (1) enables us to write:

$$\begin{aligned} \mathcal{P}(E_{t_n} = e_n, E_{t_{n-1}} = e_{n-1}, \ldots, E_{t_1} = e_1) &= \\ = \mathcal{P}(E_{t_n} = e_n \mid E_{t_{n-1}} = e_{n-1}) \mathcal{P}(E_{t_{n-1}} = e_{n-1} \mid E_{t_{n-2}} = e_{n-2}) \ldots \mathcal{P}(E_{t_1} = e_1) . \end{aligned}$$

The process is therefore perfectly defined if we know the initial distribution and the conditional distributions.

Where set T is *discrete*, we only need to know:

$$P_j(t) = \mathcal{P}[E_t = j] \qquad \forall t \in N \quad \forall j \in E$$

$$P_{jk}(t, t') = \mathcal{P}[E_t = k \mid E_{t'} = j] \qquad \forall j \in E \quad \forall t, t' \in N\ t > t'. \tag{2}$$

The set $\{P_{jk}(t, t')]$ is called the *transition probability* set of the Markov chain. The chain is termed *time homogeneous* if $P_{jk}(t, t')$ depends only on the time difference $(t - t')$.

In the case where set T is *continuous*, we only need to know:

$$P_j(t) = \mathcal{P}[E_t = j] \qquad \forall t \in R^+ \quad \forall j$$

$$P_{jk}(t, t') = \mathcal{P}[E_t = k \mid E_{t'} = j] \qquad \forall t, t' \in R^+ \quad t > t' \quad \forall j, k$$

Passing to the limit, we define, when possible, *transition intensities* or *transition rates* ρ_{jk} using the relation

$$\rho_{jk}(t) = \lim_{\substack{h \to 0 \\ \geq 0}} \frac{1}{h} \mathcal{P}[E_{t+h} = k \mid E_t = j]. \tag{3}$$

The chain is *time homogeneous* if $P_{jk}(t, t')$ depends only on $(t - t')$. In this case $\rho_{jk}(t)$ *is not dependent on* t.

The transition from state j to state k is caused by a component's change of state. Consequently, $\rho_{ij}(t)$ is a sum of system component failure or repair rates. Three cases can therefore arise:

(a) The failure and repair rates are constant. ρ_{ij} is then constant and we have a homogeneous Markov process.

(b) One or more failure or repair rates are not constant but depend solely on the elapsed time since the system was put into service; this is a non-homogeneous Markov process (at least one rate ρ_{ij} is time-dependent).

(c) One or more failure or repair rates are dependent on the elapsed time since the last transition of the component in question. This occurs, for example, when the up-times or the repair times are Weibull (but not exponentially) distributed. This is the most frequent case. However, in practice, we invariably revert to cases (a) or (b) for calculation purposes.

2.2. Chapman-Kolmogorov equations

The *Chapman-Kolmogorov* equations enable us to determine the probabilities of transition between any two states, given the probabilities of all the possible intermediate transitions:

$$P_{jk}(t, t') = \sum_{i \in E} P_{ji}(t, u)\, P_{ik}(u, t')\ t' \leq u \leq t. \tag{4}$$

Unless there is evidence to the contrary, we shall now assume that the set T is continuous and represents time, i.e. $T = R^+$.

2.3. System state equations

We put:
$P_i(t) = \mathscr{P}$ [the system is in state i at time t],
$\rho_{ij}(t)$ = instantaneous transition rate from state i to state j.

We evaluate the probability $P_i(t + \mathrm{d}t)$ of the system being in state i at time $t + \mathrm{d}t$:

$$P_i(t + \mathrm{d}t) = \mathscr{P}\,[\text{the system is in state } i \text{ at time } t \text{ and remains there in the interval } \mathrm{d}t] + \sum_{j \in E - \{i\}} \mathscr{P}\,[\text{the system is in state } j \text{ at time } t \text{ and in state } i \text{ at time } t + \mathrm{d}t].$$

In accordance with the transition rate definition (3), we therefore obtain:

$$P_i(t + \mathrm{d}t) = P_i(t)\Big[1 - \sum_{j \in E - \{i\}} \rho_{ij}(t)\,\mathrm{d}t + 0(\mathrm{d}t)\Big] + \sum_{j \in E - \{i\}} P_j(t)\,\rho_{ji}(t)\,\mathrm{d}t + 0(\mathrm{d}t)$$

hence

$$\frac{P_i(t + \mathrm{d}t) - P_i(t)}{\mathrm{d}t} = -P_i(t) \sum_{j \in E - \{i\}} \rho_{ij}(t) + \sum_{j \in E - \{i\}} \rho_{ji}(t)\,P_j(t) + \frac{0(\mathrm{d}t)}{\mathrm{d}t}.$$

Assuming that $P_i(t)$ is differentiable, by passing to the limit we obtain:

$$\frac{\mathrm{d}P_i(t)}{\mathrm{d}t} = -\Big(\sum_{j \in E - \{i\}} \rho_{ij}(t)\Big) P_i(t) + \sum_{j \in E - \{i\}} \rho_{ji}(t)\,P_j(t). \tag{5}$$

In the case of a time-homogeneous process, the terms ρ_{ij} or ρ_{ji} are constant and $P_i(t)$ is differentiable. The set of differential equations (5) constitutes the system *state equations*. Putting this in matrix form, assuming that $E = \{1, 2, \ldots, n\}$ we obtain

$$\Big[\frac{\mathrm{d}P_1}{\mathrm{d}t}(t), \frac{\mathrm{d}P_2}{\mathrm{d}t}(t), \ldots, \frac{\mathrm{d}P_n}{\mathrm{d}t}(t)\Big] = [P_1(t), P_2(t), \ldots, P_n(t)]\,A \tag{6}$$

The matrix A is called the *transition rate matrix*:

— a_{ij} is equal to the transition rate ρ_{ij} from state i to state $j (i \neq j)$,
— a_{ii} is equal to the inverse of the sum of the transition rates from state i to all the other states $\Big(a_{ii} = -\sum_{j \neq i} \rho_{ij}\Big)$.

Equation (6) is sometimes written with the matrix A transposed, which makes it possible to arrange the vectors in a column. However, Eqn (6) is the most frequently used form.

When the process is homogeneous, the elements of A are constant.

The initial property of matrix A results directly from its definition, i.e.:

$$\left.\begin{aligned} a_{ij} &\geq 0 \qquad i \neq j \\ a_{ii} &= -\sum_{j \neq i} a_{ij} \leq 0. \end{aligned}\right\} \tag{7}$$

The terms in each row sum to zero and the determinant of A is therefore zero. When a_{ii} is zero, the state is termed *absorbing*.

As the reliability is the probability of the system being in an operating state without ever having encountered a failed state, it will be necessary to eliminate the transitions from failed states to operating states in matrix A in order to compute the reliability. In other words, assuming that the operating states are numbered from 1 to l and the failed states from $l + 1$ to n:

$$i > l \quad \text{and} \quad j \leq l \Rightarrow a_{ij} = 0\,. \tag{8}$$

When this condition is satisfied, we call $P'_i(t)$ the probability of being in state i at time t.

We call A' the matrix deduced from A by applying conditions (8) and A_l the matrix $(a_{ij}, 1 < i < l, 1 < j < l)$; Eqn (6) and conditions (8) enable us to write:

$$\left[\frac{\mathrm{d}P'_1}{\mathrm{d}t}, \frac{\mathrm{d}P'_2}{\mathrm{d}t}, \ldots, \frac{\mathrm{d}P'_l}{\mathrm{d}t}\right] = [P'_1, P'_2, \ldots, P'_l]\, A_l\,. \tag{9}$$

Equation (6) can be used to compute the system availability, whereas Eqn (9) enables us to calculate the system reliability. The following notations will be subsequently retained:

$$P_i(t) = \mathscr{P}(E_t = i \mid \text{availability computation})$$

and

$$P'_i(t) = \mathscr{P}(E_t = i \mid \text{reliability computation})$$

Example: Consider a system consisting of two components in parallel (active redundancy) and with constant failure and repair rates.

Denoting a working component by i and a failed component by $\bar{i}$, we obtain the following Markov diagram for computing the availability:

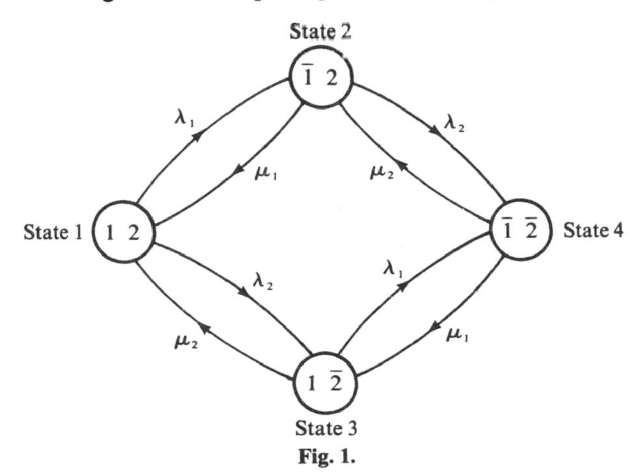

Fig. 1.

It is easy to construct the matrix A by interpreting this diagram:

$$A = \begin{bmatrix} -(\lambda_1 + \lambda_2) & \lambda_1 & \lambda_2 & 0 \\ \mu_1 & -(\mu_1 + \lambda_2) & 0 & \lambda_2 \\ \mu_2 & 0 & -(\mu_2 + \lambda_1) & \lambda_1 \\ 0 & \mu_2 & \mu_1 & -(\mu_1 + \lambda_2) \end{bmatrix}.$$

Conversely, for evaluating the reliability, it is necessary to make the failed state absorbing.

The Markov diagram then becomes:

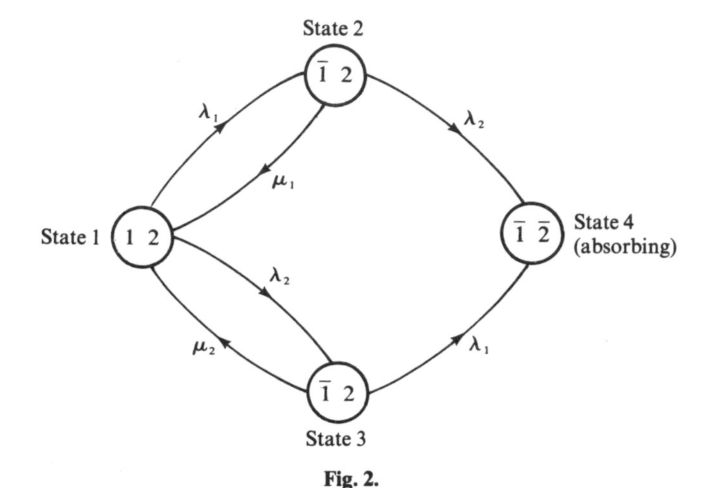

Fig. 2.

giving the transition rate matrix:

$$A' = \begin{bmatrix} -(\lambda_1+\lambda_2) & \lambda_1 & \lambda_2 & 0 \\ \mu_1 & -(\mu_1+\lambda_2) & 0 & \lambda_2 \\ \mu_2 & 0 & -(\mu_2+\lambda_1) & \lambda_1 \\ 0 & 0 & 0 & 0 \end{bmatrix}$$

If the initial distribution $P(0)$ is known, Eqns (6) and (9) can be solved by a number of methods, three of which will now be examined:

—explicit solution using the Laplace transform,
—discretization,
—computing the eigenvalues of matrix A and using the matrix exponentials.

2.4. Explicit solution of state equations using the Laplace transform

2.4.1. *Availability computation*

We denote the Laplace transforms of $P(t)$ and $P_i(t)$ by $\bar{P}(s)$ and $\bar{P}_i(s)$ respectively, and put $P_0 = P(0)$. Equation (6) can then be written as:

$$s\bar{P}(s) - P_0 = \bar{P}(s)\,A$$

from which it follows that:

$$\bar{P}(s) = P_0[sI - A]^{-1} \tag{10}$$

where I represents the nth unit matrix.

We then determine $P(t)$ by computing the inverse Laplace transform of the second member of Eqn (10). As the latter is a rational fraction in s, inversion is theoretically possible. However, as the number of states becomes moderately large, inversion is difficult. Nevertheless, Eqn (10) allows us to determine limit probabilities when t tends to infinity.

In fact, from Eqn (55), Ch. 1, we obtain:

$$P_{(\infty)} = \lim_{s\to 0} sP_0[sI - A]^{-1} = P_0 \lim_{s\to 0} \frac{s[sI - A]^*}{|sI - A|}$$

where X^* is the matrix $[X^*_{ij}]$ with:

$$X^*_{ij} = \text{cofactor of } X_{ji} \text{ in } |X|.$$

As the elements in a row of matrix A sum to zero, the sum of the elements in a row of the matrix $[sI - A]$ is s. Hence, replacing the last column of $|sI - A|$ by the sum of all the columns:

$$P_{(\infty)} = \lim_{s\to 0} \frac{sP_0[sI - A]^*}{\begin{vmatrix} s - a_{11} & -a_{12} & \cdots & s \\ & & & s \\ & & & \cdot \\ & & & \cdot \\ & & & \cdot \\ -a_{n1} & & & s \end{vmatrix}} = \frac{P_0[-A]^*}{\begin{vmatrix} -a_{11} & \cdots & -a_{1,n-1} & 1 \\ & & & 1 \\ & & & 1 \\ & & & \cdot \\ & & & \cdot \\ -a_{n1} & & -a_{n,n-1} & 1 \end{vmatrix}}$$

hence

$$P_{(\infty)} = \frac{P_0 A^*}{\Delta} \tag{11}$$

with

$$\Delta = \begin{vmatrix} a_{11} & \cdots & a_{1,n-1} & 1 \\ a_{21} & & & 1 \\ & & & 1 \\ \cdot & & & \cdot \\ \cdot & & & \cdot \\ a_{n1} & & a_{n,n-1} & 1 \end{vmatrix} \tag{12}$$

It follows from Eqn (11) that:

$$P_{j(\infty)} = \frac{1}{\Delta} \sum_{i=1}^{n} \begin{vmatrix} a_{11} & & a_{1n} \\ a_{j-1,1} & & a_{j-1,n} \\ 0-0 & P_{0i} & 0-0 \\ a_{j+1,1} & & a_{j+1,n} \\ a_{n1} & & a_{nn} \end{vmatrix} = \frac{1}{\Delta} \sum_{i=1}^{n} \begin{vmatrix} a_{11} & & a_{1,n-1} & 0 \\ a_{j-1,1} & & a_{j-1,n-1} & 0 \\ 0-0 & P_{0i} & 0-0 & \sum_{i=1}^{n} P_{0i} \\ a_{j+1,1} & & a_{j+1,n-1} & 0 \\ a_{n,1} & & a_{n,n-1} & 0 \end{vmatrix}$$

Or: as $\sum_{i=1}^{n} P_{0i} = 1$

$$P_{j(\infty)} = \frac{1}{\Delta} \begin{vmatrix} a_{11} & \cdots & a_{1,n-1} & 0 \\ & & & 0 \\ a_{j1} & & a_{j-1,1} & 1 \\ & & & \\ a_{n1} & & a_{n,n-1} & 0 \end{vmatrix} \tag{13}$$

$P_{j(\infty)}$ does not therefore depend on the initial distribution P_0.

If the states have been numbered such that the operating states are the first l, we obtain:

$$\sum_{j=1}^{l} P_{j(\infty)} = \frac{1}{\Delta} \begin{vmatrix} a_{11} & \cdots & a_{1,n-1} & 1 \\ \vdots & & & \\ a_{l1} & & a_{l,n-1} & 1 \\ a_{l+1,1} & & a_{l+1,n-1} & 0 \\ \vdots & & & \\ a_{n1} & & a_{n,n-1} & 0 \end{vmatrix} \tag{14}$$

giving the limit availability $A_{(\infty)}$:

$$A_{(\infty)} = \frac{\begin{vmatrix} a_{11} & \cdots & a_{1,n-1} & 1 \\ \vdots & & & \\ a_{l1} & & a_{l,n-1} & 1 \\ a_{l+1,1} & & a_{l+1,n-1} & 0 \\ \vdots & & & \\ a_{n,1} & & a_{n,n-1} & 0 \end{vmatrix}}{\begin{vmatrix} a_{11} & & a_{1,n-1} & 1 \\ a_{21} & & & \vdots \\ \cdot & & \cdot & \vdots \\ \cdot & & \cdot & \vdots \\ a_{n1} & & a_{n,n-1} & 1 \end{vmatrix}} \tag{15}$$

The set of non-zero values $P_{j(\infty)}$ given by Eqn (13) are termed continuous or 'stationary' availabilities. The limit unavailability is obtained by interchanging the 'zeros' and 'ones' in the last column of the numerator determinant.

2.4.2. *Reliability computation*

By replacing A by A', Eqn (10) enables us to compute the probabilities $P_i'(t)$. However, conditions (8) imply that the probabilities $P'_{i(\infty)}$ are zero when $i < l$. Equations (13)–(15) are therefore of no use.

This *a priori* obvious result prompts us to characterize the states in a different way. One method consists of calculating the mean time to failure (MTTF) or mean time before absorption.

From Eqn (77), Ch. 1, we have:

$$\mathrm{MTTF} = \bar{P}(0) = \sum_{i=1}^{l} \bar{P}_i(0)\,. \tag{16}$$

We put $P'_l = [P'_1, P'_2, \ldots, P'_l]$ and $P'_0 = [P'_1(0), P'_2(0), \ldots, P'_l(0)]$.
After applying the Laplace transform, Eqn (9) can be written as:

$$\bar{P}'_l(s) = P'_0[sI - A_l]^{-1}\,.$$

Equation (16) can therefore be written in the form:

$$\mathrm{MTTF} = P'_0[-A_l]^{-1} = -P'_0[A_l]^{-1} \tag{17}$$

then, using an identical calculation:

$$\mathrm{MTTF} = \frac{\begin{vmatrix} 0 & P'_1(0) & \ldots & P'_l(0) \\ 1 & a_{11} & & a_{1l} \\ 1 & & & \\ \cdot & & & \\ \cdot & & & \\ 1 & a_{l1} & & a_{ll} \end{vmatrix}}{|A_l|} \tag{18}$$

This relation shows that the MTTF, unlike the stationary availability, depends on the initial state.

Example: Two-component redundant system ($n = 4, l = 3$)

$$1 - A_{(\infty)} = \frac{\begin{vmatrix} -(\lambda_1+\lambda_2) & \lambda_1 & \lambda_2 & 0 \\ \mu_1 & -(\mu_1+\lambda_2) & 0 & 0 \\ \mu_2 & 0 & -(\mu_2+\lambda_1) & 0 \\ 0 & \mu_2 & \mu_1 & 1 \end{vmatrix}}{\begin{vmatrix} -(\lambda_1+\lambda_2) & \lambda_1 & \lambda_2 & 1 \\ \mu_1 & -(\mu_1+\lambda_2) & 0 & 1 \\ \mu_2 & 0 & -(\mu_2+\lambda_1) & 1 \\ 0 & \mu_2 & \mu_1 & 1 \end{vmatrix}} = \frac{\lambda_1 \lambda_2}{(\mu_1+\lambda_1)(\mu_2+\lambda_2)}$$

Assume that $P_{01} = 1$. Therefore

$$\mathrm{MTTF} = \frac{\begin{vmatrix} 0 & 1 & 0 & 0 \\ 1 & -(\lambda_1+\lambda_2) & \lambda_1 & \lambda_2 \\ 1 & \mu_1 & -(\mu_1+\lambda_2) & 0 \\ 1 & \mu_2 & 0 & -(\mu_2+\lambda_1) \end{vmatrix}}{\begin{vmatrix} -(\lambda_1+\lambda_2) & \lambda_1 & \lambda_2 \\ \mu_1 & -(\mu_1+\lambda_2) & 0 \\ \mu_2 & 0 & -(\mu_2+\lambda_1) \end{vmatrix}}$$

$$\text{MTTF} = \frac{\begin{vmatrix} 1 & \lambda_1 & \lambda_2 \\ 1 & -(\mu_1 + \lambda_2) & 0 \\ 1 & 0 & -(\mu_2 + \lambda_1) \end{vmatrix}}{\begin{vmatrix} -(\lambda_1 + \lambda_2) & \lambda_1 & \lambda_2 \\ \mu_1 & -(\mu_1 + \lambda_2) & 0 \\ \mu_2 & 0 & -(\mu_2 + \lambda_1) \end{vmatrix}}$$

$$= \frac{\lambda_1(\lambda_1 + \mu_1) + \lambda_2(\lambda_2 + \mu_2) + (\lambda_1 + \mu_1)(\lambda_2 + \mu_2)}{\lambda_1 \lambda_2(\lambda_1 + \mu_1 + \lambda_2 + \mu_2)}.$$

In the case where $\lambda_1 = \lambda_2 = \lambda$, $\mu_1 = \mu_2 = \mu$, we obtain MTTF $= (\mu + 3\lambda)/2\lambda^2$. In the case where $P_{02} = 1$, we obtain MTTF $= (\mu + 2\lambda)/2\lambda^2$.

2.4.3. *Computing the mean uptime (MUT) following repair*

After service restoration, the system is in any one of the operating states with the following probabilities:

$$P_i = \frac{\sum_{j=l+1}^{n} P_{j(\infty)} \rho_{ji}}{\sum_{i=1}^{l} \left(\sum_{j=l+1}^{n} P_{j(\infty)} \rho_{ji} \right)} \qquad i = 1, 2, \ldots, l. \tag{19}$$

The MUT is then obtained by applying Eqn (18) with the distribution given by Eqn (19) as the initial distribution.

By reversing the roles of the operating and failed states, it is easy to compute the system maintainability, the MTTR and the MDT. From this, it follows that MTBF = MUT + MDT. Note that the MTTF and the MTTR depend on the initial state and may be regarded as the mean absorption times starting from this initial state. Conversely, the MUT and the MDT do not depend on the initial state.

Example: Two-component redundant system

$$P_2 = \frac{\mu_2 \rho_4}{\mu_2 \rho_4 + \mu_1 \rho_4} = \frac{\mu_2}{\mu_1 + \mu_2}; \quad P_3 = \frac{\mu_1}{\mu_1 + \mu_2}; \quad P_1 = 0$$

hence, by applying Eqn (18):

$$\text{MUT} = \frac{\begin{vmatrix} 0 & 0 & \mu_2 & \mu_1 \\ 1 & -(\lambda_1 + \lambda_2) & \lambda_1 & \lambda_2 \\ 1 & \mu_1 & -(\mu_1 + \lambda_2) & 0 \\ 1 & \mu_2 & 0 & -(\mu_2 + \lambda_1) \end{vmatrix} \times \dfrac{1}{\mu_1 + \mu_2}}{-\lambda_1 \lambda_2(\lambda_1 + \mu_1 + \lambda_2 + \mu_2)}$$

$$= \frac{\mu_2(\lambda_1 + \mu_1)(\mu_2 + \lambda_1 + \lambda_2) + \mu_1(\lambda_2 + \mu_2)(\mu_1 + \lambda_1 + \lambda_2)}{\lambda_1 \lambda_2(\mu_1 + \mu_2)(\mu_1 + \lambda_1 + \mu_2 + \lambda_2)}.$$

In the case where $\lambda_1 = \lambda_2 = \lambda$, $\mu_1 = \mu_2 = \mu$, we obtain MUT $= (\mu + 2\lambda)/2\lambda^2$ and the diagram for calculating the maintainability is very simple:

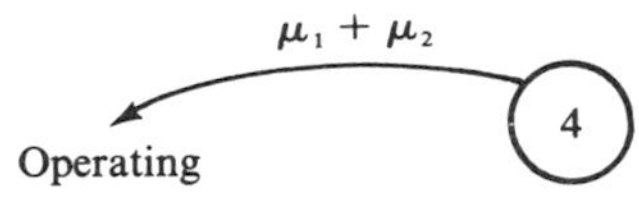

$\mu_1 + \mu_2$

Operating ← (4) $A = [-(\mu_1 + \mu_2)]$

Fig. 3.

hence, from Eqn (18):

$$\text{MTTR} = \frac{\begin{vmatrix} 0 & 1 \\ 1 & -(\mu_1 + \mu_2) \end{vmatrix}}{|-(\mu_1 + \mu_2)|} = \frac{1}{\mu_1 + \mu_2}.$$

2.4.4. *Mean state occupation frequency*

We denote the *mean residence time* in state i by d_i. Applying Eqn (18) with the following diagram:

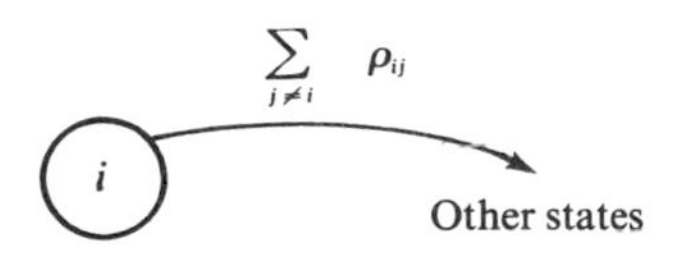

Fig. 4.

shows that $$d_i = \frac{1}{\sum_{\substack{j=1 \\ j \neq i}}^{n} \rho_{ij}} = \frac{1}{\sum_{\substack{j=1 \\ j \neq i}}^{n} a_{ij}} = \frac{-1}{a_{ii}}. \tag{20}$$

Now consider the case in which there is no absorbing state. We call d_i' the mean residence time in the other states between two encounters of state i. The *mean occupation frequency of state i*, f_i is defined by

$$f_i = \frac{1}{d_i + d_i'}.$$

It can be shown (cf. Exercise 1) that:

$$P_{i(\infty)} = \frac{d_i}{d_i + d_i'} \tag{21}$$

hence

$$f_i = P_{i(\infty)} \sum_{\substack{j=1 \\ j \neq i}}^{n} a_{ij} = -a_{ii} P_{i(\infty)}. \tag{22}$$

2.4.5. *Note on this method*

Notice that all these results relating to limit values are obtained by manipulating the determinants and can also be obtained using the flow diagram method (cf. for example [Corraza, 1975]). However, the results are difficult to calculate manually. In the next chapter, we describe a quick method for systems comprising good-quality components.

These direct methods are all limited by system size.

2.5. Solving state equations by discretization

Solving a linear system of differential equations of the first order is commonplace in numerical analysis [Lions, 1973; Richtmyer, 1967] and several programs are available. By way of example, we now give a simple algorithm which can be used when we have a program for decomposing a matrix B into the form LU:

$B = LU$, L being a lower triangular matrix and U an upper triangular matrix.

Solve:

$$\frac{\mathrm{d}P(t)}{\mathrm{d}t} = P(t)\,A\,.$$

We shall attempt to compute $P(t + \Delta t)$ as a function of $P(t)$. We put:

$$P^{n+1} = P(t + \Delta t)$$
$$P^{n} = P(t)\,.$$

The discretization of the differential system can be written as:

$$\frac{P^{n+1} - P^{n}}{\Delta t} - \frac{P^{n} + P^{n+1}}{2} A = 0\,.$$

We then put:

$$P^{n+\frac{1}{2}} = \frac{P^{n} + P^{n+1}}{2}$$

therefore

$$P^{n+1} = 2\,P^{n+\frac{1}{2}} - P^{n}$$

and

$$\frac{2}{\Delta t}\,(P^{n+\frac{1}{2}} - P^{n}) - P^{n+\frac{1}{2}} A = 0$$

or

$$P^{n+\frac{1}{2}} B = \frac{2}{\Delta t}\,P^{n}$$

with $B = (2I/\Delta t) - A$ which is decomposed into the form LU.

It is then sufficient to solve:

$$XU = \frac{2}{\Delta t} P^n, \quad (U \text{ being upper triangular})$$

$$P^{n+\frac{1}{2}} L = X, \quad (L \text{ being lower triangular})$$

$$P^{n+1} = 2\, P^{n+\frac{1}{2}} - P^n .$$

Note that the matrices L and U are fixed as long as the step remains constant. An estimation of the error in $0(\Delta t^2)$ can be found in [Lions, 1973].

2.6. Solving state equations by computing the eigenvalues of matrix A

The solution of the system of differential equations (6) is explicitly given by a matrix exponential. In fact:

$$P(t) = P(0)\, e^{At} . \tag{23}$$

If we assume that all the eigenvalues of the matrix A are different and real, there exists a regular matrix Q such that:

$$A = QDQ^{-1}$$

with

$$D = \begin{bmatrix} -\beta_1 & & 0 \\ & \ddots & \\ 0 & & -\beta_n \end{bmatrix} \qquad 0 < \beta_1 < \beta_2 \ldots < \beta_n$$

(The fact that the real parts of the eigenvalues of A are negative will be demonstrated in the following chapter.)

Therefore

$$e^{At} = Q \begin{bmatrix} e^{-\beta_1 t} & & 0 \\ & \ddots & \\ 0 & & e^{\beta_n t} \end{bmatrix} Q^{-1} \tag{24}$$

This method appears attractive. However, there is a numerical problem for very reliable systems. In fact, the largest eigenvalue $-\beta_1$ is much smaller than the others in absolute value (cf. Ch. 6) and accuracy is likely to be poor when the number of states is relatively large.

This fact is all the more troublesome in that it is this eigenvalue which determines the behaviour of the system as soon as t is relatively large. Consequently, this method is generally little used.

3. NON-MARKOVIAN SYSTEMS WITH DISCRETE STATE SPACE

The assumption that the transition rates are constant, which leads to a Markovian model, is often a valid one. However, this assumption is often incorrect as far as the repair rates are concerned, making an analytical approach much more difficult. We shall merely illustrate, by an example, the three following methods:

— supplementary variables method,
— dummy states method,
— embedded chain method.

In each of these methods we attempt a reduction to the Markovian case.

For operational calculations, refer to Ch. 4, section 6 for the availability calculation, and to Ch. 6, section 3.

3.1. Supplementary variables method [Cox, 1955]

Consider a non-Markovian system comprising n components. For each of the system states, we introduce for each component a variable measuring the elapsed time since its first change of state. A state i is therefore regarded as the juxtaposition of states $(i, x_1, x_2, \ldots, x_n)$ where x_j represents the time spent by component j in the state (operating or failed) which characterizes it in i. Consequently:

$$P_i(t) = \int_0^t \ldots \int_0^t P(i, x_1, x_2, \ldots, x_n)\, \mathrm{d}x_1\, \mathrm{d}x_2 \ldots \mathrm{d}x_n .$$

The number of supplementary variables increases very rapidly, and so this type of modelling is only possible if the number of non-constant transition rates is not excessively large. The system state equations will involve partial derivatives.

Example: Consider a system comprising two identical components in a standby redundancy configuration.

One of the components is normally operating with the other standing by. When the first fails, the other takes over while it is being repaired. We assume that the switch is not perfect: there is a probability γ of non-takeover. We assume that the component failure rate λ is constant and that the repair rate is not constant. As the repair time distribution $g(t)$ is not exponential, the repair rate can be written as:

$$\mu(t) = \frac{g(t)}{\exp\left(-\int_0^t g(u)\, \mathrm{d}u\right)} . \tag{25}$$

The state transition diagram for the system is as follows ($2a$ means 2 on standby):

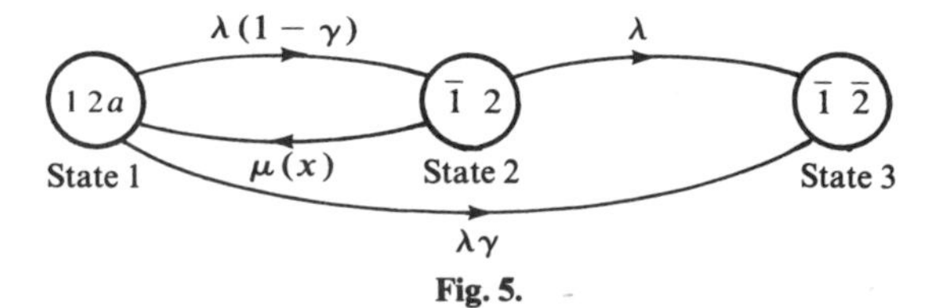

Fig. 5.

We shall denote by x the supplementary variable included to indicate how long component 1 has been under repair. We assume that $P_1(0) = 1$.

We therefore have:

$$P_2(t) = \int_0^t P_2(t, x)\,dx. \tag{26}$$

We compute the probabilities of being in the various states as time $t + \Delta t$:

$$P_1(t + \Delta t) = P_1(t)(1 - \lambda\,\Delta t) + \int_0^t P_2(t, x)\,\mu(x)\,dx\,\Delta t + 0(\Delta t). \tag{27}$$

$$P_2(t + \Delta t, x + \Delta t) = P_2(t, x)(1 - \mu(x)\,\Delta t)(1 - \lambda\,\Delta t) + 0(\Delta t). \tag{28}$$

$$P_3(t + \Delta t) = P_3(t) + \int_0^t P_1(t, x)(1 - \mu(x)\,\Delta t)\,\lambda\,\Delta t\,dx + \lambda\gamma P_1(t)\,\Delta t + 0(\Delta t). \tag{29}$$

Using a similar method to that employed by Liebowitz [Liebowitz, 1966], we then obtain by passing to the limit:

$$\frac{dP_1}{dt}(t) = -\lambda P_1(t) + \int_0^t P_2(t, x)\,\mu(x)\,dx. \tag{30}$$

$$\frac{\partial P_2}{\partial t}(t, x) + \frac{\partial P_2}{\partial x}(t, x) = -(\mu(x) + \lambda)\,P_2(t, x). \tag{31}$$

$$\frac{dP_3}{dt}(t) = \lambda \int_0^t P_2(t, x)\,dx + \lambda\gamma P_1(t). \tag{32}$$

The initial conditions are as follows:

$$\begin{aligned} &P_1(0) = 1 \quad P_2(0) = P_3(0) = 0 \quad P_2(t, x) = 0 \quad \text{if } x > t \\ &P_2(t, 0) = \lambda(1 - \gamma)\,P_1(t). \end{aligned} \tag{33}$$

Using the Laplace transform to solve Eqns (30)–(32) under the initial conditions (33), they become respectively:

$$s\bar{P}_1(s) - 1 = -\lambda\bar{P}_1(s) + \int_0^\infty \bar{P}_2(s, x)\,\mu(x)\,dx. \tag{34}$$

$$s\bar{P}_2(s, x) + \frac{\partial}{\partial x}\bar{P}_2(s, x) = -(\mu(x) + \lambda)\,\bar{P}_2(s, x). \tag{35}$$

$$s\bar{P}_3(s) = \lambda \int_0^\infty \bar{P}_2(s, x)\,dx + \lambda\gamma\bar{P}_1(s). \tag{36}$$

Equation (35) can be rearranged as:

$$\frac{1}{\bar{P}_2(s, x)}\frac{\partial}{\partial x}\bar{P}_2(s, x) = -(\mu(x) + \lambda + s)$$

hence

$$\bar{P}_2(s, x) = \bar{P}_2(s, 0)\exp\left(-(\lambda + s)x - \int_0^x \mu(u)\,du\right) \tag{37}$$

Taking account of Eqns (25), (33) and (37), Eqn (34) can then be written as:

$$(\lambda + s)\,\bar{P}_1(s) = 1 + \lambda(1 - \gamma)\,\bar{P}_1(s)\int_0^\infty g(x)\,e^{-(s+\lambda)x}\,dx$$

hence

$$\bar{P}_1(s) = \frac{1}{s + \lambda(1 - (1 - \gamma)\,\bar{g}(s + \lambda))}. \tag{38}$$

As for Eqn (36), taking account of Eqns (33) and (37), we obtain:

$$s\bar{P}_3(s) = \frac{\lambda^2(1 - \gamma)\,\bar{P}_1(s)}{\lambda + s} \int_0^\infty (\lambda + s)\exp\left(-(\lambda + s)\,x - \int_0^x \mu(u)\,du\right) dx + \lambda\gamma\bar{P}_1(s). \tag{39}$$

Integration by parts gives:

$$\bar{P}_3(s) = \frac{\lambda^2(1 - \gamma)\,\bar{P}_1(s)}{s(s + \lambda)}(1 - \bar{g}(s + \lambda)) + \frac{\lambda\gamma\bar{P}_1(s)}{s}$$

therefore, taking account of Eqn (38)

$$\bar{P}_3(s) = \frac{\lambda[\gamma + \lambda(1 - \gamma)(1 - \bar{g}(s + \lambda))]}{s(s + \lambda)[s + \lambda - \lambda(1 - \gamma)\,\bar{g}(s + \lambda)]} \tag{40}$$

then, from Eqn (36)

$$\bar{P}_2(s) = \frac{s}{\lambda}\,\bar{P}_3(s) - \gamma\bar{P}_1(s)$$

hence

$$\bar{P}_2(s) = \frac{\lambda(1 - \gamma)(1 - \bar{g}(s + \lambda))}{(s + \lambda)[s + \lambda - \lambda(1 - \gamma)\,\bar{g}(s + \lambda)]}. \tag{41}$$

Inversion of Eqns (38), (40) and (41) is generally impossible explicitly, except where $\bar{g}(s + \lambda)$ is rational (Erlangian distributions). In the other cases, we can attempt numerical inversion (cf., for example [Abate and Dubner, 1968]).

However, it is possible to compute the MTTF of the system:

$$\text{MTTF} = \bar{P}_1(0) + \bar{P}_2(0) = \frac{1 + (1 - \gamma)(1 - \bar{g}(\lambda))}{\lambda[1 - (1 - \gamma)\,\bar{g}(\lambda)]}.$$

3.2. Dummy states method

An Erlangian distribution with integer-valued parameter k may be regarded as the convolution of k exponential distributions, i.e. a transition (failure or repair) corresponding to an Erlangian distribution may be decomposed into k transitions corresponding to the same exponential law.

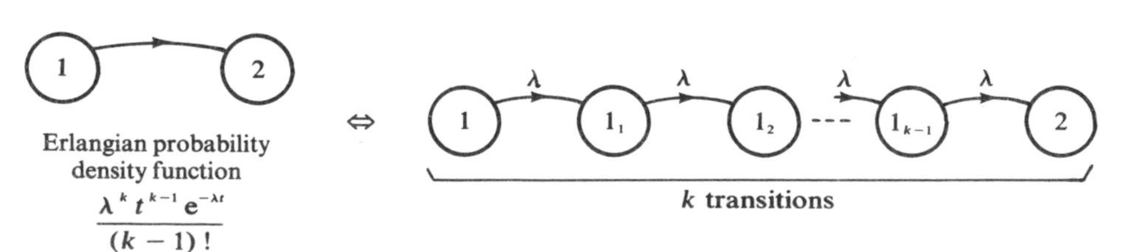

Fig. 6.

The $1_1, 1_2, \ldots, 1_{k-1}$ are dummy states but enable us to convert the process into a Markovian one. This method is quite useful, as approximations for a large number of distributions can be obtained via an Erlangian distribution. However, the multiplication of states is very problematic.

The use of dummy states can be extended to more general situations (cf., for example [Billington and Singh, 1972; Marguin, 1978; Costes et al., 1976]); initially, the transition rates may be different. We then have the following transition diagram:

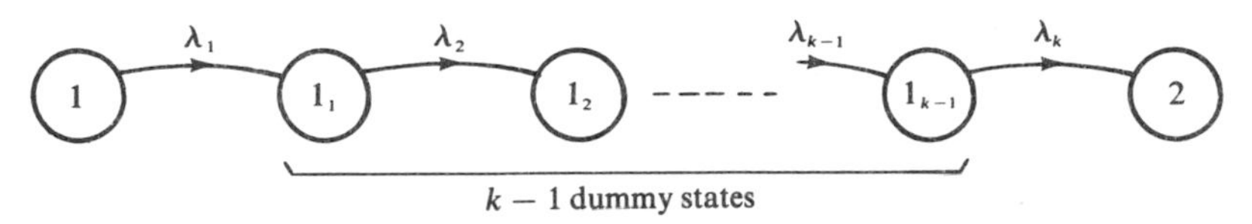

Fig. 7.

The dummy states are termed 'in series'.

Equally conceivable is a transition from state 1 to state 2 such that the residence time in state 1 is exponentially distributed with parameter λ_i having probability α_i:

$$\mathcal{P}(X = X_i) = \alpha_i \; : \; \alpha_i > 0, \qquad \sum_{i=1}^{k} \alpha_i = 1 \, .$$

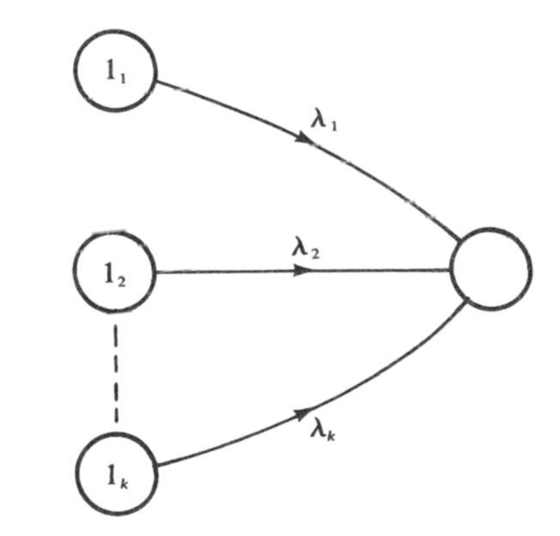

Fig. 8.

The dummy states are termed 'in parallel'. Of course, it is possible to use both representations simultaneously. This method yields distributions for residence in state 1 known as *generalized Erlangian distributions* which provide a good approximation for a large number of distributions ([Marguin, 1978] and Exercise 4).

Example: Compute the availability of a component whose failure rate is constant and whose repair time distribution $g(t)$ is Erlangian with parameter 2

$$g(t) = \mu^2 \, t \, \mathrm{e}^{-\mu t} \qquad \tau = \int_0^\infty t g(t) \, \mathrm{d}t = \frac{2}{\mu} \, .$$

The state transition diagram is as follows:

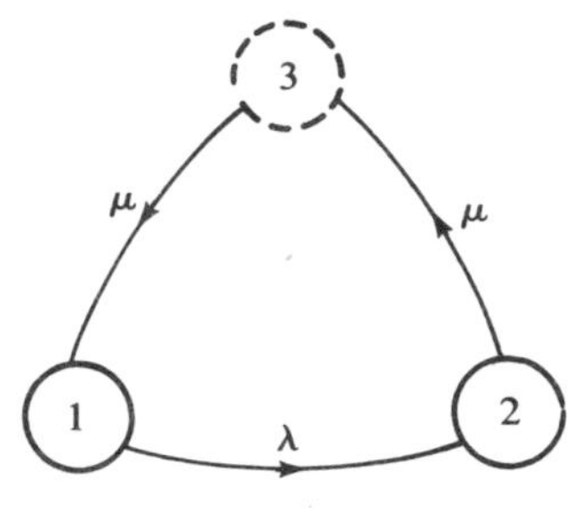

Fig. 9.

State 3 is a dummy state;
hence:

$$A = \begin{bmatrix} -\lambda & \lambda & 0 \\ 0 & -\mu & \mu \\ \mu & 0 & -\mu \end{bmatrix}$$

$$\bar{P}(s) = P(0) \begin{bmatrix} s+\lambda & -\lambda & 0 \\ 0 & s+\mu & -\mu \\ -\mu & 0 & s+\mu \end{bmatrix}^{-1}.$$

Taking, for example, $P_1(0) = 1$, then:

$$\bar{P}(s) = \frac{[(s+\mu)^2, \lambda(s+\mu), \lambda\mu]}{s(s^2 + s(\lambda + 2\mu) + \mu^2 + 2\lambda\mu)}$$

hence the availability $\bar{A}(s)$ is given by:

$$\bar{A}(s) = \frac{(s+\mu)^2}{s(s^2 + s(\lambda + 2\mu) + \mu^2 + 2\lambda\mu)}.$$

Note that when $(\lambda - 4\mu)$ is negative (as is usually the case), the denominator of $\bar{A}(s)$ has two complex conjugate roots.

We again obtain the well-known formula:

$$A(\infty) = \frac{\mu^2}{\mu^2 + 2\lambda\mu} = \frac{1}{1 + \dfrac{2\lambda}{\mu}} = \frac{1}{1 + \lambda\tau}.$$

3.3. Embedded chain method

3.3.1. *Definitions*

Certain non-Markovian systems are such that there are times for which the system state encapsulates its entire past history. These are called regeneration points. If we take the set of these regeneration times as the set of indices of the process describing the system, we define a discrete Markov chain known as an *embedded chain.* Analysis of this process gives an idea of the behaviour of the system. In practice, an important set of processes having this property is constituted by the semi-Markov processes. For such processes, the interstate transition probabilities depend on a single variable: *the elapsed time since arrival in that state.*

These processes were introduced simultaneously by Levy and by Smith [Levy, 1954; Smith, 1955]. We shall consider only those processes with a finite and regular number of states (the number of visits to each state is finite over a finite interval). Consider a system S consisting of p repairable components: we denote the operating states by $1, 2, \ldots, l$ and the failed states by $l+1, l+2, \ldots, n$.

We shall call B the transition probability matrix in the embedded chain associated with the semi-Markov process. Consequently

$$\left.\begin{aligned} b_{ii} &= 0 \\ \sum_{j=1}^{n} b_{ij} &= 1 \quad \text{for a non-absorbing state} \\ b_{ij} &= 0 \quad \text{for an absorbing state} \end{aligned}\right\} \tag{42}$$

Notice that a Markov process is also semi-Markovian. In the case of a time-homogeneous Markov process, for a non-absorbing state we obtain:

$$b_{ij} = \frac{\rho_{ij}}{\sum_{j \neq i} \rho_{ij}} . \tag{43}$$

We call the *transition distribution matrix* the matrix Q, each element $Q_{ij}(t)$ of which represents the probability of leaving state i during $[0, t]$ and entering state j in a single transition. This matrix is very important because, as we shall see, being associated with the initial state it determines the system completely.

The following properties follow from the definition of this matrix:

(a) If the transition $i \to j$ is created by the change of state of component l between u and $u + du$, then:

$$Q_{ij}(t) = \int_0^t \mathscr{P}\begin{pmatrix} \text{the components apart from } l \\ \text{do not change state during } [0, u] \end{pmatrix} f(u)\,du \tag{44}$$

where $f(u)$ is the time to failure density or repair time density of component l.

(b) From the definition of B, it follows that:

$$\left.\begin{aligned} Q_{ij}(0) &= 0 \\ Q_{ij}(\infty) &= b_{ij} \end{aligned}\right\} \tag{45}$$

(c) $H_i(t) = \sum_{j=1}^{n} Q_{ij}(t)$ represents the distribution function of the time spent in state i. When the state is absorbing, this distribution function is degenerate ($H_i(t) = 0$).

3.3.2. *System availability and reliability computation*

To calculate reliability or availability, it is sufficient to calculate the probability of being in a given state j. We therefore put:

$$P_{ij}(t) = \mathscr{P}\,(\text{being in state } j \text{ at } t/\text{the system is in state } i \text{ at } t = 0).$$

The functions $P_{ij}(t)$ and $Q_{ij}(t)$ are linked by the following relations [Pyke, 1961]:

$$P_{ij}(t) = \sum_k \frac{\mathrm{d}Q_{ik}}{\mathrm{d}t} * P_{kj}(t) \qquad i \neq j \tag{46}$$

$$P_{jj}(t) = 1 - H_j(t) + \sum_k \frac{\mathrm{d}Q_{jk}}{\mathrm{d}t} * P_{kj}(t)\,. \tag{47}$$

This gives the expression for computing the Laplace transform $\bar{P}_{ij}(s)$ of $P_{ij}(t)$:

$$\bar{P}_{ij}(s) = [I - s\bar{Q}(s)]_{ij}^{-1} \left(\frac{1}{s} - \bar{H}_j(s)\right). \tag{48}$$

Availability computation (no absorbing states)

In this case:

$$P_{ij}(\infty) = \lim_{s \to 0} \; [I - s\bar{Q}(s)]_{ij}^{-1} \, (1 - s\bar{H}_j(s))$$

we put:

$$d_j = \int_0^\infty t \, \mathrm{d}H_j = \text{mean residence time in state } j,$$

then, from Eqn (45) and taking account of the fact that $1 - s\bar{H}_j(s) \underset{s \to 0}{\simeq} sd_j$

$$P_{ij}(\infty) = \lim_{s \to 0} \; [I - s\bar{Q}(s)]_{ij}^{-1} \, sd_j$$

The matrix $[I - s\bar{Q}(s)]$ tends to the matrix $[I - B]$ as s tends to zero, which is unusual since the sum of the elements of the same row is zero. We can then put s as a factor in the determinant $|I - s\bar{Q}(s)|$ by replacing the last column by the sum of the columns. Ultimately:

$$P_{ij}(\infty) = \frac{\begin{vmatrix} 1 & -b_{12} & \ldots & -b_{1,i-1} & 0 & -b_{1,i+1} & \ldots & -b_{1n} \\ -b_{21} & 1 \ldots & & & 0 \ldots & & & -b_{2n} \\ & & & & & & & \\ -b_{j1} & & & -b_{j,i-1} & d_j & -b_{j,i+1} & & -b_{jn} \\ & & & & 0 & & & \\ & & & & \cdot & & & \\ -b_{n1} & & & & 0 & & & 1 \end{vmatrix}}{\begin{vmatrix} 1 & -b_{12} & \ldots & -b_{1,i-1} & d_1 & -b_{1,i+1} & \ldots & -b_{1n} \\ -b_{21} & 1 & \ldots & -b_{2,i-1} & d_2 & -b_{2,i+1} & \ldots & -b_{2n} \\ & & & & \cdot & & & \\ & & & & \cdot & & & \\ & & & & \cdot & & & \\ -b_{n1} \ldots & & & & d_n & & & 1 \end{vmatrix}} \tag{49}$$

Moreover, this expression is identical to Eqn (12), noting that from Eqns (6), (20) and (43):

$$A = \begin{bmatrix} d_1 & & 0 \\ & d_2 & \\ 0 & & \ddots \; d_n \end{bmatrix}^{-1} [B - I] \tag{50}$$

Reliability computation

Assume that the operating states are numbered from 1 to l. Denoting by Q^l the matrix formed by the first l rows and the first l columns of Q, since the failed states are absorbing, Eqn (48) can be written as:

$$\bar{P}_{ij}(s) = [I - s\bar{Q}^l(s)]_{ij}^{-1} \left(\frac{1}{s} - \bar{H}_j(s)\right) \tag{51}$$

hence, the reliability $\bar{R}_i(s) = \sum_{j=1}^{l} \bar{P}_{ij}(s)$. It is also possible to calculate the MTTF when the initial state is i. We denote this MTTF by μ_i. From Eqn (77), Ch. 1, we obtain:

$$\mu_i = \sum_{j=1}^{l} [I - B^l]_{ij}^{-1} d_j \quad \forall i = 1, \ldots, l.$$

Or putting:

$$\mu = \begin{bmatrix} \mu_1 \\ \mu_2 \\ \cdot \\ \cdot \\ \cdot \\ \mu_l \end{bmatrix} \quad \text{and} \quad d = \begin{bmatrix} d_1 \\ d_2 \\ \cdot \\ \cdot \\ \cdot \\ d_l \end{bmatrix}$$

$$\mu = [I - B^l]^{-1} d. \tag{52}$$

3.3.3. *Examples*

With the exception of Markovian systems (which are obviously semi-Markovian), large systems which lend themselves to analysis by this method are very rare as it is generally difficult to find regeneration points. (An example is given in Exercise 2.) In this chapter we will examine two very simple examples: one for computing the availability of a single-component system which will enable us to see the results of renewal theory (Ch. 1, section 6.3), and a typical reliability calculation.

Example 1: consider a single-component system for which $F(t)$ and $G(t)$ are the uptime and repair time distribution functions respectively.

Firstly, this system is in fact semi-Markovian as the transition probabilities between the operating and failed states only depend on the elapsed time since the last transition. We shall denote the operating state by 1 and the failed state by 2.

All the calculations in this section depend on the matrix Q. We shall therefore calculate this matrix. For each state there is only one transition out of it; Eqn (44) can therefore be written quite simply as:

$$Q = \begin{bmatrix} 0 & F(t) \\ G(t) & 0 \end{bmatrix}$$

hence, from the properties of Q:

$$B = \begin{bmatrix} 0 & 1 \\ 1 & 0 \end{bmatrix} \quad ; H_1(t) = F(t),\ H_2(t) = G(t).$$

Applying formulae (48) yields the expressions of the Laplace transforms of $P_{ij}(t)$:

$$[I - s\bar{Q}(s)]^{-1} = \frac{1}{1 - s^2 \bar{F}(s)\bar{G}(s)} \begin{bmatrix} 1 & s\bar{F}(s) \\ s\bar{G}(s) & 1 \end{bmatrix}$$

hence:

$$\bar{P}_{11}(s) = \frac{\frac{1}{s} - \bar{F}}{1 - s^2 \bar{F}\bar{G}} \qquad \bar{P}_{21}(s) = \frac{s\bar{G}\left(\frac{1}{s} - \bar{F}\right)}{1 - s^2 \bar{F}\bar{G}}$$

$$\bar{P}_{12}(s) = \frac{s\bar{F}\left(\frac{1}{s} - \bar{G}\right)}{1 - s^2 \bar{F}\bar{G}} \qquad \bar{P}_{22}(s) = \frac{\frac{1}{s} - \bar{G}}{1 - s^2 \bar{F}\bar{G}}.$$

Note that the expression of $\bar{P}_{11}(s)$ is none other than Eqn (71) of Ch. 1 and the expression of $\bar{P}_{12}(s)$ that of Eqn (73). We therefore find the renewal theory results described in Ch. 1.

The availability is equal to $P_{11}(t)$ when the initial state is state 1 and equal to $P_{21}(t)$ when the initial state is state 2. Inversion of the Laplace transforms is often impossible. However, formula (49) enables us to determine the stationary (*a priori* obvious) probabilities:

$$P_{11}(\infty) = \frac{\begin{vmatrix} d_1 & -1 \\ 0 & 1 \end{vmatrix}}{\begin{vmatrix} d_1 & -1 \\ d_2 & 1 \end{vmatrix}} = \frac{d_1}{d_1 + d_2}$$

$$P_{21}(\infty) = \frac{\begin{vmatrix} 1 & d_1 \\ -1 & 0 \end{vmatrix}}{\begin{vmatrix} 1 & d_1 \\ -1 & d_2 \end{vmatrix}} = \frac{d_1}{d_1 + d_2}.$$

Example 2: Consider a system comprising two identical components which are independent and in an active redundancy configuration. Assume that the failure rate λ is constant and that the repair time distribution density is $g(t)$. The aim is to determine the reliability of the system, which is indeed semi-Markovian (for computing the availability, we must assume, for example, that there is only one repair facility operating on a first-failed-first-repaired basis for the system to be semi-Markovian). Let 1 be the state in which both components are operating, 2 the state where only one of them is operating and 3 the failed state.

Matrix Q can be determined using Eqn (44):

$$Q_{12}(t) = 2 \int_0^t e^{-\lambda u} \lambda e^{-\lambda u} du = 1 - e^{-2\lambda t}$$

$$Q_{21}(t) = \int_0^t e^{-\lambda u} g(u) du$$

$$Q_{23}(t) = \int_0^t (1 - G(u)) \lambda e^{-\lambda u} du$$

hence:

$$\bar{Q}_{(s)} = \begin{vmatrix} 0 & \dfrac{2\lambda}{s(s+2\lambda)} & 0 \\ \dfrac{\bar{g}(s+\lambda)}{s} & 0 & \dfrac{\lambda}{s(s+\lambda)}(1 - \bar{g}(s+\lambda)) \\ 0 & 0 & 0 \end{vmatrix}$$

$$\bar{H}_1(s) = \frac{2\lambda}{s(s+2\lambda)}, \quad \frac{1}{s} - \bar{H}_1(s) = \frac{1}{s+2\lambda}, \quad d_1 = \frac{1}{2\lambda}$$

$$\bar{H}_2(s) = \frac{\lambda}{s(s+\lambda)} + \frac{\bar{g}(s+\lambda)}{s+\lambda}, \quad \frac{1}{s} - \bar{H}_2(s) = \frac{1 - \bar{g}(s+\lambda)}{s+\lambda}, \quad d_2 = \frac{1 - \bar{g}(\lambda)}{\lambda}.$$

Then ($l = 2$):

$$[I - s\bar{Q}'(s)]^{-1} = \frac{s+2\lambda}{s+2\lambda - 2\lambda\bar{g}(s+\lambda)} \begin{bmatrix} 1 & \dfrac{2\lambda}{s+2\lambda} \\ \bar{g}(s+\lambda) & 1 \end{bmatrix}$$

hence, assuming that the initial state is state 1:

$$\bar{R}(s) = \bar{P}_{11}(s) + \bar{P}_{12}(s) = \frac{s + 3\lambda - 2\lambda\bar{g}(s+\lambda)}{(s+\lambda)(s+2\lambda - 2\lambda\bar{g}(s+\lambda))}$$

$$\text{MTTF} = \mu_1 = \frac{1}{\lambda} \frac{3 - 2\bar{g}(\lambda)}{2(1 - \bar{g}(\lambda))}.$$

We thus find the same results as those obtained by Liebowitz using the supplementary variables method [Liebowitz, 1966].

4. MONTE CARLO SIMULATION

The term 'simulation' often gives rise to confusion as it is generally applied to two types of situation: situations where we wish to determine the evolution of a system in time, which is termed *dynamic* or operational simulation, and situations where the aim is to define the probability of certain events occurring, in which case it is known as *Monte Carlo simulation.*

The confusion therefore stems from the fact that, in the case of operational simulation, we may take into account a number of scenarios each with a certain probability of occurring.

In addition, when the mathematical model describing the system is stochastic, operational simulation becomes very similar to Monte Carlo simulation.

4.1. Principle of Monte Carlo simulation

We shall discuss Monte Carlo techniques from the point of view of computing system reliability or availability. It will therefore be assumed that the aim is to estimate a probability or a mean uptime.

Consider, then, a random variable depending on a certain number $X_1, X_2, \ldots, X_n$ of random variables whose distributions are known.

We shall attempt to determine the distribution of this random variable or to estimate some of its parameters.

Example: We know the distributions X_i of the various system components as well as the system logic, and we wish to estimate the system MTTF.

This parameter is a random variable written as:

$$\varphi = \varphi(X_1, X_2, \ldots, X_n).$$

We estimate the MTTF as follows:

(a) We place ourselves at time $t_0 = 0$.

—For each of the components i, we then sample the time they take to fail. (The sampling procedure is described in section 4.3.) Let t_1 be the smallest of these values defined for a particular $i\,(i_1)$.

—We place ourselves at time t_k. At this point in time, X is in a certain position and the system logic then tells us whether or not the system has failed. If so, the realization of the uptime is t_k.

—If not, we consider variable X_{i_k} and randomly sample the time it takes to change state. Let t_{k+1} be the nearest time corresponding to the change of state of one of the variables and i_{k+1} the index of that variable. We then go back to the previous step.

(b) A certain number of sample runs of the above type give realizations T_i of the uptime. The average of these values gives an estimate μ of the MTTF

$$\mu = \frac{1}{n} = \sum_{i=1}^{n} T_i .$$

This method appears attractive by virtue of the great variety of assumptions which it can take into account; however, it has a serious drawback. Let us assume, in fact, that we have ten components each with an MTTF of one year and a mean repair time of the order of a week and that the system has an MTTF of the order of 10^7 years.

We therefore have, on average, some ten accidents per year and consequently some 10^8 events in a failure sample in order to reach system failure.

The various realizations T_i are realizations of the same random variable T with mean MTTF and standard deviation σ.

According to the central limit theorem (Ch. 1, section 4.2.3), μ is asymptotically normally distributed $N(\text{MTTF}, \sigma/\sqrt{n})$. $\sigma/\sqrt{n}$ is called the *standard error* [Hammersley and Handscomb, 1967].

If we wish to determine the MTTF with a relative error ε at symmetrical confidence level α, it is therefore necessary to select n (cf. Ch. 1, section 4.2.3) such that:

$$\frac{u_{\alpha/2}}{\text{MTTF}} \frac{\sigma}{\sqrt{n}} = \varepsilon .$$

In the special case where the random variable T is exponential, since $\sigma = \text{MTTF}$:

$$n = \left(\frac{u_{\alpha/2}}{\varepsilon}\right)^2 .$$

Example: $\alpha = 0.1, u_{\alpha/2} = 1.645$

Table 1.

ε	0.3	0.2	0.1	0.05
n	30	68	270	~ 1000

In order to estimate our MTTF we need, for example, some 30 sample runs which will give 3×10^9 events to be taken into account.

If each event (random sampling, verification of system operation or failure) takes of the order of 100 microseconds (10^{-4} s), we need 3×10^5 s $\simeq$ 90 h computer time to estimate this parameter.

The difficulty of applying the Monte Carlo technique lies in this prohibitive computing time whenever a *very rare event* has to be shown. The time required *for a single sample run* is therefore prohibitive (obviously if the MTTF is of the order of 10^5 years, then the computer time reduces to some tens of minutes and simulation therefore becomes possible).

In the following section, we will attempt to get round this difficulty.

Shooman gives examples of the Monte Carlo technique applied to reliability analysis [Shooman, 1968].

4.2. Variance reduction

Variance reducing techniques have a dual purpose: to reduce the length of a sample run and to increase accuracy using the same number of runs.

We shall concentrate mainly on the first.

4.2.1. *Conditional sampling (variate control* [Hammersley and Handscomb, 1967])

This is performed by representing an auxiliary random variable Z such that φ can be written as:

$$\varphi(X_1, X_2, \ldots, X_n) = f(X_1, X_2, \ldots, X_k, Z)$$

with

$$Z = \Psi(X_{k+1}, X_{k+2}, \ldots, X_n).$$

φ is then estimated in two stages:

Stage 1
Determination of the distribution of random variable Z by simulation or by an analytical method.

Stage 2
Simulation (or analytical method) for the function

$$f(X_1, \ldots, X_k, Z).$$

This sampling can be performed sequentially a number of times.

Example:

Consider the following reliability block diagram:

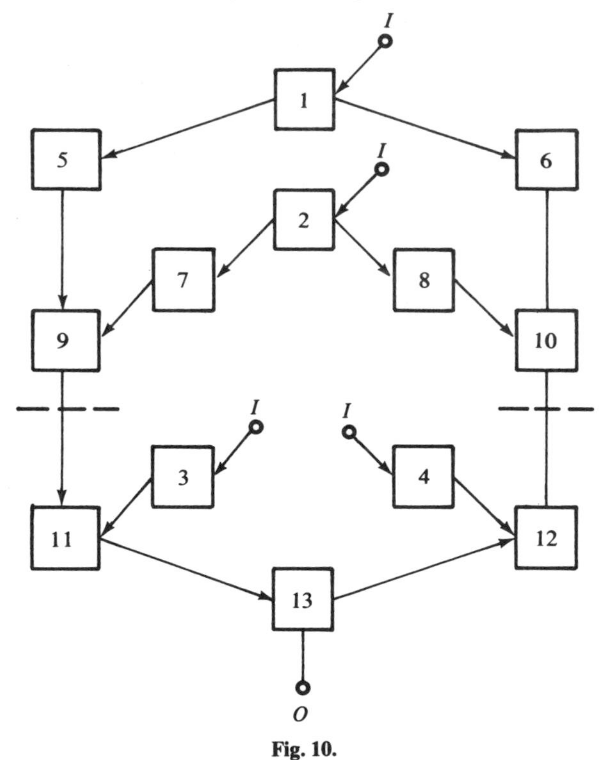

Fig. 10.

1 and 2 represent electric power systems, 3 and 4 diesel engines. The other components are circuit breakers or busbars. The aim is to determine the probability of supply failure in 13. Components 1 to 12 have independent and known operating distributions.

We can divide the system at the dotted lines; we then define a random variable $Z = \psi(X_1, X_2, X_5, X_6, X_7, X_8, X_9$ and $X_{10})$ having four modalities (no supply from

$\boxed{1}$ or $\boxed{2}$ to $\boxed{11}$ and $\boxed{12}$, supply from $\boxed{1}$ or $\boxed{2}$ to $\boxed{11}$ only, supply from $\boxed{1}$ or $\boxed{2}$ to $\boxed{12}$ only, supply from $\boxed{1}$ or $\boxed{2}$ to $\boxed{11}$ and $\boxed{12}$).

This provides a sampling in time which will subsequently enable us to define the sequence for the entire system.

4.2.2. *Importance sampling* [Hammersley and Handscomb, 1967]

Generally known as stratification, importance sampling consists of increasing the frequency of rare and important events.

More precisely, consider the random variable:

$$K = \Psi(X_1, X_2, \ldots, X_n)$$

whose distribution is well known.

For example, K may be the number of null variables ($K = \sum_i 1_{xi} = 0$) and the $\mathscr{P}(K = k)$, $k = 0, 1, \ldots, n$ are known.

Consider N realizations of the variable

$$X = (X_1, X_2, \ldots, X_n).$$

Let X^j be a realization of this kind.

A conventional estimator of the mean is:

$$\hat{\varphi} = \frac{1}{N} \sum_j \varphi(X^j).$$

Another unbiased estimator is:

$$\varphi^* = \sum_k \mathscr{P}(K = k) \frac{1}{N_k} \sum_{\Psi(X^j)=k} \varphi(X^j)$$

where N_k is the number of X^j such that $\psi(X^j) = k$.

Then, if $V(\hat{\varphi})$ and $V(\varphi^*)$ are the variances of the estimators $\hat{\varphi}$ and φ^* respectively:

$$\frac{V(\varphi^*)}{V(\hat{\varphi})} \text{ tends to } 1 - \mu_{K,\varphi}^2 \text{ as } N \text{ tends to infinity}$$

where $\mu_{K,\varphi}$ is the correlation coefficient of K and φ.

Therefore, if K and φ are highly correlated, φ^* is a better estimator than $\hat{\varphi}$. In addition, if we can sample under conditionality (e.g. by fixed k), we make *a priori* stratification (we fix the N_k and then sample for fixed k) with $N_k = N\mathscr{P}(K = k)$ which is the optimum.

This method unfortunately appears difficult to apply to reliability computations, as the definition of a sampling sequence takes into account a series of realizations rather than a single realization.

4.2.3. *Other methods*

There are many other methods, all of which appear somewhat difficult to apply to reliability:

—The *Russian roulette* or separation process [Agard, 1968] consists of reducing the simulation time by eliminating any trial which would end in failure, without pursuing it to its normal fixed conclusion.

—Using *antithetic variables* and *regression methods* [Hammersley and Handscomb, 1967] gives an estimation by overlapping processes, which greatly reduces the variance of the result.

—The *comparison method* leads to a simplification by partially neutralizing the effect of the random variates [Hammersley and Handscomb, 1967]. Thus, in order to compare two different policies, it is sufficient to have an individual uniform sequence for each type of random variable. The two policies are then simulated using the same pseudo-random numbers. Consequently, the randomness encountered will be the same in each policy. The differences found will therefore be much less attributable to the variance of the sampling runs.

4.3. Random number generation

Random sampling of operating or repair times is necessary for Monte Carlo simulation. Random number generation is based on generating uniformly distributed random numbers.

At the present time, the most frequently used method is the *congruential method* which can be summarized by the formula:

$$x_{i+1} = Ax_i + M(\text{modulo } B)$$

where x_i and x_{i+1} are two successive realizations, and A, B and M are constants. Constant B is the largest representable integer number. Hamming has shown that the theoretical period of a generator of this kind is equal to $B/4$ if $A = (8k \pm 3)^l$ and equal to $B/6$ if $A = (8k \pm 1)^l$, exponent l being odd [Hamming, 1962]. Deltour showed that the numbers generated were indeed random and uniformly distributed for a value of A equal to the largest representable odd-numbered power of 5 [Deltour, 1967].

There are a large number of tests for ensuring the quality of the numbers generated, known as *pseudo-random numbers* [Knuth, 1969]. Non-uniformly distributed random numbers can then be generated using uniformly distributed pseudo-random numbers by inverting the distribution function or by a rejection method (cf. Exercise 3).

5. SUMMARY

System modelling by means of stochastic processes is particularly useful if the system is Markovian. For small systems, explicit solution is possible using the Laplace transform. The following formulae are used:

$\mathrm{d}P/\mathrm{d}t = PA$, $P = [P_1, P_2, \ldots, P_n]$ being the line vector of the system state probabilities and A the transition rate matrix:

$$a_{ij} = \text{transition rate from state } i \text{ to state } j$$

$$a_{ii} = -\sum_{j \neq i} a_{ij}$$

$$A(\infty) = \frac{\begin{vmatrix} a_{11} & \cdots & a_{1,n-1} & 1 \\ & & & 1 \\ & & & 1 \\ a_{l1} & \cdots & a_{l,n-1} & 1 \\ & & & 0 \\ & & & 0 \\ a_{n1} & \cdots & a_{n,n-1} & 0 \end{vmatrix}}{\begin{vmatrix} a_{11} & \cdots & a_{1,n-1} & 1 \\ & & & \vdots \\ a_{n1} & \cdots & a_{n,n-1} & 1 \end{vmatrix}} \quad \left.\right\} \text{operating states}$$

$A(\infty)$ does not depend on the initial state P_0

$$\text{MTTF} = \frac{\begin{vmatrix} 0 & P_1(0) & \ldots & P_l(0) \\ 1 & a_{11} & \ldots & a_{1l} \\ \vdots & \vdots & & \vdots \\ 1 & a_{l1} & \ldots & a_{ll} \end{vmatrix}}{\begin{vmatrix} a_{11} & a_{1l} \\ \vdots & \vdots \\ a_{l1} & a_{ll} \end{vmatrix}}$$

For average-sized systems, it is necessary to consider a numerical treatment, discretization methods being preferable to calculating eigenvalues. However, in this case we no longer have formulae for evaluating the effects of the various parameters.

Direct use of stochastic processes does not enable us to deal with large systems, as the size of the matrix A increases exponentially with the number of system components.

As far as Monte Carlo simulation is concerned, we find that this method enables us to take account of a large number of hypotheses, at the cost of mediocre accuracy and very considerable computing time.

EXERCISES

Exercise 1 — Demonstration of Eqn (21)

(a) Make state i absorbing and calculate the mean absorption time, taking as the initial distribution:

$$P_i(0) = 0$$

$$P_j(0) = \frac{\rho_{ij}}{\sum_{j \neq i} \rho_{ij}} \quad j \neq i$$

deduce:

$$\frac{d_i'}{d_i}.$$

(b) Calculate $\dfrac{1 - \rho_{i\infty}}{\rho_{i\infty}}$.

Deduce Eqn (21).

Exercise 2—Calculate the reliability and availability of the following semi-Markovian system:

Two identical components in parallel: one is in service, the other on standby in an operable condition. The operating failure rate λ_1 and the standby failure rate λ_2 are constant. Failures while on standby are detected immediately.

The distribution of the repair time after failure during operation has a density $q_1(t)$ and that of the repair time following standby failure has a density $g_2(t)$. There is a single repair facility, the first failed component being repaired on a priority basis.

The system has five states:

—State 1: two components down after having operated.

—State 2: two components down, one after having operated, the other having failed while on standby.

—State 3: one component operating, the other having failed while on standby.

—State 4: one component operating, the other having failed while operating.

—State 5: one component operating, the other on standby.

The interstate transitions are as follows:

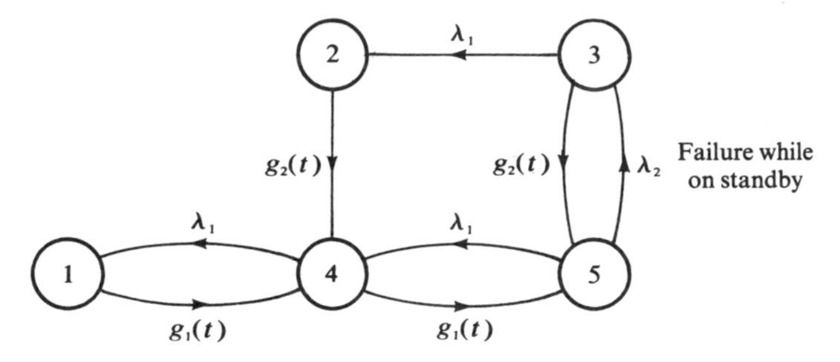

Calculate the MTTF and the MTTR.

Select exponential distributions for g and examine the effect of the distribution on the result for a fixed MTTF:MTTR ratio.

Exercise 3— Random number generation

We have seen in section 4.3 how to generate uniformly distributed pseudo-random numbers. The aim of this exercise is to show how to use these pseudo-random numbers to generate random numbers distributed according to normal, lognormal, exponential, Weibull, Erlangian and gamma laws.

(1) *Generation of normally or lognormally distributed random numbers* [Box and Muller, 1958].

Let X_1 and X_2 be two independent random variables with a truncated normal distribution $N(0, 1)$. Show that the random variables Γ and θ defined by:

$$X_1 = \Gamma \cos \theta$$

$$X_2 = \Gamma \sin \theta$$

$$\Gamma \geqslant 0$$

$$0 \leqslant \theta \leqslant 2\pi$$

are independent. Show that θ is uniformly distributed over $[0, 2\pi]$ and that Γ^2 is exponentially distributed.

Deduce that if U_1 and U_2 are two random variables uniformly distributed over $[0, 1]$, then

$$Y_1 = \sqrt{-2 \log U_1}\, \mathrm{Cos}\,(2\pi U_2)$$

$$Y_2 = \sqrt{-2 \log U_1}\, \mathrm{Sin}\,(2\pi U_2)$$

are two independent random variables distributed according to $N(0, 1)$.

Application: give a method of generating random variables having a normal distribution $N(m, \sigma)$ and a lognormal distribution (a, b).

(2) *Generation of random numbers with an exponential or Weibull distribution.*

Consider a random variable X with a continuous distribution function $F(X)$.
Show that $Y = F(X)$ if uniformly distributed over $[0, 1]$.
Deduce a method of generating random variables with an exponential, or more generally, a Weibull distribution.

(3) *Generation of random numbers having an Erlangian distribution with parameters* (λ, k).

Show that it is possible to generate a random variable of this type using k random variables uniformly distributed over $[0, 1]$.

(4) *Generation of gamma-distributed random numbers using a rejection method.*

As the inversion of the distribution function of the gamma distribution is very difficult to obtain, we will examine another method known as the rejection method.
Let $f(x)$ be the probability density of a random variable for which we want to generate realizations $(x \in \Omega)$.
Let $h(x, \alpha)$ be another density such that:

$$\exists K \geq 1 : \begin{cases} f(x) \leq Kh(x, \beta), & \in \Omega \\ \beta \text{ is a parameter which can be selected.} \end{cases}$$

Show, using Baye's theorem, that we can generate random numbers having f as probability density using the following method:

(a) Generate a random number with probability density h.
(b) Generate a number Γ uniformly distributed over $[0, 1]$.
(c) If $\Gamma \leq \dfrac{f(x)}{Kh(x, \beta)}$, accept the number drawn in (a).

If not, return to (a).
For this method to provide good results, it must be easy to generate random numbers having h as probability density and coefficient K must be as small as possible.
Application: Show that we can use this gamma distribution technique by taking an exponential or Erlangian distribution for h and that an optimal parameter β therefore exists [Tadikamalla, 1978]

Exercise 4—Generalized Erlangian distributions and the dummy states method

(1) *Dummy states 'in series'.*

The transition $1 \to 2$ is, in reality, a sequence of transitions $1 \to 1_1 \to 1_2 \ldots \to 1_{k-1} \to 2$ with respective constant transition rates λ_i.
Let X be the random variable representing the residence time in states $1, 1_1, 1_2, \ldots 1_{k-1}$.

Calculate the probability density of X, then $m = E[X]$ and $\sigma^2 = E[X - m]^2$. Show that $\dfrac{1}{\sqrt{k}} \leqslant \dfrac{\sigma}{m} \leqslant 1$.

Examine the transition rate $\Lambda(t)$.

(2) *Dummy states 'in parallel'.*

The random variable X representing the residence time in state i is X_i with probability α_i.

$$X = \sum_{i=1}^{k} \alpha_i X_i \quad \sum_{i=1}^{k} \alpha_i = 1, \alpha_i > 0 .$$

Calculate the probability density of X, then m and σ.
Show that $\sigma/m \geqslant 1$.
Examine the transition rate $\Lambda(t)$.

SOLUTIONS TO EXERCISES

Exercise 1

d'_i may be regarded as the mean residence time in states other than i with an initial state proportional to the a_{ij}, hence

$$P_j(0) = \frac{a_{ij}}{\sum_{j \neq i} a_{ij}} = -\frac{a_{ij}}{a_{ii}}.$$

We call A^i the matrix resulting from the transition rate matrix by deleting row and column i. Hence, by applying Eqn (18):

$$d'_i = \frac{\begin{vmatrix} 0 & P_j(0) \\ 1 & \\ 1 & A^i \\ 1 & \end{vmatrix}}{|A^i|} = -\frac{1}{a_{ii}} \frac{\begin{vmatrix} 0 & a_{i1} \ \dots \ a_{in} \\ 1 & \\ 1 & A^i \\ 1 & \end{vmatrix}}{|A^i|}$$

$$d'_i = -\frac{1}{a_{ii}} \frac{\begin{vmatrix} a_{11} & \dots & a_{1,i-1} & 1 & a_{1,i+1} & \dots & a_{1,n} \\ \cdot & & & 1 & & & \\ \cdot & & & 1 & & & \\ a_{i1} & \dots & a_{i,i-1} & 0 & a_{i,i+1} & \dots & a_{i,n} \\ \cdot & & & 1 & & & \\ \cdot & & & & & & \\ a_{n1} & \dots & a_{n,i-1} & 1 & a_{n,i+1} & \dots & a_{nn} \end{vmatrix}}{|A^i|}$$

On the other hand, from Eqn (12), $P_i(\infty) = |A^i|$. Using Eqn (14) we can therefore write:

$$\frac{1 - P_i(\infty)}{P_i(\infty)} = \frac{\begin{vmatrix} a_{11} & \dots & a_{1,i-1} & 1 & a_{1,i+1} & \dots & a_{1n} \\ \cdot & & & 1 & & & \\ \cdot & & & \cdot & & & \\ \cdot & & & 1 & & & \\ a_{n1} & \dots & a_{n,i-1} & 1 & a_{n,i+1} & \dots & a_{nn} \end{vmatrix} - |A^i|}{|A^i|}$$

hence the desired result, using Eqn (20).

Exercise 2 — see [Bharat Sham and Branson, 1971]

Exercise 3

(1) From Eqn (22), Ch. 1, the probability density of the pair X_1, X_2 is written as

$$f(X_1, X_2) = \frac{1}{2\pi} \exp\left(-\frac{X_1^2 + X_2^2}{2}\right)$$

hence the density of the pair (Γ, θ) is:

$$g(\Gamma, \theta) = \frac{1}{2\pi} \exp\left(-\frac{\Gamma^2}{2}\right) \Gamma$$

i.e.:

$$g(\Gamma, \theta)\, \mathrm{d}\Gamma\, \mathrm{d}\theta = \exp\left(-\frac{\Gamma^2}{2}\right) d\left(\frac{\Gamma^2}{2}\right) \frac{\mathrm{d}\theta}{2\pi}.$$

Therefore, θ is uniformly distributed over $[0, 2\pi]$ and Γ^2 is exponentially distributed with parameter 1/2. In addition, they are independent. Consequently, if U_1 and U_2 are uniformly distributed over $[0, 1]$ and are independent, $(-2 \log U_1)^{1/2}$ and $2\pi U_2$ are distributed as Γ and θ respectively. Hence the result, since $X_1 = \Gamma \cos\theta$ and $X_2 = \Gamma \sin\theta$.

Application: After generating X_1 and X_2, it follows that:

$$X_1' = \sigma X_1 + m$$
$$X_2' = \sigma X_2 + m$$

are independent and distributed according to a law $N(m, \sigma)$.

And, from Table 1, Ch. 1:

$$X_1'' = e^{bX_1 + a}$$
$$X_2'' = e^{bX_2 + a}.$$

are independent and lognormally distributed (a, b).

(2) $y = F(x) = \mathscr{P}(X \leqslant x) = \mathscr{P}(F(X) \leqslant F(x))$ since F is monotonic. Hence

$$y = \mathscr{P}(Y \leq y) ;$$

therefore Y is uniformly distributed over $[0, 1]$.

Application: $F(x) = 1 - e^{-\lambda x}$

$$x = -\frac{1}{\lambda}\log(1 - y)$$

$$F(x) = 1 - \exp\left(-\left(\frac{x - \gamma}{\eta}\right)^{\beta}\right)$$

$$x = \gamma + \eta - [\log(1 - y)]^{1/\beta}.$$

(3) An Erlangian distribution may be regarded as the convolution of k exponential distributions with parameter λ. Consequently, if the x_i ($i = 1$ to k) are distributed exponentially with parameter λ, $\sum_{i=1}^{k} x_i$ has an Erlangian distribution with parameters (λ, k).

(4) Put $g(x) = \dfrac{f(x)}{Kh(x, \beta)}$.

From Bayes' theorem:

$$\mathscr{P}(x \leq X < x + dx \mid \Gamma \leq g(x)) = \frac{\mathscr{P}(\Gamma \leq g(X) \mid x \leq X \leq x + dx)\, h(x, \beta)\, dx}{\mathscr{P}(\Gamma \leq g(x))}.$$

On the other hand:

$$\Gamma \text{ being uniform over } [0, 1], \mathscr{P}(\Gamma \leq g(x) \mid x \leq X < x + dx) = g(x)$$

and

$$\mathscr{P}(\Gamma \leq g(x)) = \int_{\Omega} \mathscr{P}(\Gamma \leq g(X))\, h(x, \beta)\, dx = \frac{1}{K}\int_{\Omega} f(x)\, dx = \frac{1}{K}.$$

Therefore:

$$\mathscr{P}(x \leq X < x + dx \mid \Gamma \leq g(x)) = \frac{g(x)\, h(x, \beta)\, dx}{\frac{1}{K}} = f(x)\, dx.$$

On the other hand, the acceptance rate of x will be equal to $1/K \leqslant 1$. It is therefore advisable that K should be as small as possible.

Application:

$$f'(t) = \frac{\lambda^{\alpha} t^{\alpha - 1} e^{-\lambda t}}{\Gamma(\alpha)}, \alpha > 1.$$

putting $x = \lambda t$:

$$f(x) = \frac{x^{\alpha-1} e^{-x}}{\Gamma(\alpha)}$$

we put $h_1(x, \beta) = \frac{1}{\beta} \exp\left(-\frac{x}{\beta}\right)$ (exponential distribution).

We can then determine β^* as follows:

Let $y(x, \beta) = \frac{f(x)}{h_1(x, \beta)}$

determine x_0 such that $y(x, \beta) \leq y(x_0, \beta)$
determine β^* such that $y(x_0, \beta) \geq y(x_0, \beta^*)$
hence $\beta^* = \alpha$.

Put $h_2(x, \beta) = \frac{x^{m-1} \exp\left(-\frac{x}{\beta}\right)}{\beta^m (m-1)!}$ (Erlangian distribution).

Using the same method we obtain:

$$\beta^* = \frac{\alpha}{m^*} \quad m^* = E(\alpha).$$

A discussion on the effectiveness of these two functions h is given in [Tadikamalla, 1978]. A general method valid for $\alpha > 0$ is given in [Wallace, 1974].

Exercise 4

(1) $X = \sum_{i=1}^{k} X_i$, hence $f(x) = f_1 * f_2 \dots * f_k$

and

$$\bar{f}(s) = \prod_{i=1}^{k} \bar{f}_i(s) = \prod_{i=1}^{k} \frac{\lambda_i}{\lambda_i + s} = \prod_{i=1}^{k} \frac{1}{1 + \frac{s}{\lambda_i}}.$$

Denote by $\lambda_1, \lambda_2, \dots, \lambda_h$ the h distinct values of the λ_i ($h \leqslant k$).
Hence

$$\bar{f}(s) = \prod_{i=1}^{h} \left(\frac{\lambda_i}{\lambda_i + s}\right)^{\alpha_i} \quad \text{with} \quad \sum_{i=1}^{h} \alpha_i = k \quad \alpha_i \in N^+$$

α_i being the multiplicity of λ_i.
Formula (54) of Ch. 1 enables inversion to be carried out.
In the case where all the λ_i are different, we obtain:

$$f(t) = \sum_{i=1}^{k} \lambda_i \rho_i e^{-\lambda_i t} \quad \text{with} \quad \rho_i = \prod_{\substack{j=1 \\ j \neq i}}^{k} \frac{\lambda_j}{\lambda_j - \lambda_i}.$$

In the case where all the λ_i are equal, we obtain:

$$f(t) = \frac{\lambda^k t^{k-1} e^{-\lambda t}}{(k-1)!} \quad \text{(Erlangian distribution)}$$

On the other hand, expanding $\bar{f}(s)$ in series, we obtain:

$$\bar{f}(s) = \prod_{i=1}^{k} \left(1 - \frac{s}{\lambda_i} + \frac{s^2}{\lambda_i^2} + \cdots\right) = 1 - s\left(\sum_{i=1}^{k} \frac{1}{\lambda_i}\right) + s^2\left(\sum_{i=1}^{k} \frac{1}{\lambda_i^2} + \sum_{i=1}^{k} \sum_{j>i} \frac{1}{\lambda_i \lambda_j}\right) + \cdots$$

hence $E[X] = \sum_{i=1}^{k} \frac{1}{\lambda_i}$

$$E[X^2] = 2\left(\sum_{i=1}^{k} \frac{1}{\lambda_i^2} + \sum_{i=1}^{k} \sum_{j>i} \frac{1}{\lambda_i \lambda_j}\right)$$

and

$$\sigma^2 = \sum_{i=1}^{k} \frac{1}{\lambda_i^2}, \quad \frac{\sigma}{m} = \frac{\sqrt{\sum_{i=1}^{k} \frac{1}{\lambda_i^2}}}{\sum_{i=1}^{k} \frac{1}{\lambda_i}} \leq 1 .$$

On the other hand, it is easy to see that

$$y = k\alpha^2 - 2\alpha \sum_{i=1}^{k} \frac{1}{\lambda_i} + \sum_{i=1}^{k} \frac{1}{\lambda_i^2} = \sum_{i=1}^{k} \left(\alpha - \frac{1}{\lambda_i}\right)^2 \text{ is positive or zero.}$$

Consequently, the discriminant of the second degree expression in α is negative or zero, hence $\frac{\sigma}{m} \geq \frac{1}{\sqrt{k}}$.

The expression for $f(t)$ shows that when t tends to infinity, $\Lambda(t)$ tends to $\min_i (\lambda_i)$.

When t tends to zero, we obtain:

$$\bar{f}(s) \underset{s \to \infty}{\simeq} \frac{\prod_{i=1}^{k} \lambda_i}{s^k}$$

hence, since $1 - F(t) \to 1 : \Lambda(t) \sim f(t) \sim \prod_{i=1}^{k} \lambda_i \frac{t^{k-1}}{(k-1)!}$.

(2)

$$f(t)\,dt = \mathcal{P}(t \leq X < t + dt) = \sum_{i=1}^{k} \alpha_i \, \mathcal{P}(t \leq X_i < t + dt) = \sum_i f_i(t)\,dt$$

hence

$$f(t) = \sum_{i=1}^{k} \alpha_i \lambda_i \, e^{-\lambda_i t}$$

and

$$F(t) = 1 - \sum_{i=1}^{k} \alpha_i \, e^{-\lambda_i t}$$

hence $m = \sum_{i=1}^{k} \frac{\alpha_i}{\lambda_i}, \; \sigma^2 = 2 \sum_{i=1}^{k} \frac{\alpha_i}{\lambda_i^2} - \left(\sum_{i=1}^{k} \frac{\alpha_i}{\lambda_i}\right)^2$

$$\frac{\sigma}{m} = \sqrt{\frac{2 \sum_{i=1}^{k} \frac{\alpha_i}{\lambda_i^2}}{\left(\sum_{i=1}^{k} \frac{\alpha_i}{\lambda_i}\right)^2} - 1} \geq 1 \quad \text{as} \quad \left(\sum_{i=1}^{k} \alpha_i\right)\left(\sum_{i=1}^{k} \frac{\alpha_i}{\lambda_i^2}\right) \geq \left(\sum_{i=1}^{k} \frac{\alpha_i}{\lambda_i}\right)^2 .$$

$\Lambda(0) = \sum_{i=1}^{k} \alpha_i \lambda_i$.

When $t \to \infty$, $\Lambda(t) \to \min_i (\lambda_i)$.

$\Lambda(t)$ is a decreasing function, since $\Lambda'(t) \leqslant 0$.

REFERENCES

ABATE J. and DUBNER H. (1968): Numerical inversion of Laplace transforms by relating them to the finite Fourier cosine transform; *ACM Journal*, **15**, no. 1, pp. 115–123.

AGARD J. (1968): *Les méthodes de simulation*. AFCET Monograph 7, Dunod, Paris.

BHARAT SHAM and BRANSON M. H. (1971): Reliability analysis of systems composed of units with arbitrary repair-time distributions; *IEEE Transactions on Reliability*, **R20**, no. 4.

BILLINGTON R. and SINGH C. (1972): Reliability in systems with non exponential down times; NATO Conference on *Reliability Testing and Reliability Evaluation*.

BOX G. E. P. and MULLER M. E. (1958): A note on the generation of random normal deviates; *Ann. Math. Stat.*, **28**, pp. 610–611.

CORRAZA M. (1975): *Techniques mathématiques de la fiabilité prévisionnelle*; Cepadues Editions, France.

COSTES A., LANDRAULT C. and LAPRIE J. C. (1976): *The Use of Stochastic Processes for Reliability Prediction of Maintainable Structures*; LAAS.

COX D. R. (1955): The analysis of non-markovian stochastic processes by the inclusion of supplementary variables; *Proc. Comb. Publ. Soc.*, **51**, pp. 433–441.

DELTOUR J. (1967): Etude d'une distribution de nombres pseudo-aléatoires; *Bulletin de Recherche Agronomique*, Gembloux 3, pp. 450–460.

HAMMERSLEY J. M. and HANDSCOMB D. C. (1967): *Les méthodes de Monte Carlo*; Dunod, Paris.

HAMMING R. W. (1962) : *Numerical Methods for Scientists and Engineers*, McGraw-Hill, New York.

KNUTH D. E. (1969): *The Art of Computer Programming*, Vol. 2; Addison-Wesley, London.

KOZLOV B. A. (1966): Determination of reliability indices of systems with repair; *Engineering Cybernetics*.

LEVY P. (1954): Processus semi-markoviens; *Proc. Int. Cong. Math. (Amsterdam)*, pp. 416–426.

LIEBOWITZ B. H. (1966): Reliability considerations for a two elements redundant system with generalized repair times; *Operational Research*, **14**, pp. 233–241.

LIONS J. L. (1973): *Cours d'analyse numérique de l'Ecole Polytechnique*; Hermann, Paris.

MARGUIN J. (1978): Intérêt de la classe des lois d'Erlang en fiabilité; *Colloque International sur la Fiabilité et la Maintenabilité*, Paris.

PYKE R. (1961): Markov renewal processes with finitely many states; *Ann. Math. Stat.*, **32**, no. 4, pp, 1243–1259.

RICHTMYER R. D. (1967): *Difference Methods for Initial Value Problem*; Wiley, Chichester.

SHOOMAN M. L. (1968): *Probabilistic Reliability: an Engineering Approach*; McGraw-Hill, New York.

SMITH W. L. (1955): Regenerative stochastic process. *Proc. Roy Soc. Series A*, **232**, pp. 6–31.

TADIKAMALLA P. R. (1978): Computer generation of gamma random variables; *Communications of the ACM*, **21**, no. 5.

WALLACE N. D. (1974): Computer generation of gamma random variates with non integral shape parameters; *Communications of the ACM*, **17**, no. 12.

CHAPTER 6

RELIABILITY EVALUATION USING THE CRITICAL OPERATING STATES METHOD

For analysing large systems, the methods discussed in the previous chapter are found wanting. In most cases, there is no alternative. However, the systems analysed are very often of 'good quality', i.e. they consist of components whose mean time to repair is small compared with the mean uptime. Of course, if a system consists of n components in series with a stationary availability of 10^{-2}, this system will not be of good quality if n runs into tens, even though each component is of good quality. However, we intend to show that if certain relationships between the various failure and repair rate values are satisfied (as is very often the case), it is possible to obtain a good reliability approximation for large systems.

1. PRINCIPLE OF CRITICAL OPERATING STATES METHOD FOR COHERENT, HOMOGENEOUS MARKOVIAN SYSTEMS WITH DISCRETE STATE SPACE

We intend to compute an approximation for the reliability $R(t)$. To do this, we shall attempt to find an approximation for the system failure rate $\Lambda(t)$.

1.1. Nature of chosen approximation

In order to determine the system failure rate $\Lambda(t)$ defined by

$$\Lambda(t) = \lim_{\Delta t \to 0} \frac{1}{\Delta t} \mathscr{P}\begin{pmatrix} \text{the system fails between } t \text{ and } t + \Delta t \\ \text{given that it has not failed during } [0, t] \end{pmatrix}$$

we have to compute the probability of system failure during the interval $[t, t + \Delta t]$ (given that there has been no failure during $[0, t]$).

For the system to fail between t and $t + \Delta t$, it is necessary that, at time t, it should occupy a state possessing at least one transition to the failed states.

We therefore have to consider two types of operating states:

—*critical operating states* having at least one transition to the failed states,
—*non-critical operating states* having no transition to the failed states.

For a given critical operating state, we shall denote a component whose failure causes system failure by the term *critical component*.

For simplicity, we shall adopt the following notations:

M: set of operating states (critical or non-critical),
M_c: set of critical operating states,
P: set of failed states,
$\mathscr{C}_i$: set of the critical components of critical operating state i (this set is empty for the non-critical operating states),
$\Lambda_i (i \in M)$: sum of the transition rates from a state to the failed states. For a non-critical operating state, Λ_i is zero. For a critical operating state Λ_i represents the sum of the failure rates of the critical components of this state: $\Lambda_i = \sum_{j \in \mathscr{C}_i} \lambda_j$,
MTTF_i: system MTTF when the initial state is state i,
State 1: system state for which all the components are available.

We can now begin to calculate the system failure rate.

Given that the system is in state i at time t without ever having failed, the probability of it failing between t and $t + \Delta t$ is equal to $\Lambda_i \Delta t$.

Applying the conditional probabilities definition (Eqn (6), Ch. 1), we therefore obtain:

$$\Lambda(t) = \frac{\sum_{i \in M_c} \Lambda_i P'_i(t)}{\sum_{i \in M} P'_i(t)} \tag{1}$$

$P'_i(t)$ representing the probability of being in state i when the failed states are absorbing. (We shall adopt the same convention as in the previous chapter and consequently $P_i(t)$ will represent the probability of being in state i when there is no absorbing state, i.e. when computing the availability.)

Now assume that this system is repaired following a failure; the subsequent evolution of the system can then be modelled by an alternating *renewal process*. The *failure density* $v(t)$ in this renewal process is defined by: (cf. Chapter 1, Section 6.3)

$$v(t) = \lim_{\Delta t \to 0} \frac{1}{\Delta t} \mathscr{P}(\text{the system fails during } [t, t + \Delta t]).$$

Using similar reasoning to that behind Eqn (1) we obtain:

$$v(t) = \sum_{i \in M_c} \Lambda_i P_i(t) \tag{2}$$

By analogy with Eqn (1), consider the *pseudo-rate of failure* $\tilde{\Lambda}(t)$ defined by:

$$\tilde{\Lambda}(t) = \frac{\sum_{i \in M_c} \Lambda_i P_i(t)}{\sum_{i \in M} P_i(t)} \tag{3}$$

Using the conditional probabilities theorem, we can write:

$$\tilde{\Lambda}(t) = \lim_{\Delta t \to 0} \frac{1}{\Delta t} \mathscr{P}\begin{pmatrix} S \text{ fails between } t \text{ and } t + \Delta t \text{ given that} \\ \text{it is operating at time } t \end{pmatrix}.$$

We can see therefore that the difference between this and the failure rate definition (Ch. 1, Eqn (3)) is finely drawn. Vesely was the first to propose computing $\tilde{\Lambda}(t)$ [Vesely, 1970], but without demonstrating the relationships between $\Lambda(t)$ and $\tilde{\Lambda}(t)$. It was Gondran [Gondran, 1975] who seized upon this difference and demonstrated the first equivalences.

In this chapter, we intend to show that under unrestrictive hypotheses, $\tilde{\Lambda}(t)$ is a good approximation for $\Lambda(t)$. Note that this property is not *a priori* obvious, since, as we saw in the previous chapter:

$$\lim_{t \to \infty} P_i'(t) = 0$$

$$\lim_{t \to \infty} P_i(t) = P_i(\infty) > 0.$$

1.2. Example

In order to understand the difference between Eqns (1) and (3), we shall examine the simple example of a system comprising two identical components in parallel (active redundancy).

In Ch. 3 we computed the reliability of a system of this type using the following Markov diagram:

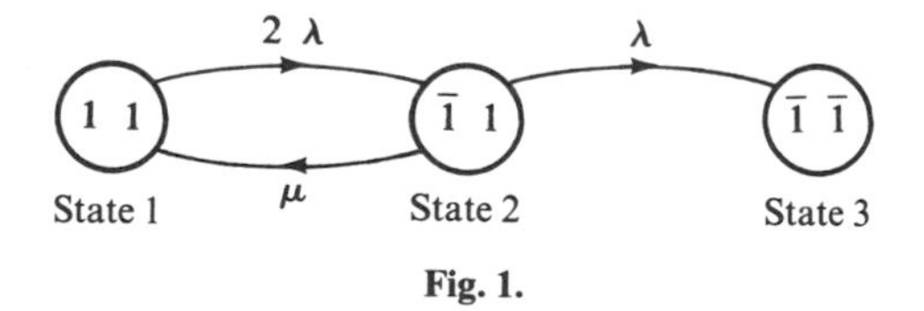

Fig. 1.

We assume that the system is in state 1 at time $t = 0$. It had been shown that:

$$P_1'(t) = \frac{s_1 + \lambda + \mu}{s_1 - s_2} e^{s_1 t} + \frac{s_2 + \lambda + \mu}{s_2 - s_1} e^{s_2 t}$$

$$P_2'(t) = \frac{2\lambda}{s_1 - s_2} e^{s_1 t} + \frac{2\lambda}{s_2 - s_1} e^{s_2 t}$$

with s_1 and s_2 solutions of:

$$S^2 + (3\lambda + \mu) S + 2\lambda^2 = 0 \tag{4}$$

$$s_2 < s_1 < 0.$$

The system has a single critical operating state, state 2 and

$$\Lambda_1 = \lambda.$$

From this, we deduce the system failure rate by applying Eqn (1):

$$\Lambda(t) = \frac{2\lambda^2(e^{s_1 t} - e^{s_2 t})}{-s_2 e^{s_1 t} + s_1 e^{s_2 t}} \tag{5}$$

We naturally obtain the same value as in Ch. 3.

Let us now evaluate the pseudo-rate of failure $\tilde{\Lambda}(t)$ of the system. The Markov diagram for computing the availability, assuming that sufficient repair facilities are available, looks like this:

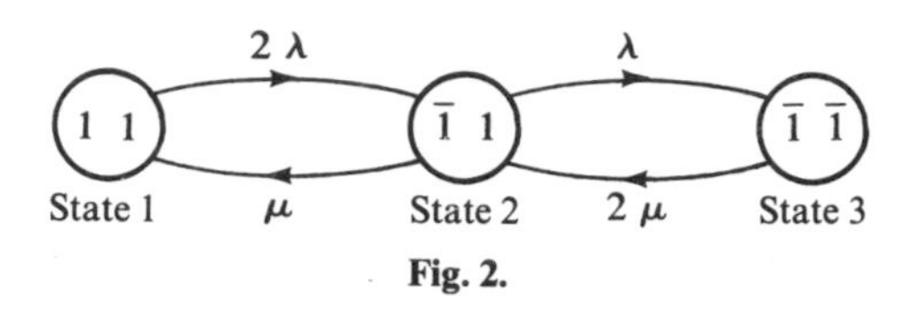

Fig. 2.

As the components are independent, if we call $q(t)$ the availability of the component, it follows that:

$$P_1(t) = q^2(t)$$
$$P_2(t) = 2\,q(t)\quad(1 - q(t)).$$

Hence, by applying Eqn (3):

$$\tilde{\Lambda}(t) = \frac{2\lambda q(t)(1 - q(t))}{q(t)[2 - q(t)]} = \frac{2\lambda(1 - q(t))}{2 - q(t)}.$$

Or, in view of Eqn (11), Ch. 3

$$\tilde{\Lambda}(t) = \frac{2\lambda^2(1 - e^{-(\lambda+\mu)t})}{\mu + 2\lambda - \lambda e^{-(\lambda+\mu)t}}. \tag{6}$$

Let us now assume that the mean uptime of the components is large compared with their mean time to repair; i.e.:

$$\lambda \ll \mu.$$

Therefore, by expanding with respect to λ/μ, the roots of Eqn (4) can be written:

$$s_2 = \frac{-(3\lambda + \mu) - \sqrt{\mu^2 + 6\lambda\mu + \lambda^2}}{2} = -(\mu + 3\lambda)\left(1 - 2\left(\frac{\lambda}{\mu + 3\lambda}\right)^2 + \ldots\right)$$

$$s_1 = \frac{-(3\lambda + \mu) + \sqrt{\mu^2 + 6\lambda\mu + \lambda^2}}{2} = -\frac{2\lambda^2}{\mu + 3\lambda}\left(1 + 2\left(\frac{\lambda}{\mu + 3\lambda}\right)^2 + \cdots\right)$$

i.e.

$$s_2 \simeq -(\mu + 3\lambda)$$

$$s_1 \simeq -\frac{2\lambda^2}{\mu + 3\lambda}.$$

Examine the behaviour of functions $\Lambda(t)$ and $\tilde{\Lambda}(t)$ when t tends to infinity.

From Eqns (5) and (6) we obtain respectively:

$$\Lambda_\infty = \lim_{t\to\infty} \Lambda(t) = \frac{2\lambda^2}{-s_2} = -s_1 \simeq \frac{2\lambda^2}{\mu + 3\lambda}$$

$$\tilde{\Lambda}_\infty = \lim_{t\to\infty} \tilde{\Lambda}(t) = \frac{2\lambda^2}{\mu + 2\lambda}$$

hence

$$\Lambda_\infty = \tilde{\Lambda}_\infty\left(1 - \frac{\lambda}{\mu + 3\lambda}\right).$$

Now examine the behaviour of functions $\Lambda(t)$ and $\tilde{\Lambda}(t)$ when t tends to zero

$$\Lambda(t) \underset{t\to 0}{\simeq} 2\lambda^2 t$$

$$\tilde{\Lambda}(t) \underset{t\to 0}{\simeq} 2\lambda^2 t .$$

The limit Λ_∞ is in fact very rapidly attained. In fact $s_2 \simeq -(\mu + 3\lambda)$. Therefore, when $(\mu + 3\lambda)t$ is greater than a few units, the term $\exp(s_2 t)$ becomes negligible and $\Lambda(t) \sim \Lambda_\infty$. From Eqns (5) and (6) we can verify that $\tilde{\Lambda}(t)$ is a good approximation for $\Lambda(t)$ for all values of t.

1.3. Properties of the transition rate matrix

Let us look at the reliability calculation; there are l operating states and the failed states are combined to form a single absorbing state denoted by $l + 1$.

The transition matrix can then be written as:

$$A' = \begin{bmatrix} a_{11} & a_{12} & \dots & a_{1l} & \Lambda_1 \\ a_{21} & a_{22} & & a_{2l} & \Lambda_2 \\ \cdot & & & & \\ \cdot & & & & \\ \cdot & & & & \\ a_{l1} & & & a_{ll} & \Lambda_l \\ 0 & & & 0 & 0 \end{bmatrix}$$

1.3.1. *Properties of eigenvalues*

We shall call A_l the matrix obtained from A' by eliminating the last column and row. This matrix has very useful properties.

The a_{ij} satisfy Eqn (7), Ch. 5, which can be rewritten as:

$$\left.\begin{aligned} a_{ij} &\geqslant 0 \quad \text{if} \quad i \neq j \\ a_{ii} &= -\sum_{\substack{j=1\\ j\neq i}}^{l} a_{ij} - \Lambda_i < 0 \,\forall i = 1, 2, \dots, l \end{aligned}\right\} \tag{7}$$

Property 1 – *Matrix A_l is non-singular and irreducible*

A matrix is *irreducible* if its associated Markov diagram is strongly connected, i.e. if there exists, for any ordered pair of states (i, j), a path connecting i to j. As all the components in our system are repairable, matrix A_l is therefore irreducible. Note that the same does not apply if there is a non-repairable component (cf. section 3.2).

On the other hand, matrix A_l has a *dominant diagonal* since

$$|a_{ii}| - \sum_{j \neq i} |a_{ij}| = \Lambda_i \geqslant 0 . \tag{8}$$

In addition, there is at least one transition to the failed state. There is therefore an index i for which Λ_i is strictly positive. Consequently (cf. [Varga, 1962, theorem 1.8]), matrix A_l is non-singular and $|A_l| \neq 0$.

Property 2 – *Matrix A has a negative real eigenvalue s_1 such that the real parts of the other eigenvalues are less than this value. When t tends to infinity, system reliability is equivalent to $\alpha_1 \exp(s_1 t)$*

Demonstration: Let a_0 be a positive number such that $a_0 + a_{ii}$ is positive for all $i = 1, \ldots, l$.

The matrix $a_0 I + A_l$ has positive terms and is irreducible. From Frobenius' theorem [Varga, 1962] it can be stated that the matrix $a_0 I + A_l$ possesses a simple positive eigenvalue ρ. The moduli of all the eigenvalues are less than or equal to ρ; if there are $(k - 1)$ other eigenvalues with moduli ρ, they are roots of the equation $x^k = \rho^k$.

Consequently $\mathcal{R}(x) < \rho \; \forall x \neq \rho$.

It follows from this that matrix A_l has a simple real eigenvalue $s_1 = \rho - a_0$ such that

$$s_1 > \mathcal{R}(s_i) \; \forall \; s_i \text{ eigenvalue of } A \text{ distinct from } s_1. \tag{9}$$

The system reliability can therefore be expressed in the form:

$$R(t) = \alpha_1 \, e^{s_1 t} + \sum_{i=2}^{h} f_i(t) \, e^{[\mathcal{R}(s_i) + i \mathcal{I}(s_i)]t} \tag{10}$$

$s_1, s_2, \ldots, s_h$ being the eigenvalues distinct from $A_l . f_i(t)$ is a polynomial of degree lower than or equal to $n_i - 1$, n_i being the order of multiplicity of the eigenvalue s_i. $\left(\sum_{i=1}^{h} n_i = l \right)$.

It therefore follows from Eqns (9) and (10) that:

$$R(t) \underset{t \to \infty}{\sim} \alpha_1 \, e^{s_1 t} . \tag{11}$$

On the other hand, Gerschgorin's theorem [Varga, 1962; Parodi, 1959] indicates that:

$$|a_{ii} - s_i| \leqslant \sum_{j \neq i} a_{ij} . \tag{12}$$

Consequently, from Eqn (7), the eigenvalues s_i are zero or have negative real parts.

Since A_l is regular according to property 1, it follows that all the eigenvalues have negative real parts.

The double inequality [Varga R. S., 1962, lemma 2.5] can also be demonstrated.

$$-\max_{i=1,l} \Lambda_i < s_1 < -\min_{i=1,l} \Lambda_i .$$

Generally speaking, $\min_i \Lambda_i$ is zero as there are non-critical operating states. However, if there is a component in series with the rest of the system, then Eqn (8) shows that the absolute value of s_1 is greater than or equal to the failure rate of this component. Experience shows that s_1 is much smaller in modulus than the other eigenvalues, which causes numerical difficulties in the computation of the eigenvalues of A_l.

1.3.2. *Application to MTTF*

Differentiating Eqn (10), we obtain:

$$\Lambda_\infty = -s_1. \tag{13}$$

The coefficient α_1 of Eqn (10) can be computed as follows:

$$\alpha_1 = \lim_{t \to \infty} R(t)\, e^{-s_1 t} .$$

In practice, we compute $\alpha_1(t_0)$ such that

$$\alpha_1(t_0) = \exp\left(-\int_0^{t_0} \Lambda(u)\, du \right) e^{-s_1 t_0} . \tag{14}$$

We then find that $\alpha_1(t_0)$ rapidly converges to α_1, the limit being reached for a value of t_0 equal to a few times the largest mean time to repair of the system components. In addition, when no component is failed in the initial state, α_1 is approximately 1.

We can proceed in the same way to compute the system MTTF. In fact:

Let t_0 be such that $\left| \dfrac{\Lambda_\infty - \Lambda(t_0)}{\Lambda_\infty} \right| \leqslant \varepsilon$. It can be shown (cf. Exercise 5) that

$$\left| \text{MTTF} - \int_0^{t_0} R(t)\, dt - \frac{R(t_0}{\Lambda_\infty} \right| \leqslant \frac{\varepsilon}{\Lambda_\infty (1 - \varepsilon)} .$$

Experimental results indicate that for reliable systems, $R(t_0)$ is virtually 1 and that t_0 is small compared with $R(t_0)/\Lambda_\infty$, which amounts to saying that the reciprocal of the stationary failure rate is a good approximation for the MTTF. In fact, the MTTF of a system depends on the initial state (cf. Ch. 5) and we find that the best approximation is obtained when, in the initial state, no component has failed (State 1).

This phenomenon may be explained as follows: the distribution of the lifetime of a very reliable system is asymptotically exponential and for an exponential distribution $\Lambda_\infty = 1/\text{MTTF}$.

Precise demonstration of this fact is the object of Exercise 1 due to P. Oger. It may be summarized as follows: Consider a system state i; each time the system leaves state i, it has a probability q of returning to it (without having encountered a failed state) and a probability p of being absorbed by the failed state.

As the system is homogeneous Markovian, this probability p is not time-dependent. If the experiment is repeated n times, the number of times when it is absorbed follows a binomial distribution $B(n, p)$.

For reliable systems, the probability p is very small and the binomial distribution is very close to a Poisson distribution (cf. Ch. 1, section 4.1.1). The smaller the parameter p, the better the approximation. For a given coherent system, the probability p will be minimal if, in the initial state, no element has failed. Consequently:

$$\text{MTTF}_1 \simeq \frac{1}{\Lambda_\infty} = \frac{-1}{s_1}. \tag{15}$$

1.4. Analysis of the approximation for large values of t

We shall calculate the pseudo-rate of failure when the system has reached its equilibrium state. By partitioning the system states between operating and failed states, using the flow conservation formula (cf. Ch. 4, section 5.1) we can write:

$$\sum_{i \in M} \sum_{j \in P} a_{ij} P_i(\infty) = \sum_{j \in P} \sum_{i \in M} a_{ji} P_j(\infty). \tag{16}$$

If we reduce the set of failed states to a single state denoted by $l + 1$, the transition rates from this state to the critical operating states i have a value:

$$\gamma_i = \sum_{j \in P} \frac{a_{ji} P_j(\infty)}{\sum_{j \in P} P_j(\infty)}. \tag{17}$$

The limit values $P_i(\infty)$ of the operating states are not affected by this transformation. We put:

$$\gamma_{l+1} = -\sum_{i=1}^{l} \gamma_i. \tag{18}$$

The system transition rate matrix is then as follows:

$$A'' = \left[\begin{array}{ccc|c} & & & \Lambda_1 \\ & & & \Lambda_2 \\ & & & \cdot \\ & A_l & & \cdot \\ & & & \cdot \\ & & & \Lambda_l \\ \hline \gamma_1 & \cdots & \gamma_i & \gamma_{l+1} \end{array}\right] \tag{19}$$

Applying Eqn (12), Ch. 5, we compute:

$$\sum_{i=1}^{l} \Lambda_i P_i(\infty) = \frac{\left|\begin{array}{ccc|c} & & & \Lambda_1 \\ & A_l & & \vdots \\ & & & \Lambda_l \\ \hline \gamma_1 & \cdots & \gamma_l & 0 \end{array}\right|}{\left|\begin{array}{ccc|c} & & & 1 \\ & & & 1 \\ & A_l & & 1 \\ & & & \vdots \\ \hline \gamma_1 & \cdots & \gamma_l & 1 \end{array}\right|}$$

and, from Eqn (14), Ch. 5:

$$\frac{\sum_{i=1}^{l} \Lambda_i P_i(\infty)}{\sum_{i=1}^{l} P_i(\infty)} = \frac{\left|\begin{array}{ccc|c} & & & \Lambda_1 \\ & A_l & & \vdots \\ & & & \Lambda_l \\ \hline \gamma_1 & \cdots & \gamma_l & 0 \end{array}\right|}{\left|\begin{array}{ccc|c} & & & 1 \\ & & & 1 \\ & A_l & & 1 \\ & & & \vdots \\ & & & 1 \\ \hline \gamma_1 & \cdots & \gamma_l & 0 \end{array}\right|}$$

Then, as the sum of the elements of row i of A_l are equal to $-\Lambda_i$:

$$\tilde{\Lambda}_\infty = \frac{-\gamma_{l+1} \, | A_l |}{\left|\begin{array}{ccc|c} & & & 1 \\ & & & 1 \\ & A_l & & 1 \\ & & & \vdots \\ \hline \gamma_1 & \cdots & \gamma_l & 0 \end{array}\right|} \tag{20}$$

By comparing Eqn (20) with Eqn (18) in Ch. 5, we deduce that $1/\tilde{\Lambda}_\infty$ is equal to the system MTTF when the initial state of the system is

$$\left(-\frac{\gamma_1}{\gamma_{l+1}}, -\frac{\gamma_2}{\gamma_{l+1}}, \ldots, -\frac{\gamma_l}{\gamma_{l+1}}\right);$$

the values $P_i'(0)$ of Eqn (18) in Ch. 5 are therefore:

$$P_i'(0) = \frac{\sum_{j \in P} a_{ji} P_j(\infty)}{\sum_{i \in M} \sum_{j \in P} a_{ji} P_j(\infty)}$$

consequently, from Eqn (19), Ch. 5:

Theorem 1

$$\tilde{\Lambda}_\infty = \frac{1}{\text{MUT}} \tag{21}$$

$$\tilde{\Lambda}_\infty = \frac{1}{\sum_{i=1}^{l} \left(\frac{-\gamma_i}{\gamma_{l+1}}\right) \text{MTTF}_i} \tag{22}$$

1.5. Analysis of the approximation for small values of t

Intuitively, we realize that when t tends to zero, repairs play no further part and the system behaves as if it were non-repairable and, for a non-repairable system, reliability equals availability. Consequently (cf. Exercise 2)

$$\tilde{\Lambda}(t) \underset{t \to 0}{\sim} \Lambda(t). \tag{23}$$

1.6. Characterization of systems for which $\tilde{\Lambda}(t)$ is a good approximation for the instantaneous system failure rate

Equation (23) shows that the approximation is valid for all systems when t tends to zero.

For large values of t, Eqn (22) shows that $1/\tilde{\Lambda}_\infty$ is a linear combination of the mean uptimes obtained by taking as the initial state the various critical operating states. From Eqn (15), the approximation $\tilde{\Lambda}_\infty$ for Λ_∞ will therefore be correct if the various mean uptimes differ little from the MTTF_1. For this to be so, it is sufficient that when the system is in a critical operating state, the probability q of it failing is small compared with the probability $p = 1 - q$ of it still being in an operating state each time it leaves that state, i.e.

$$\nu_i = \frac{\Lambda_i}{\sum\limits_{\substack{j \in M \\ j \neq i}} a_{ij}} \ll 1 \qquad \forall_i \in M_c . \tag{24}$$

We can therefore choose as a criterion $\max\limits_i \nu_i \ll \nu_0$ (e.g. 0.1).

We want the MUT to be a good approximation to the MTTF_1. Now, the MUT is a linear combination of the MTTF_i for all the critical operating states i, the linear combination coefficients being equal to $\frac{-\gamma_i}{\gamma_{l+1}}$ (cf. Eqn (22)). We therefore obtain a global condition by weighting the various conditions (24) relating to the critical operating states using the same weighting system. Hence, from Eqn (18):

$$\sum_{i \in M_c} \left(\frac{\gamma_i}{-\gamma_{l+1}} \right) \frac{\Lambda_i}{\sum\limits_{j \in M} a_{ij}} \ll 1 . \tag{25}$$

Example: Let us return to the example given in section 1.2.

Equation (25) can be written: $\lambda/\mu \ll 1$ and we obtain the result already stated. More generally, Eqn (25) shows that the more failed components that are involved in the critical operating states, the more flexible are the conditions for the approximation being correct. In fact, in this case the sum $\sum\limits_{j \in M} a_{ij}$ is made up of several repair rates. Therefore, the more redundant the system logic, the better the approximation. Systems comprising n identical independent components in a redundancy configuration constitute a particularly favourable case, since Eqn (25) can be written as:

$$\frac{\lambda}{(n-1)\,\mu} \ll 1 .$$

This formula has been well tested experimentally.

On the other hand, Eqn (24) appears stricter when state 1 is a critical operating state. However, as the components are independent, the system can be broken down into two subsystems in series: the critical components for state 1 and the others. Analysis of the first poses no difficulties since $\Lambda = \tilde{\Lambda}$. Equation (25) can therefore be satisfied on the second subsystem, as state 1 is no longer critical.

Several experiments have shown that Eqn (25) is indeed a sufficient condition. Good results are obtained when the first member of Eqn (25) is less than 1/10. However, there are systems for which the approximation holds good for values greater than 1/10. For practical reasons, we usually make do with satisfying Eqn (24), as the γ_i are not always easy to compute. When Eqn (25) is satisfied, $\tilde{\Lambda}(t)$ is a good approximation to $\Lambda(t)$ for small and large values of t. In practice, the limits $\tilde{\Lambda}_\infty$ and Λ_∞ are reached rapidly, without oscillation, and the approximation holds good for all time values. We can then compute a reliability approximation by taking:

$$R(t) \simeq \exp\left(- \int_0^t \tilde{\Lambda}(u)\,\mathrm{d}u \right).$$

2. APPLICATIONS OF THE CRITICAL OPERATING STATES METHOD

2.1. Automatic evaluation of the failure rate of large repairable systems

To use Eqn (3) we need to know the probabilities $P_i(t)$.

The case where all the components are *independent* is a particularly simple one, since:

$$P_i(t) = \prod_{j \in \mathcal{M}_i} q_j(t) \prod_{j \in \mathcal{P}_i} (1 - q_j(t)) \tag{26}$$

where $\mathcal{M}_i$: set of components operating in state *i*,
$\mathcal{P}_i$: set of components failed in state *i*,
$q_j(t)$: availability of component *j*.

The numerator of Eqn (3) can then be calculated easily as soon as we have determined the critical operating states. The latter can easily be enumerated by finding the critical operating states with 0, 1, 2, . . . , failed components and disregarding those whose probability is too low. Another quick method of finding a good approximation for these critical operating states is by means of paths and cut sets (cf. Exercise 3).

As for the denominator, this is merely the system availability, which can be calculated, for example, using minimal cut sets (cf. Ch. 4). In many cases we can even make this value equal to one.

In order to determine the minimal cut sets and the critical operating states, it is necessary to know the system logic. A large number of methods are available. The FIABC program [Gondran and Pagès, 1976] determines these sets by enumeration using successful paths. The latter can be obtained from a reliability block diagram [Batts et al., 1970], from a fault tree (see Ch. 4) or manually.

Note that we can calculate $\tilde{\Lambda}(t)$ using only the minimal cut sets, since from these we can generate the critical operating states by identifying the critical components. However, from two different minimal cut sets we can generate the same critical operating state. Consequently, as we have to compute the probability of the union of a state set, we have to apply a generally time-consuming formula of the Poincaré type. This method was described by Vesely [Vesely, 1970] and improved upon by Caldarola [Caldarola and Wickenhauser, 1977]. However, when it comes to computing the function $\tilde{\Lambda}(t)$ for a large number of points or several times for different numerical values of the parameters, computing the critical operating states represents a worthwhile investment.

Unlike the Vesely method, the critical operating states method makes it possible to take account of dependences between the components of a critical operating state.

2.2. Evaluating the repair rate of large repairable systems

We have seen that the critical operating states play a crucial part in analysing the failure rate of a system. As far as the repair rate is concerned, it is the *critical failed states* which are important. (A critical failed state is a failed state having at least one transition to the operating states in the Markov graph for computing the availability; we shall denote the set of these states by P_c.)

By analogy with the approximation used for the failure rate, we can show that the *pseudo-rate of repair* $\tilde{\mathcal{M}}(t)$ defined by:

$$\tilde{\mathcal{M}}(t) = \frac{\sum_{i \in P_c} \mathcal{M}_i P_i}{1 - A(t)} \tag{27}$$

$$\mathcal{M}_i = \sum_{j \in M} a_{ij}$$

is a good approximation for the system repair rate $\mathcal{M}(t)$.

It follows from Eqns (3) and (27) that the three functions $A(t)$, $\tilde{\Lambda}(t)$ and $\tilde{\mathcal{M}}(t)$ are linked by the relation:

$$\frac{\mathrm{d}A(t)}{\mathrm{d}t} = -\tilde{\Lambda}(t)\,A(t) + \tilde{\mathcal{M}}(t)(1 - A(t))\,. \tag{28}$$

Consequently, for the reliability calculation, the system is equivalent to a single-component system with failure rate $\tilde{\Lambda}(t)$ and repair rate $\tilde{\mathcal{M}}(t)$. This equivalent system is non-homogeneous Markovian and its Markov diagram is as follows:

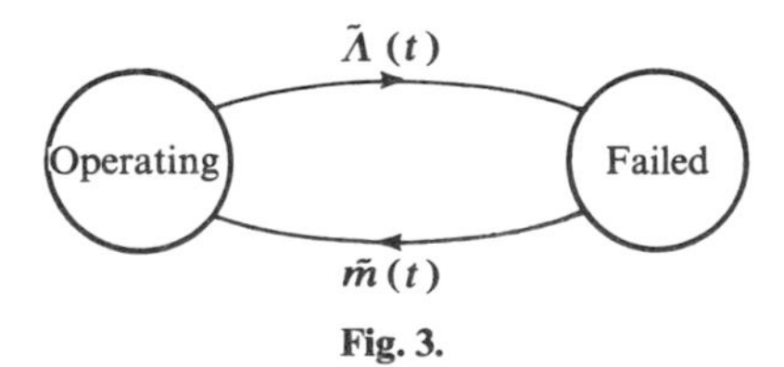

Fig. 3.

Equation (28) enables us to compute the pseudo-rate of repair:

$$\tilde{\mathcal{M}}(t) = \frac{\dfrac{\mathrm{d}A(t)}{\mathrm{d}t} + \tilde{\Lambda}A(t)}{1 - A(t)}\,. \tag{29}$$

In general, we merely compute the limit $\tilde{\mathcal{M}}_\infty$ which is very quickly attained in practice:

$$\tilde{\mathcal{M}}_\infty = \frac{\tilde{\Lambda}_\infty A(\infty)}{1 - A(\infty)}\,. \tag{30}$$

The condition equivalent to Eqn (25) is written as:

$$\left.\begin{aligned} &\sum_{i \in P_c} \gamma'_i \frac{\mathcal{M}_i}{\sum_{j \in M} a_{ij}} \gg \sum_{i \in P_c} \gamma'_i \\ &\gamma'_i = \frac{\sum_{j \in M} a_{ij} P_j(\infty)}{A(\infty)} \end{aligned}\right\} \tag{31}$$

When Eqn (25) is satisfied, Eqn (31) is of course satisfied except for series configuration systems. In fact, in this case the pseudo-rate of failure is identical to the failure rate. Therefore, apart from this special case, there is no need to test the validity of hypothesis (31).

2.3. Manual method for solving small systems

2.3.1. *Independent components*

The independence of the components enables us to compute the probability of occupation of the various states, using Eqn (26).

Example 1: A system consists of n components in series. There is a single operating state, which is critical. Hence:

$$\tilde{\Lambda} = \sum_{i=1}^{n} \lambda_i = \Lambda . \tag{32}$$

There are n critical failed states, each of them characterized by a single component being failed. Therefore:

$$\tilde{\mathcal{M}}(t) = \frac{\sum_{i=1}^{n} \mu_i (1 - q_i(t)) \prod_{j \neq i} q_j(t)}{1 - \prod_{j=1}^{n} q_j(t)} = \frac{A(t)}{1 - A(t)} \sum_{i=1}^{n} \mu_i \frac{1 - q_i(t)}{q_i(t)} \tag{33}$$

$$\tilde{\mathcal{M}}_\infty = \frac{\prod_{j=1}^{n} \mu_j}{\prod_{j=1}^{n} (\lambda_j + \mu_j) - \prod_{j=1}^{n} \mu_j} \sum_{j=1}^{n} \lambda_j . \tag{34}$$

Equation (31) can therefore be written as:

$$\sum_{i=1}^{n} \lambda_i \frac{\mu_i}{\sum_{j} \lambda_j - \lambda_i} \gg \sum_{i=1}^{n} \lambda_i .$$

If this condition is fulfilled, $\tilde{\mathcal{M}}(t)$ is a good approximation to $\mathcal{M}(t)$. For systems with good availability, Eqn (34) may be written in the form:

$$\tilde{\mathcal{M}}_\infty \simeq \frac{\sum_{j=1}^{n} \lambda_j}{\sum_{j=1}^{n} \frac{\lambda_j}{\mu_j}} .$$

Example 2: n components in an active redundancy configuration.

The system has n critical operating states, each of them characterized by a single component being operable. Hence:

$$\tilde{\Lambda}(t) = \frac{\sum_{i=1}^{n} \lambda_i q_i(t) \prod_{j \neq i} (1 - q_j(t))}{1 - \prod_{j=1}^{n} (1 - q_j(t))} \tag{35}$$

$$\tilde{\Lambda}_\infty = \frac{\prod_{i=1}^{n} \lambda_i \sum_{i=1}^{n} \mu_i}{\prod_{i=1}^{n} (\lambda_i + \mu_i) - \prod_{i=1}^{n} \lambda_i}. \tag{36}$$

From Eqn (17), the rates γ_i are equal to the repair times of the components μ_i. Equation (25) can therefore be written as:

$$\sum_{i=1}^{n} \frac{\mu_i}{\sum_{j=1}^{n} \mu_j} \frac{\lambda_i}{\left(\sum_{j=1}^{n} \mu_j\right) - \mu_i} \ll 1 .$$

In the case where the system unavailability $\prod_{i=1}^{n} \left(\frac{\lambda_i}{\lambda_i + \mu_i}\right)$ is small, Eqn (36) assumes the form:

$$\tilde{\Lambda}_\infty \simeq \prod_{i=1}^{n} \left(\frac{\lambda_i}{\lambda_i + \mu_i}\right) \sum_{i=1}^{n} \mu_i . \tag{37}$$

The repair rate is easily computed as there is only one failed state:

$$\tilde{\mathcal{M}} = \mathcal{M} = \sum_{i=1}^{n} \mu_i . \tag{38}$$

Example 3: n identical components in an r/n redundancy configuration.

The critical operating states are constituted by r operable components and $(n - r)$ failed components. Hence, if we call $S_k(t)$ the probability of k components being up at time t:

$$\tilde{\Lambda}(t) = \frac{r\lambda S_r(t)}{\sum_{k=r}^{n} S_k(t)} \tag{39}$$

$$\tilde{\mathcal{M}}(t) = \frac{(n - r + 1)\,\mu S_{r-1}(t)}{\sum_{k=0}^{r-1} S_k(t)} \tag{40}$$

with $S_k(t) = C_n^k q^k(t)(1 - q(t))^{n-k}$.

Hence, the stationary values

$$\tilde{\Lambda}_\infty = \frac{r\lambda\, C_n^r \mu^r \lambda^{n-r}}{\sum_{k=r}^{n} C_n^k \mu^k \lambda^{n-k}} \tag{41}$$

$$\tilde{\mathcal{M}}_\infty = \frac{(n - r + 1)\,\mu C_n^{r-1} \mu^{r-1} \lambda^{n-r+1}}{\sum_{k=0}^{r-1} C_n^k \mu^k \lambda^{n-k}}. \tag{42}$$

The coefficients γ_i of the critical operating states are all identical. Consequently, Eqn (25) is written as:

$$\frac{r\lambda}{(n-r)\,\mu} \ll 1\,. \tag{43}$$

2.3.2. *Functional dependence*

By *functional dependence* we mean an interaction between components which alters the Markov diagram such as Eqn (26) is no longer applicable.

Example: Components on standby and only switched on in certain system configurations; if the failure rates in operation and on standby are not identical or there is a probability of non-start-up, Eqn (26) is no longer applicable.

Interpreting the Markov diagram

As we saw in section 1.3.2, the reciprocal of the MTTF (taking state 1, in which all the components are available, as the initial state) is a good approximation for the stationary failure rate of reliable systems. We shall therefore evaluate $1/\mathrm{MTTF}_1$ using the matrix formulation (18) from Ch. 5.

However, experience shows that hand-calculation of determinants above the 4th or 5th order is laborious. *We shall propose a method of computing the determinants by simple interpretation of the Markov diagram.*

Assume, therefore, that state 1 is the initial state of the system. In this case, noting that $\sum_{j=1}^{l} a_{ij} = -\Lambda_i$, Eqn (18) of Ch. 5, can be written as:

$$\frac{1}{\mathrm{MTTF}_1} = \frac{\begin{vmatrix} \Lambda_1 & a_{12} & a_{1l} \\ \Lambda_2 & a_{22} & a_{2l} \\ \cdot & & \\ \cdot & & \\ \cdot & & \\ \Lambda_l & a_{l2} & a_{ll} \end{vmatrix}}{\begin{vmatrix} 1 & a_{12} & a_{1l} \\ 1 & a_{22} & a_{2l} \\ \cdot & & \\ \cdot & & \\ \cdot & & \\ 1 & a_{l2} & a_{ll} \end{vmatrix}} = \frac{N}{\Delta} \tag{44}$$

Computation of the numerator N of Eqn (44).

By the definition of a determinant, any element in the numerator N has the form:

$$(-1)^{s_\sigma} \prod_{i=1}^{l} a_{i,\bar{\sigma}(i)} \tag{45}$$

where $\sigma(.)$ is a permutation of the set $(1, 2, \ldots, l)$ and

$$\bar{\sigma}(.) = \sigma(.) \quad \text{if} \quad \sigma(.) \neq 1$$
$$\bar{\sigma}(.) = l + 1 \quad \text{if} \quad \sigma(.) = 1$$

s_σ is the signature of the permutation σ.

Equation (45) is zero if one of the transitions $i \to \bar{\sigma}(i)$ does not exist in the Markov diagram. Consequently, it is only necessary to take the following sets into account:

The paths between 1 *and* $l + 1$ *associated with the circuits passing via elements not belonging to the path under consideration. Each state must be taken only once as the origin of a transition and only once as the destination of a transition* (state 1 cannot be the destination of a transition since the terms a_{j1} do not exist, and state $l + 1$ can only be the destination of a transition of the type Λ_j). The absolute value of Eqn (45) is easy to compute as it is equal to a product of transition rates. As far as the signature of the permutation of a circuit involving k elements is concerned, it is equal to $(-1)^{k-1}$ (cf. [Berge, 1968] or Exercise 2), and the path between 1 and $l + 1$ may be regarded as a circuit in the wide sense by merging the states 1 and $l + 1$.

In the total, we disregard all the negligible circuits. Most of the time, the latter are made up, for a fixed path $1 \to l + 1$, solely of circuits of the type $\circlearrowleft_{a_{ii}}^{i}$. In fact they are of order μ, since $i \geqslant 2$ and any other circuit contains at least one term in λ.

Hence:

$$N \simeq \sum_{\text{paths from 1 to } l+1} \left(\Pi \text{ transition rate of path}\right)\left(-1\right)^{\text{number of intermediate states of path}} \left(\prod_{i \notin \text{path}} a_{ii}\right) \tag{46}$$

Example: Consider the following path:

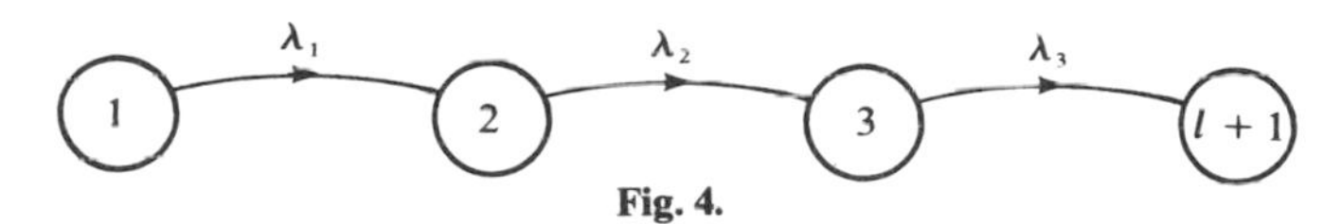

Fig. 4.

The contribution of this path to N is:

$$\lambda_1 \lambda_2 \lambda_3 (-1)^2 \prod_{i=4}^{l} a_{ii}.$$

As for the denominator Δ, note that as the failure rate tends to zero, it tends to $\prod_{i=2}^{l} a_{ii}$.

More precisely, putting $\lambda = \max_i \lambda_i$

$$\lim_{\lambda \to 0} \Delta = \prod_{i=2}^{l} a_{ii}. \tag{47}$$

In practice, this asymptotic value is a good approximation for the determinant when the probability of no component being failed is high (e.g. greater than 0.9). This experimental condition is once again sufficient but not necessary, as the approximation may be very good even though it is not satisfied.

Hence the following theorem, since a_{ii} is negative:

Theorem 2 when $\Delta \simeq \prod_{i=2}^{l} a_{ii}$, then:

$$\Lambda_\infty \simeq \sum_{\text{paths from state 1 to failed state}} \frac{\Pi \text{ transition rate of path}}{\prod_{i \in \text{set of intermediate states of path}} |a_{ii}|} \tag{48}$$

with $|a_{ii}| = -a_{ii} = \sum_{j \neq i} a_{ij}$ = sum of transition rates out of i.

Note that this formula is only correct under one condition$\left(\Delta \simeq \prod_{i=2}^{l} a_{ii}\right)$which is more stringent than condition (25).

Consequently, $\tilde{\Lambda}(t)$ can be a good approximation to $\Lambda(t)$ without Eqn (48) being applicable. However, this equation can only be used manually for small systems, in which case the condition is relatively unrestrictive. In all events, it is always possible to use this equation and to compare it with the discretization method described in Ch. 5. Equation (48), if it is satisfied, therefore has the advantage of giving an analytical expression of the result, providing a simple means of measuring the importance of the various parameters. It is possible to establish an equation similar to Eqn (48) for computing the availability (cf. Ch. 4, section 5.1, theorem 2).

Examples

Example 1: Consider a system comprising three redundant components. Component 1 is normally operating, with the other two (identical) components on standby. When component 1 fails, the other two are switched on simultaneously. A component's failure rate on being called upon to operate is equal to γ. Assume that the standby failure rates are zero, the operating failure rates of components 2 and 3 are λ, the repair rates of components 2 and 3 are μ, the failure rate of component 1 is λ_1 and the repair rate of component 1 is μ_1.

The Markov diagram is then as follows, the standby status of component i being denoted by i_a:

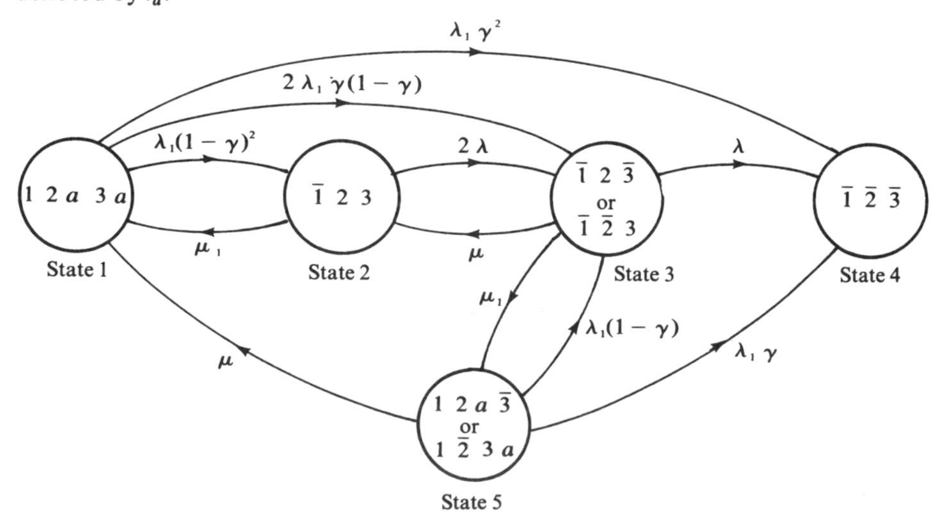

Fig. 5.

● denotes the contribution of the paths

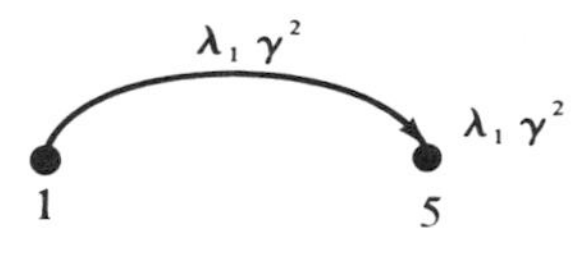

Fig. 6.

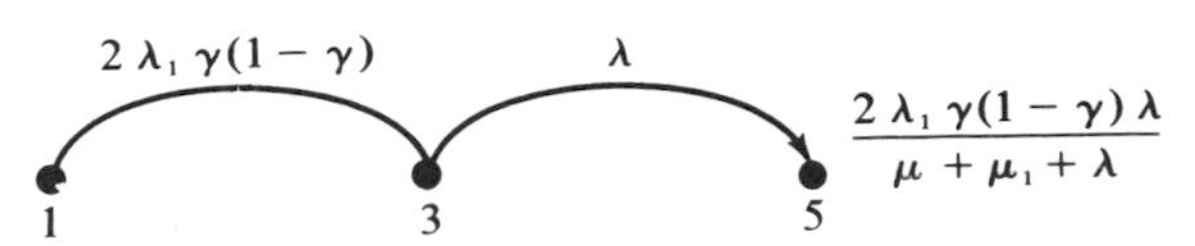

Fig. 7.

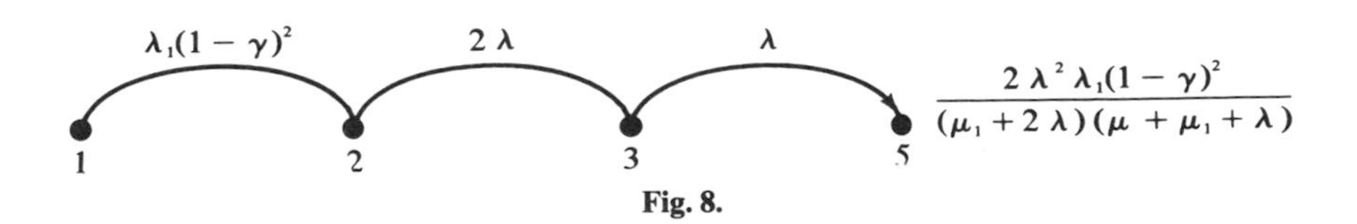

Fig. 8.

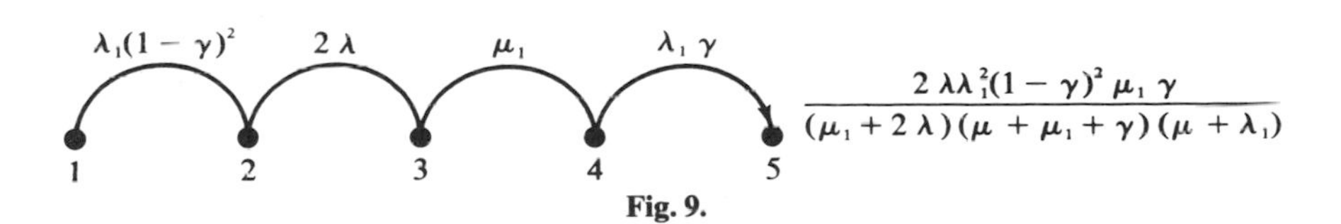

Fig. 9.

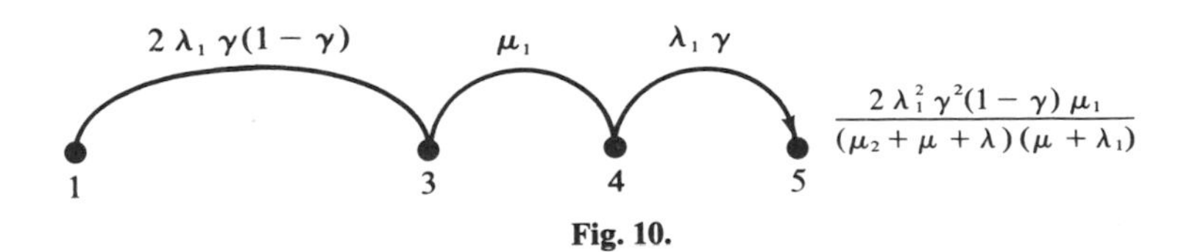

Fig. 10.

As the contributions of the last two paths are negligible, it follows that:

$$\Lambda_\infty \simeq \lambda_1 \gamma^2 + 2\lambda(1-\gamma)\frac{\lambda_1}{\mu_1+2\lambda}\left[\frac{\gamma(\mu_1+2\lambda)+\lambda(1-\gamma)}{\mu+\mu_1+\lambda}\right]. \tag{49}$$

Example 2: The reader may wish to verify that, in the case where component 2 is switched on and component 3 is left inoperative, we obtain:

$$\Lambda_\infty \simeq \lambda_1\gamma^2 + \lambda(1-\gamma)\frac{\lambda_1}{\mu_1+\lambda}\,\frac{\gamma(\mu_1+\lambda)+\lambda}{\mu+\mu_1+\lambda}. \tag{50}$$

Example 3: The reader may also consider the case where the repair rate μ' is different when a failure occurs when a component is called upon to operate (generally $\mu' > \mu$).

For example, in the case of simultaneous switch-on, we obtain:

$$\Lambda_\infty \simeq \lambda_1 \gamma^2 + 2\lambda(1-\gamma)\frac{\lambda_1}{\mu_1 + 2\lambda}\left[\frac{\lambda(1-\gamma)}{\mu + \mu_1 + \lambda} + \frac{\gamma(\mu_1 + 2\lambda)}{\mu_1 + \mu' + \lambda}\right] \tag{51}$$

and when switch-on is sequential:

$$\Lambda_\infty \simeq \lambda_1 \gamma^2 + \lambda(1-\gamma)\frac{\lambda_1}{\mu_1 + \lambda}\left|\frac{\lambda}{\mu_1 + \mu + \lambda} + \frac{\lambda(\mu_1 + \lambda)}{\mu_1 + \mu' + \lambda}\right| . \tag{52}$$

2.4. Aid to system design. Importance factors

Computing the reliability or availability of a system is not generally an end in itself. For example, at the design stage, it is very important to be able to answer such questions as:

α_1—which component must be improved as a matter of priority to increase system reliability?

α_2—which parameters have most bearing on the outcome?

In operating terms, we may wish to resolve the following question:

α_3—given that the system is down, which component must receive priority repair?

In order to attempt to answer these questions, we shall assign to each component a set of functions in order to understand their importance in the system. No economic considerations will be introduced at this stage (optimization problems will be examined in Ch. 8).

The notion of an *importance factor* was introduced by Birnbaum [Birnbaum, 1969] and taken further by several authors; Lambert has compiled a fairly comprehensive list [Lambert, 1976].

The importance factor $B_i(t)$ of component i due to Birnbaum was the partial derivative of the system unavailability $\bar{A}(t)$ with respect to the unavailability $\bar{q}_i(t)$ of component i.

$$B_i(t) = \frac{\partial \bar{A}(t)}{\partial \bar{q}_i} . \tag{53}$$

$\bar{A}(t)$ being regarded as a function of the $\bar{q}_i(t)$.

This function enables us to evaluate the variation of $\bar{A}(t)$ as a function of $\bar{q}_i(t)$ (sensitivity). Computation of $B_i(t)$ therefore requires computation of $\partial\bar{A}(t)/\partial\bar{q}_i$. Murchland [Murchland, 1973] has shown that the renewal density of a coherent system with independent components can be written as:

$$v(t) = \sum_{i=1}^{n} \lambda_i q_i(t) \frac{\partial \bar{A}}{\partial \bar{q}_i}(t) . \tag{54}$$

Consequently:

$$\tilde{\Lambda}(t) = \frac{v(t)}{A(t)} = \frac{\sum_{i=1}^{n} \lambda_i q_i(t) \frac{\partial \bar{A}}{\partial \bar{q}_i}(t)}{A(t)} \tag{55}$$

On the other hand, we can incorporate in Eqn (3) all the terms containing λ_i and write:

$$\tilde{\Lambda}(t) = \frac{\sum_{i=1}^{n} \lambda_i f_i(t)}{A(t)} \tag{56}$$

with

$$f_i(t) = \sum_{j \in J_i} P_j(t) \tag{57}$$

where J_i represents the set of the critical operating states having i as a critical component. For computing $\tilde{\Lambda}(t)$ it is very easy to compute the various functions $f_i(t)$. Comparing Eqns (55) and (56), it follows that:

$$\frac{\partial \bar{A}}{\partial \bar{q}_i}(t) = \frac{f_i(t)}{q_i(t)}. \tag{58}$$

We can now examine some importance factors.

2.4.1. *Birnbaum's importance factor or marginal importance factor ($B_i(t)$)*

In view of Eqn (58), definition (53) of this importance factor allows us to write:

$$B_i(t) = \frac{f_i(t)}{q_i(t)}. \tag{59}$$

This means that $B_i(t)$ can be interpreted as the probability of the system being in an operating state with i as critical component, given that i is operating.

When the system comprises independent components, $\bar{A}$ is linear in $q_i(t)$ and $B_i(t)$ does not depend on $q_i(t)$.

In accordance with the properties of the partial derivatives, we can write:

$$B_i(t) = \frac{\partial \bar{A}}{\partial \bar{S}} \frac{\partial \bar{S}}{\partial q_i}(t), \tag{60}$$

$\bar{S}$ representing the unavailability of a subsystem. The importance factor of the component in the system is the product of the importance factor in the subsystem multiplied by the importance factor of the subsystem.

2.4.2. *Critical importance factor ($C_i(t)$)*

The critical importance factor is the probability of a component i failing, given that the system has failed. We can naturally compute $C_i(t)$ using the minimal cut sets, but we can also use the $f_i(t)$. In fact, it can be shown that when the components are independent:

$$C_i(t) = \frac{\bar{q}_i(t)}{\bar{A}(t)} \frac{\partial \bar{A}}{\partial \bar{q}_i}(t) \tag{61}$$

hence

$$C_i(t) = \frac{\bar{q}_i(t)\, f_i(t)}{q_i(t)\, \bar{A}(t)} \tag{62}$$

Equation (61) shows that $\Delta\bar{A}$ and $\Delta\bar{q}_i$ are linked by the relation:

$$\frac{\Delta\bar{A}}{\bar{A}} = C_i \frac{\overline{\Delta q_i}}{\bar{q}_i} + 0(\overline{\Delta q_i^2}) \tag{63}$$

This importance factor seems well suited to answering questions α_1 and α_2 when system availability is of interest.

2.4.3. *Vesely–Fussell importance factor ($VF_i(t)$)*

When a system has failed, there is generally more than one component down and it is preferable to repair a component whose recommittal will restore the system to an operating condition. Consider, for example, the system whose reliability block diagram is given below:

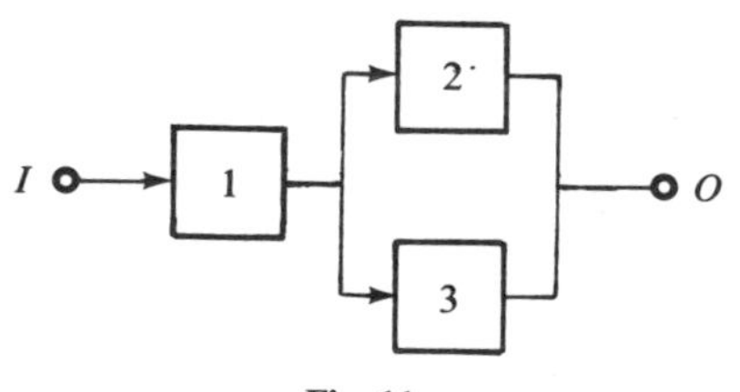

Fig. 11.

Assume that component 2 fails and then component 1 fails, before the repair facility has been able to take action. If component 2 is repaired first, system operation will not be restored since component 1 is still down. In fact, this problem arises either during fault diagnosis or where the number of repair facilities is limited.

The Vesely–Fussell importance factor is defined as the probability of a component contributing to system failure, given that the latter has failed. It can be shown that

$$VF_i(t) = \frac{\bar{q}_i(t)}{\bar{A}(t)}\ \bar{A}\ (t/\text{component } i \text{ has failed}) \tag{64}$$

In the above example, we obtain

$$VF_1(t) = \frac{\bar{q}_1}{\bar{A}(t)}$$

$$VF_2(t) = \frac{\bar{q}_2(\bar{q}_1 + \bar{q}_3 - \bar{q}_1\,\bar{q}_3)}{\bar{A}(t)}\,.$$

When the components are of good quality, $VF_1(t) \gg VF_2(t)$, which is to be expected.

This importance factor provides an answer to question α_3.

2.4.4. *Barlow and Proschan importance factor ($BP_i(t)$)*

This importance factor is defined as the probability of a component having caused system failure at time t (it is the failure of this component at time t which causes system failure). Hence:

$$BP_i(t) = \frac{\lambda_i f_i(t)}{\sum_{i=1}^{n} \lambda_i f_i(t)} = \frac{\lambda_i f_i(t)}{A(t)\,\Lambda(t)}. \tag{65}$$

Naturally

$$\sum_{i=1}^{n} BP_i(t) = 1\,.$$

2.4.5. *Summary*

We have just described a few importance factors. A more detailed study can be found in [Lambert, 1975]. The importance factors are functions of time but in practice they are only computed for a single value of t (mission duration, where t = infinity gives the stationary value).

In this section, we have referred to component importance factors. In fact, we know (cf. Ch. 2) that in a reliability block diagram or a fault tree, the events do not always correspond to the failure of a component. It is therefore more appropriate to speak of the importance factors of components or events.

In conclusion, we shall give the values of the various importance factors for two systems comprising two components (λ_1, μ_1), (λ_2, μ_2) in series or in parallel.

We can compute the various values for $t = \infty$ and for *component* 1

$$\bar{q}_1 = \frac{\lambda_1}{\lambda_1 + \mu_1} \qquad \bar{q}_2 = \frac{\lambda_2}{\lambda_2 + \mu_2}.$$

Table 1.

	Series configuration	Parallel configuration
Unavailability	$\bar{q}_1 + \bar{q}_2 - \bar{q}_1\,\bar{q}_2$	$\bar{q}_1\,\bar{q}_2$
Birnbaum	$1 - \bar{q}_2$	$\bar{q}_2$
Critical	$\dfrac{\bar{q}_1(1-\bar{q}_2)}{\bar{q}_1 + \bar{q}_2 - \bar{q}_1\,\bar{q}_2}$	1
Vesely–Fussell	$\dfrac{\bar{q}_1}{\bar{q}_1 + \bar{q}_2 - \bar{q}_1\,\bar{q}_2}$	1
Barlow–Proschan	$\dfrac{\lambda_1}{\lambda_1 + \lambda_2}$	$\dfrac{\mu_1}{\mu_1 + \mu_2}$

3. EXTENSION OF THE CRITICAL OPERATING STATES METHOD TO MORE GENERAL SITUATIONS

3.1. Non-Markovian systems

In the case of non-Markovian systems, direct application of Eqn (3) is no longer possible as the failure rates are no longer constant. In this case [Vesely, 1970], we may be interested in the following function:

$$\tilde{\Lambda}(t) = \frac{\sum_{i \in M_c} \Omega_i(t)\, P_i(t)}{\sum_{i \in M} P_i(t)} \tag{66}$$

with

$$\Omega_i(t) = \sum_{j \in \mathscr{C}_i} \frac{h_j(t)}{q_j(t)} \tag{67}$$

$h_j(t)$ being the failure density of the renewal process of component j and $q_j(t)$ the availability of component j. It is no longer possible to show that $\tilde{\Lambda}(t)$ is a good approximation, but the checks which we have carried out on a number of systems (cf., for example [Pagès, 1977]) give grounds for thinking that the property holds good, assuming that the MUT is not too different from the MTTF (computed when no element has failed in the initial state).

Analysis of these non-Markovian systems shows that they often behave like Markovian systems when t is large, i.e. when their failure rate tends to a limit.

For this, it is sufficient that a very loose hypothesis is satisfied. Consider a non-Markovian system having at least one regeneration point reached periodically (e.g. times at which the system arrives in the state where all components are available). Each time the system leaves this state, there is therefore a constant probability q of returning to it and a probability $p = 1 - q$ of being absorbed by the failed state.

Exercise 1 shows that when p tends to zero (very reliable system) the lifetime tends to an exponential distribution and the failure rate tends to the reciprocal of the mean uptime.

In practice, we can extend the principal of 'interpreting' the Markov diagram to non-Markovian cases for which the failure rates λ_i and repair times τ_i are constant, and the products $\lambda_i \tau_i$ are small compared with one. This is the approach which we discussed for the case of availability in Ch. 4, section 6 [Gondran and Laleuf, 1981].

In the system state transition diagram, the changes of state therefore depend on the residence time in each state. We shall again denote by α the vertex for which all the components are available and by B the set of vertices β for which the system is regarded as failed.

We then consider all the *elementary paths* Γ from α to β. Such a path is called an *accidental sequence.* We denote by $\Lambda(\Gamma)$ the contribution of the accidental sequence Γ to the system's asymptotic failure rate (which exists as shown in Exercise 1 of this chapter).

We can therefore demonstrate the following theorem (cf. [Gondran and Laleuf, 1981; Laleuf, 1983] and Ch. 4, section 6):

Theorem 2: *If all the system components are repairable with $\lambda_i \tau_i \leqslant 1$, an excellent approximation for the asymptotic failure rate is given by*:

$$\Lambda_\infty \simeq \sum_{\substack{\text{paths } \Gamma \\ \text{from } \alpha \text{ to } \beta}} \Lambda(\Gamma) \tag{68}$$

If the sequence Γ corresponds to the successive failures of components $1, 2, \ldots, n$, *using the notations from Ch. 4, section 3.2,* we have:

$$\Lambda(\Gamma) = \left(\prod_{i=1}^{n-1} \lambda_i \tau_i' \right) \lambda_n \tag{69}$$

with

$$\tau_{i+1}' = \min\left(\tau_{i+1}, \frac{\tau_i'}{2} \right). \tag{70}$$

Compared with the Markovian case, we obtain differences in the failure rate result, which is relatively small for systems with few components.

Example: Consider the system made up of a component 1 (failure rate λ_1, repair time τ_1) and two components in a passive (standby) redundancy configuration (failure rate λ, repair time τ, probability of starting γ), only one of which performs the system function.

If we consider the non-repairable components, the state transition diagram is as follows:

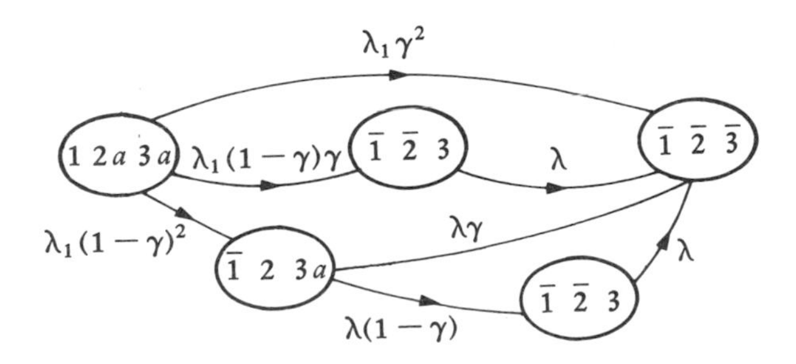

which reveals four accidental sequences:

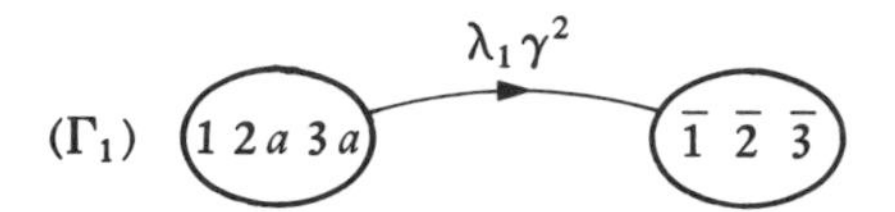

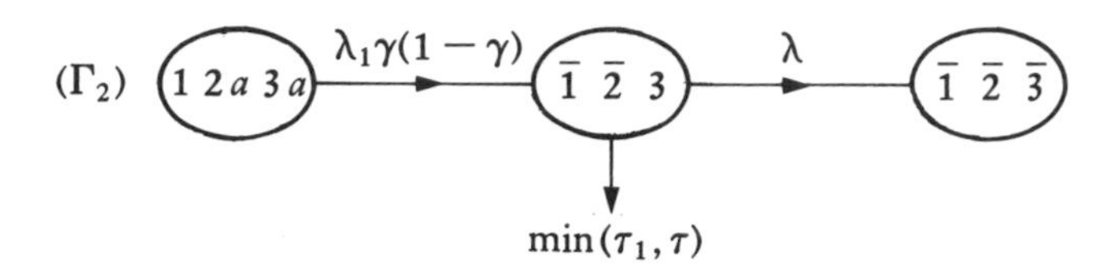

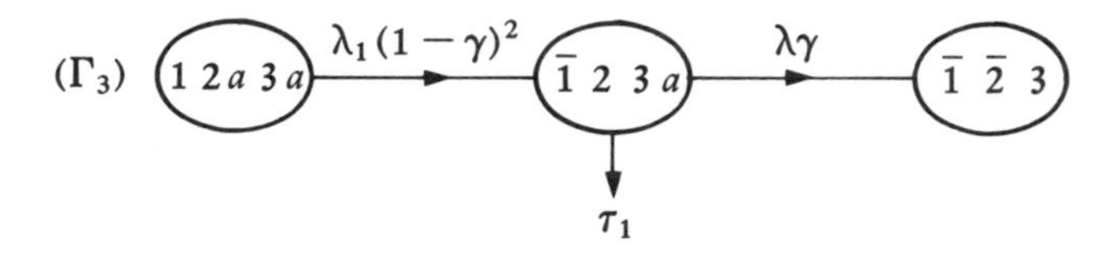

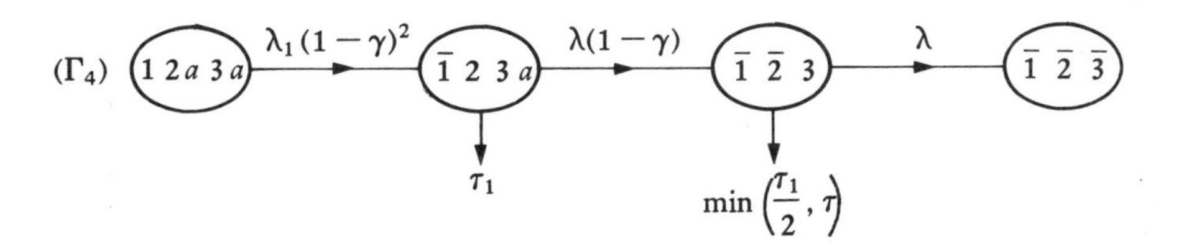

hence

$$\Lambda_\infty \simeq \lambda_1\gamma^2 + \lambda_1\lambda\gamma\min(\tau_1,\tau) + \lambda_1\lambda\gamma\tau_1 + \lambda_1\lambda^2\tau_1\min\left(\frac{\tau_1}{2},\tau\right) \tag{71}$$

a formula comparable with relation (50), section 2.3.

3.2. Partially repairable systems

We call a system comprising both repairable and non-repairable components 'partially repairable'. We assume that it is impossible to break down the system into two subsystems in series, one consisting solely of repairable components and the other of non-repairable components. In fact, in this case, the reliability would be the product of the reliability of the two subsystems, which can be analysed.

3.2.1. *Difficulties due to the non-repairability of certain components*

It is still possible and useful to compute $\tilde{\Lambda}(t)$. *However, the A_l matrix used in the reliability computation is no longer irreducible and the Frobenius theorem cannot now be applied, which means that $\tilde{\Lambda}(t)$ ceases to be an approximation for $\Lambda(t)$.*

Consider, for example, the system whose reliability block diagram is represented by Fig. 11 but in which component number 2 is not repairable.

The Markov diagram is as follows:

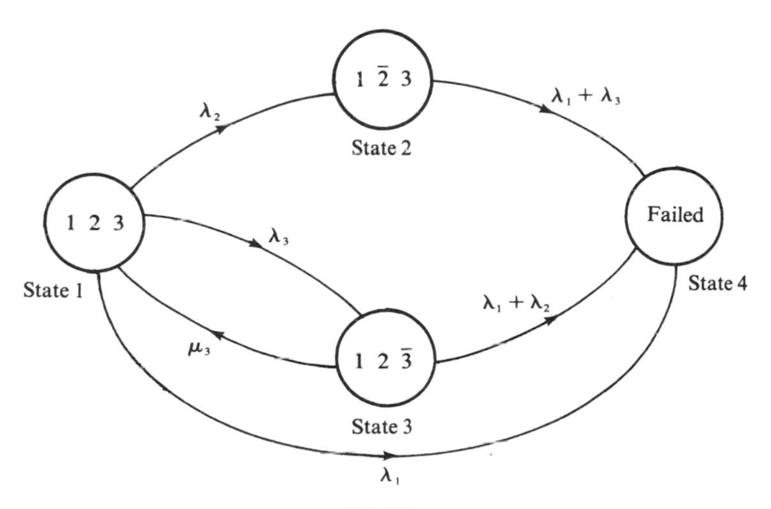

Fig. 12.

The system state equations are:

$$\left.\begin{aligned}
\frac{dP_1(t)}{dt} &= -(\lambda_1+\lambda_2+\lambda_3)\,P_1(t) + \mu_3\,P_3(t)\\
\frac{dP_2(t)}{dt} &= \lambda_2\,P_1(t) - (\lambda_1+\lambda_3)\,P_2(t)\\
\frac{dP_3(t)}{dt} &= \lambda_3\,P_1(t) - (\lambda_1+\lambda_2+\mu_3)\,P_3(t)
\end{aligned}\right\} \qquad (72)$$

Solving this system, we obtain:

$$\left.\begin{aligned}
\Lambda_\infty &= \lambda_1+\lambda_2 \quad \text{if} \quad \lambda_3 > \lambda_2\\
\Lambda_\infty &= \lambda_1+\lambda_3 \quad \text{if} \quad \lambda_3 < \lambda_2\\
\tilde{\Lambda}_\infty &= \lambda_1+\lambda_3\\
\text{MTTF}_1 &= \frac{1}{\lambda_1+\lambda_2} + \frac{\lambda_2(\mu_3+\lambda_1+\lambda_2)}{(\lambda_1+\lambda_2)(\lambda_1+\lambda_3)(\mu_3+\lambda_1+\lambda_2+\lambda_3)}
\end{aligned}\right\} \qquad (73)$$

We shall show how to attempt to simplify system (72) by breaking down the Markov diagram into strongly connected components (here 1–3, 2, 4). Each strongly connected component consists of a set of states whose configuration of non-repairable components is identical for each state. *We can then apply the critical operating states method* to these strongly connected components by eliminating the failed non-repairable components and considering the operating non-repairable components as having *an availability of* 1 *and retaining the same failure rate.*

We obtain, for example, for the first strongly connected component:

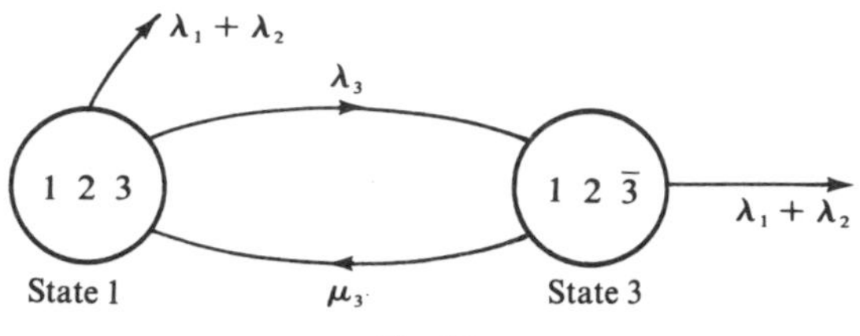

Fig. 13.

Hence $P_1(t) + P_3(t) = \exp(-(\lambda_1 + \lambda_2)t)$.

We can then calculate the instantaneous transition rates of this strongly connected component to the others or to the failed state. We then obtain a reduced Markov diagram with only failures as transition.

In the previous example we obtain:

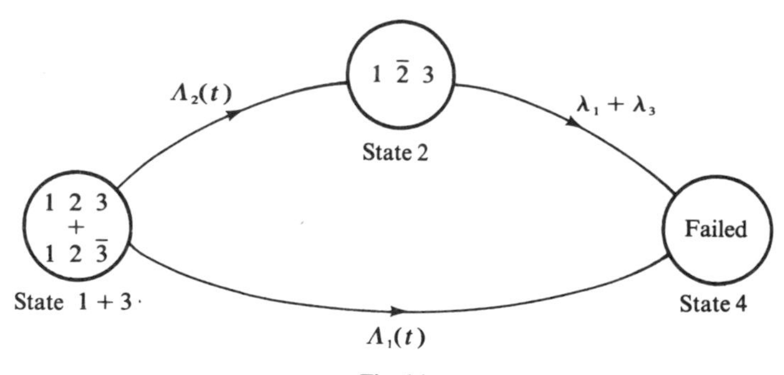

Fig. 14.

with

$$\Lambda_1(t) = \lambda_1 + \lambda_2\,\bar{q}_3(t)$$
$$\Lambda_2(t) = \lambda_2\,q_3(t)$$
$$q_3(t) = \text{availability of component 3.}$$

System (72) then reduces to the following system:

$$\frac{\mathrm{d}P_2(t)}{\mathrm{d}t} = \Lambda_2(t)\,\mathrm{e}^{-(\lambda_1+\lambda_2)t} - (\lambda_1 + \lambda_3)\,P_2(t)\,. \tag{74}$$

We introduce a number of connected components which may increase exponentially with the number of non-repairable components. This is why the following method is to be preferred when the system mission time is not excessively long.

3.2.2. *Reliability analysis over a reduced time interval*

When t tends to zero, we know that $\tilde{\Lambda}(t)$ is a good approximation for $\Lambda(t)$, whatever the system.

Suppose we want to analyse the reliability of a partially repairable system over the interval $[0, T]$. We can then replace the non-repairable components by repairable components whose *repair time is constant and equal to T.*

If condition (25) is satisfied by taking $\mu = 1/T$ as the repair rate for these components, the critical operating states method is applicable over $[0, T]$.

If (25) is not satisfied, we can determine the value T_{max} of the time t until which condition (25) is satisfied. Experiments show that this condition is not always necessary, in reality the approximation is valid for larger time values (in practice we can often accept ν_0 values of 0.5 or 0.6).

3.3. Dependent components

Equation (3) is equally valid when the components are dependent. However, it is then difficult to compute the $P_i(t)$. Each type of dependence is a special case and must be treated as such.

Here are a few examples:

— Standby components switched on under certain conditions (cf. [Gondran and Pagès, 1976]).

— Admissible short-duration failures (cf. [Pagès, 1976] and Ch. 3, section 3.1.4).

— Limited repair facilities.

The reader is also referred to Ch. 4, section 5.5 and Exercises 1, 5, 6 and 7.

In practice, if all the system components are repairable with $\lambda_i \tau_i \ll 1$ (assumptions of theorems 2 and 3), the asymptotic failure rate will be obtained directly from Eqn (48) or Eqn (68).

4. SUMMARY

• When computing the reliability of repairable Markovian systems by conventional methods, the problem is that the volume of computations increases exponentially as the component count rises. The *critical operating states* method gets round this problem by providing a good approximation for the system failure rate $\Lambda(t)$.

Using the notation:

$P_i'(t)$, the probability of state i in the Markov diagram associated with the reliability computation;

$P_i(t)$, the probability of state i in the Markov diagram associated with the availability calculation;

M, the set of operating states;

M_c, the set of critical operating states;

Λ_i, the sum of the transition rates from an operating state i to the failed states. $\Lambda(t)$ is written as:

$$\Lambda(t) = \frac{\sum_{i \in M_c} \Lambda_i P_i'(t)}{\sum_{i \in M} P_i'(t)}.$$

We then define the function $\tilde{\Lambda}(t)$ by analogy:

$$\tilde{\Lambda}(t) = \frac{\sum_{i \in M_c} \Lambda_i P_i(t)}{\sum_{i \in M} P_i(t)}.$$

The numerator of $\tilde{\Lambda}(t)$ represents the system failure density in the alternating renewal process associated with the Markov diagram for computing the availability, and the denominator represents the system availability (which can very often be taken as being 1).

For small time values, $\tilde{\Lambda}(t)$ is a very good approximation to $\Lambda(t)$ for any system. On the other hand, when t tends to infinity, $\tilde{\Lambda}_\infty$ is only a good approximation to Λ_∞ if condition (25) is satisfied. In fact, this condition is virtually unrestrictive and the critical operating states method is therefore applicable to most repairable systems.

Enumeration of the critical operating states is simple for small systems. Exercise 3 contains an algorithm for generating the critical operating states for large systems. Note that, in general, the number of critical operating states is much smaller than the number of operating states, especially if we only take into account those whose numerical contribution to the computation of $\Lambda(t)$ is significant.

• When the system components are independent, $P_i(t)$ is simple to compute, since:

$$P_i(t) = \prod_{j \in \mathcal{M}_i} q_j(t) \prod_{j \in \mathcal{P}_i} (1 - q_j(t))$$

where $\mathcal{M}_i$ = set of components operating in state i,
$\mathcal{P}_i$ = set of components failed in state i,
$q_j(t)$ = availability of component j.

The critical operating states method therefore provides an easy way of hand-calculating the failure rates of small systems. For example:
n independent components in an active redundancy configuration:

$$\tilde{\Lambda}_\infty = \frac{\prod_{i=1}^{n} \lambda_i \sum_{i=1}^{n} \mu_i}{\prod_{i=1}^{n} (\lambda_i + \mu_i) - \prod_{i=1}^{n} \lambda_i}$$

n identical independent components in an r/n redundancy configuration:

$$\tilde{\Lambda}_\infty = \frac{r\lambda C_n^r \mu^r \lambda^{n-r}}{\sum_{k=r}^{n} C_n^k \mu^k \lambda^{n-k}}.$$

For more complex systems, it is necessary to use a program (e.g. FIABC [Mulet Marquis and Dubreuil-Chambardel, 1984]).

• When the components are not independent, the $P_i(t)$ are more difficult to compute.

For small systems of good quality (e.g. probability of no component failing is higher than 0.9), we can obtain a good approximation for the stationary failure rate of the system merely by interpreting the Markov diagram.

$$\Lambda_\infty \simeq \sum_{\substack{\text{paths from state 1} \\ \text{to the failed} \\ \text{state of the system}}} \frac{\Pi \text{ path transition rate}}{\prod\limits_{\substack{i \in \text{set of operating states} \\ \text{of the path except} \\ \text{state 1 and the} \\ \text{failed state}}} |a_{ii}|} \tag{48}$$

with $|a_{ii}|$ = sum of the transition rates out of state i.

Typical applications of this formula can be found in section 2.3.2.

• The critical operating states method can be generalized to other situations:

— Non-Markovian systems.
— Partially repairable systems.
— Computing the $P_i(t)$ in certain dependence cases.
— Computing the propagation of data uncertainties (cf. Ch. 7).

In particular, for systems comprising repairable components with constant repair times such that $\lambda_i \tau_i \ll 1$, an excellent approximation for the asymptotic failure rate is given by

$$\Lambda_\infty \simeq \sum_{\substack{\text{paths } \Gamma \\ \text{from } \alpha \text{ to } \beta}} \Lambda(\Gamma)$$

where the $\Lambda(\Gamma)$ are given by relations (69) and (70).

Finally, a *very important point* should be made. In the case of good-quality repairable systems, the methodology used to calculate the asymptotic availability of a large system can be used to calculate the asymptotic failure rate. Let C_j, j from 1 to m, be the set of minimal cut sets of the system.

Denoting the asymptotic failure rate of the minimal cut set by $\Lambda_\infty(\bar{C}_j)$, an excellent approximation for the asymptotic failure rate of the system is

$$\Lambda_\infty \simeq \sum_{j=1}^{m} \Lambda_\infty(\bar{C}_j). \tag{75}$$

This is of great practical importance as it largely limits the calculations in the case of intercomponent dependences to the dependent components in each minimal cut set. It also makes it possible to measure the contribution of each minimal cut set to the system failure rate.

Analysis of the critical operating states combined with examination of the minimal cut sets, provides useful qualitative analysis of a system. This analysis is facilitated by computing importance factors.

EXERCISES

Exercise 1 — Behaviour of very reliable systems (P. Oger)

Consider a system having at least one regeneration point (e.g. one where no component has failed) and one absorbing state.

Let q be the probability of the system returning to this regeneration state after having left it. In our hypothesis, q is constant.

We put $p = 1 - q$; p represents the probability of being absorbed.

X = random variable measuring the time between two passages via the regeneration point.

Y = random variable representing the absorption time (without returning via the regeneration point).

(1) Compute, as a function of X and Y, the random variable T representing the system uptime. Deduce $E[T]$, then the Laplace transform of T.

(2) Compute the Laplace transform of $\mathscr{T} = \dfrac{T}{E[T]}$.

(3) Now assume that the system is very reliable. Its failure is a rare event and p is very small.

Calculate the limit, when p tends to zero, of $\bar{\mathscr{T}}(s)$.

Deduce that T tends to an exponential distribution with parameter $1/E[T]$.

Exercise 2 — Behaviour of systems for small t

(1) Consider a system comprising repairable components possessing l operating states. Assume that the failure and repair rates are constant.

Evaluate the matrix Q of the transition distributions using Eqn (44) from Ch. 5. Compute the Laplace transforms $\bar{Q}_{ij}(s)$ of the functions $Q_{ij}(t)$. Deduce the behaviour of these functions as t tends to zero.

(2) Consider the subset of the permutations $\bar{\sigma}_k$ (on a set of components of power k) such that there is only one circuit of the type $i \to \bar{\sigma}(i) \to \bar{\sigma}(\bar{\sigma}(i)) \to \ldots \to i$. Show that the sign of the permutations of this set is equal to $(-1)^{k-1}$.

(3) Show that the Laplace transform of the reliability, given that the initial state is state j, can be expressed in the form:

$$\underset{j,l+1}{\bar{P}(s)} = \frac{(-1)^{l+1+j}}{s\,|I - s\bar{Q}(s)|} \begin{vmatrix} 1 & -s\bar{Q}^{(s)}_{1,2} & \cdots & -s\bar{Q}^{(s)}_{1,j-1} & 0 & -s\bar{Q}^{(s)}_{1,j+1} & \cdots & -s\bar{Q}^{(s)}_{1,l+1} \\ -s\bar{Q}^{(s)}_{2,1} & 1 & & & \vdots & & & \\ & & \ddots & & \vdots & & & \\ & & & & \vdots & \ddots & 1 & -s\bar{Q}^{(s)}_{l,l+1} \\ & & & & 0 & & & \\ 0 & \cdots & \cdots & 0 & 1 & 0 & \cdots & 0 \end{vmatrix}$$

From which it follows, as was *a priori* obvious, that $\underset{j,l+1}{P(t)} \underset{t=0}{\to} 0$.

(4) Consider the behaviour of $\underset{j,l+1}{P(t)}$ when t tends to zero.

Show that the absolute value of the smallest degree terms in s^{-1} in the expansion of $\underset{j,l+1}{\bar{P}(s)}$ may be expressed in the form

$$\frac{1}{s^{p+1}} \Pi \rho_{jj_1} \rho_{j_1 j_2} \cdots \rho_{j_{p-2} j_{p-1}} \rho_{j_{p-1} l}$$

where p represents the minimum size of the paths linking state j to state $l + 1$.

Deduce from question 2 that all these terms are of the same sign and, consequently:

$$\underset{j,l+1}{\bar{P}(s)} \underset{s \to \infty}{\simeq} \frac{\sum\limits_{\text{paths of size } p} \rho_{jj_1} \rho_{j_1, j_2} \cdots \rho_{j_{p-1} l}}{s^{p+1}} = \frac{\alpha}{s^{p+1}}.$$

Deduce that:

$$P(t)_{j,l+1} \underset{t\to 0}{\simeq} \alpha t^p$$

where p represents the minimum number of components having to fail in order to cause system failure.

(5) Deduce that α does not depend on the repair times for coherent systems and that $\Lambda(t) \underset{t\to 0}{\sim} \bar{\Lambda}(t)$.

Exercise 3—Algorithm for generating the critical operating states

Consider a system whose set of minimal paths and set of minimal cut sets is known. We shall use the following notations:

$\mathscr{P} = \{P_i, i = 1, n\} =$ set of minimal paths.

$\mathscr{C} = \{C_i, i = 1, p\} =$ set of minimal cut sets.

$P_i = \{e_{ij}, j = 1, p_i\} =$ set of components constituting path P_i (if none of the components e_{ij} has failed, the system has not failed).

$C_i = \{\bar{e}_{ij}, j = 1, c_i\} =$ set of components constituting the cut set C_i (if the set of the components $\bar{e}_{ij}$ have failed, the system has failed).

(1) Consider a path P_i and a cut set C_j having one and only one component in common. What can we say of the set of the states made up as follows?:

—the components belonging to P_i have not failed,
—the components belonging to C_j have failed, except component r, also belonging to P_i which has not failed,
—the other components are in an indeterminate state.

(2) A set of this type is called a critical operating subset. The component common to P_i and C_j is called the critical component.

Show that two critical operating subsets are such that the failed components of one cannot be included in the failed components of the other.

(3) Deduce an algorithm to determine the critical operating states having k failed components.

Apply this algorithm to determine the critical operating states of a bridge system.

Show that in order to determine these critical operating sets, it is necessary to know all the minimal cut sets whose order is less than or equal to $k + 1$.

(4) Show that the failure density of a critical operating subset may be expressed in the form:

$$\lambda_r \prod_{l \in P_i} q_l \prod_{\substack{m \in C_j \\ m \neq r}} \bar{q}_m$$

where

λ_r represents the failure rate of the critical component,
q_l represents the availability of a component of path P_i,
$\bar{q}_m$ represents the unavailability of a component of cut set C_j (except for component r).

(5) Consider the set of the critical operating subsets having a given critical component γ and a list of given failed components. Denote by $\bar{P}_j$ the set of components common to all the P_i. P_j is not empty since it contains at least the common critical component.

Show that such a set is defined by:

—set of operating components $= \bar{P}_j$,
—set of failed components $= C_j - \{r\}$,
—set of the indeterminate components = Union of indeterminate-state components of the critical operating subsets constituting the set.

We shall call this set the pseudo-critical operating state. Such a set consists both of operating and failed states. Show that the failure density of the operating states of a pseudo-critical operating state can be written as:

$$\lambda_r \prod_{l \in \bar{P}_j} q_l \prod_{m \in C_j - \{r\}} \bar{q}_m A$$

with $A = 1$
or A = availability of a subsystem made up of components not belonging to either $\bar{P}_j$ or C_j.

(6) Show that the subsystem whose availability is given by A contains no first-order cut set.
Deduce that when the components comprising the system are of good quality, we can take $A = 1$.

(7) Show that for a given critical component, the intersection of the two pseudo-critical operating states, if it exists, is obtained by declaring failed at least one of the non-critical components of each pseudo-critical operating state; deduce a good approximate for $\tilde{\Lambda}(t)$; application: determine the pseudo-states of critical operation for a bridge system.

(8) Numerical application of questions 3 and 7:

$$\lambda_1 = 0.2 \quad \lambda_2 = 0.2 \quad \lambda_3 = 0.2 \quad \lambda_4 = 0.1 \quad \lambda_5 = 0.1$$
$$\mu_1 = 10 \quad \mu_2 = 10 \quad \mu_3 = 20 \quad \mu_4 = 2 \quad \mu_5 = 2$$

(in h^{-1}).

Exercise 4 — Example of a system with standby redundancy

(1) Consider the following two-component system. One component is normally operating while the other is on standby. When the operating component fails, the other component takes over, but the switch is not perfect; there is a probability γ of the standby component refusing to take over. Assume that the transition rates are constant and that we have two repair facilities (the mean repair time of a component is the same if failure occurs during operation or when the component is called upon to operate). Plot the Markov diagrams for computing system availability and reliability.

(2) Compute the stationary unavailability of the system and its stationary failure rate Λ_∞ by computing the eigenvalues of the matrix A_l. Take $\lambda \ll \mu$, $\gamma < 0.05$.

(3) Obtain an approximation for Λ_∞ using the method described in section 2.2.2. Assume $\lambda \ll \mu$, $\gamma < 0.05$. Do the same for the unavailability.

(4) Compute the system MTTFs for the various possible initial states.

(5) Compute $\tilde{\Lambda}_\infty$.

(6) Repeat the exercise assuming that the repair times are different depending on the failure mode (in operation or when load is applied).

Exercise 5 — Compute the mean and the variance of the uptimes of repairable systems

(1) Consider a Markovian system such that

$$\lim_{t \to \infty} \Lambda(t) = \Lambda_\infty .$$

Assume that for $\quad t \geqslant t_0, \dfrac{|\Lambda_\infty - \Lambda(t)|}{\Lambda_\infty} \leqslant \varepsilon.$

Show that

$$\text{MTTF} = \lim_{t \to \infty} \int_0^t R(u)\,du + \frac{R(t)}{\Lambda_\infty}.$$

Deduce that $\rho_1 = \int_0^{t_0} R(t)\,dt + \dfrac{R(t_0)}{\Lambda_\infty}$ is a good approximation for the MTTF, more precisely that:

$$|\,\text{MTTF} - \rho_1\,| \leqslant \frac{\varepsilon}{\Lambda_\infty(1-\varepsilon)}.$$

Deduce that if t_0 is small compared with $1/\Lambda_\infty$, then 1/MTTF is a good approximation for Λ_∞.

(2) Show that the 2nd-order moment of the uptime T satisfies:

$$E[T^2] = \lim_{t\to\infty} 2\int_0^t uR(u)\,du + \frac{2\,tR(t)}{\Lambda_\infty} + \frac{2\,R(t)}{\Lambda_\infty^2}.$$

Deduce that $\rho_2 = 2\int_0^{t_0} uR(u)\,du + \frac{2\,t_0\,R(t_0)}{\Lambda_\infty} + \frac{2\,R(T_0)}{\Lambda_\infty^2}$ is a good approximation for $E[T^2]$, more precisely that:

$$|E[T^2] - \rho_2| \leq \frac{2\,\varepsilon R(t_0)}{\Lambda_\infty(1-\varepsilon)}\left(t_0 + \frac{2}{\Lambda_\infty(1-\varepsilon)}\right).$$

(3) Deduce a practical method of computing the mean and the variance of the uptime.

SOLUTIONS TO EXERCISES

Exercise 1

(1) $T = Y + \sum_{i=1}^{N} X$ with $\mathcal{P}(N = n) = q^n(1-q)$

hence $E[T] = E[Y] + \frac{q}{p}E[X] = \frac{\alpha}{p}$

with $\alpha = qE[X] + pE[Y]$

$$\overline{T}(s) = \sum_{n=0}^{\infty} (\overline{X(s)})^n\,\overline{Y(s)})\,pq^n$$

$$\overline{T}(s) = p\overline{Y(s)}\sum_{n=0}^{\infty}(q\overline{X(s)})^n$$

$$\overline{T}(s) = \frac{p\overline{Y(s)}}{1 - q\overline{X(s)}}.$$

(2) $\mathcal{T} = \frac{T}{E[T]} = p\frac{T}{\alpha}$ $\qquad T = \frac{\alpha}{p}\mathcal{T}$.

This amounts to taking $u = E[T]$ as the time unit. Thus, by change of variables

$$T(t)\,dt = \mathcal{T}(u)\,du$$

hence $\overline{\mathcal{T}(s)} = \int_0^\infty \mathcal{T}(u)e^{-su}\,du = \int_0^\infty T(t)\exp\left(-\frac{sp}{\alpha}t\right)dt = \overline{T}\left(\frac{sp}{\alpha}\right)$

hence $\overline{\mathcal{T}(s)} = \dfrac{\overline{Y}\left(\frac{sp}{\alpha}\right)p}{1 - q\overline{X}\left(\frac{sp}{\alpha}\right)}.$

(3) $\lim_{p\to 0}\alpha = E[X]$ since $q = 1 - p \to 1$

$$\lim_{p\to 0}\overline{\mathcal{T}(s)} = \lim_{p\to 0}\frac{p\left(1 - \frac{sp}{\alpha}E[Y]\right)}{1 - q\left(1 - \frac{sp}{\alpha}E[X]\right)}$$

$$= \lim_{p\to 0}\frac{p}{p\left(1 + \frac{sq}{\alpha}E[X]\right)} = \frac{1}{1+s}$$

consequently, T tends to an exponential distribution with parameter $1/E[T]$ as p tends to zero.

Exercise 2

(1)

$$Q_{ij}(t) = \int_0^t \exp\Big(-\sum_{\substack{k\neq i\\ k\neq j}} \rho_{ik}\, u\Big)\, \rho_{ij}\, e^{-\rho_{ij} u}\, du = \frac{\rho_{ij}}{\sum\limits_{k\neq i} \rho_{ij}} \Big(1 - \exp\Big(-\sum_{k\neq i} \rho_{ik}\, t\Big)\Big)$$

$$\overline{Q_{ij}}(s) = \frac{\rho_{ij}}{s\Big(s + \sum\limits_{k\neq i} \rho_{ik}\Big)}$$

$$\overline{Q_{ij}}(s) \underset{s\to\infty}{\simeq} \frac{\rho_{ij}}{s^2}.$$

(2) The formula is satisfied for $k = 1$ and $k = 2$. Reason by recurrence and assume the formula is valid for $k - 1$.

Let $\{i_1, i_2, \ldots, i_k\}$ be the set of k components. Assume $\bar{\sigma}(i_m) = i_1$. m is necessarily different from 1 and $\bar{\sigma}(i_1)$ is necessarily different from i_m according to the definition of $\bar{\sigma}$. Interchanging i_1 and $\bar{\sigma}_{i1}$ gives a conclusive result.

(3) From Eqn (48), Ch. 5, we have:

$$\underset{j,l+1}{\overline{P}(s)} = [I - s\overline{Q}(s)]^{-1}_{j,l+1} \Big(\frac{1}{s} - \underset{l+1}{\overline{H}(s)}\Big).$$

As state $l + 1$ is absorbing, it follows that $\underset{l+1}{\bar{H}(s)}$ is zero. Hence the formula. Note that $(I - s\bar{Q}(s)) \to \overset{l+1}{I}$ when $s \to \infty$ hence $|I - s\bar{Q}(s)| \to 1$ when $s \to \infty$.

(4) Expansion of the determinant appearing in the numerator shows that each term is, in absolute value, equal to the product of the components $s\bar{Q}_{ij}(s)$ of a path connecting j to $l + 1$ and of the similar components of cycles formed with states not belonging to the path. The minima terms in $1/s$ of these cycles are obtained when the cycle comprises a single component, since the component then has a value of 1. In order to obtain the minimum degree terms in $1/s$ it is then necessary to select the shortest paths between j and $l + 1$.

(5) The path of minimum size, between j and $l + 1$, cannot therefore contain transitions corresponding to a repair if the system is coherent. When t tends to zero, the system is therefore equivalent to the system constituted by the same components but which is not repairable. For a non-repairable system, the reliability is equal to the availability. Consequently, $A(t)$ and $\tilde{A}(t)$ are equivalent when t tends to zero.

Exercise 3

(1) This set is a set of 2^l critical operating states, where l is the number of components whose state is indeterminate.

(2) Let there be critical operating subsets associated with cut sets C_1 and C_2.
Assume that $C_1 - \{r\} \subset C_2 - \{r\}$. Therefore, $C_1 \subset C_2$ and C_2 is not a minimal cut set.

(3) This is the set of the critical operating subsets associated with cut sets C_i such that:

- card $(C_i) \leqslant k + 1$,
- there are at least $k_i = k + 1 - \text{card}\,(C_i)$ components whose state is indeterminate.

For such a cut set, there are therefore $C_{k'}^{k_i}$ critical operating states having k failed components (k' being the number of components whose state is indeterminate). Note that this algorithm may give identical critical operating states which must be eliminated.

In order to determine these critical operating states, it is therefore necessary to know all the minimal cut sets of order less than or equal to $k + 1$.

Example of a bridge system

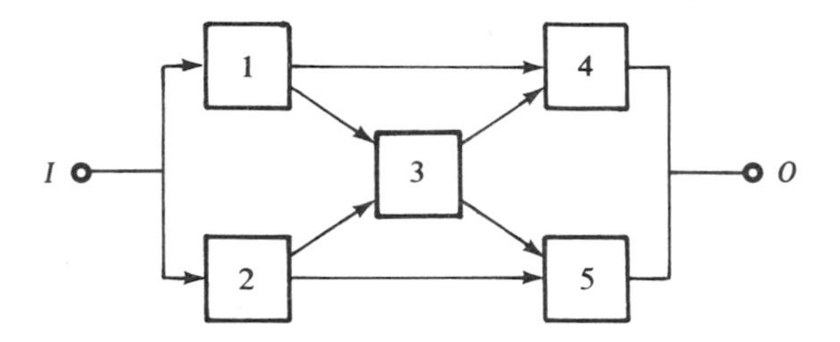

minimal paths: 1-4, 2-5, 1-3-5, 2-3-4
minimal cut sets: 1-2, 4-5, 1-3-5, 2-3-4 .

The algorithm gives the following combinations:

paths	cut sets	critical component	failed components	non-failed components	non-critical components
1-4	1-2	1	2	4	3-5
	4-5	4	5	1	2-3
	1-3-5	1	3-5	4	2
	2-3-4	4	2-3	1	5
2-5	1-2	2	1	5	3-4
	4-5	5	4	2	1-3
	1-3-5	5	1-3	2	4
	2-3-4	2	3-4	5	1
1-3-5	1-2	1	2	3-5	4
	4-5	5	4	1-3	2
	2-3-4	3	2-4	1-5	—
2-3-4	1-2	2	1	3-4	5
	4-5	4	5	2-3	1
	1-3-5	3	1-5	2-4	—

Therefore, the 15 critical operating states (the critical components are ringed) are:

$\bar{1}$ ② 3 4 5 ; $\bar{1}$ ② $\bar{3}$ 4 5 ; ① $\bar{2}$ ③ $\bar{4}$ ⑤ ; ① $\bar{2}$ 3 ④ $\bar{5}$;
① $\bar{2}$ 3 4 5 ; $\bar{1}$ ② 3 $\bar{4}$ ⑤ ; ① $\bar{2}$ $\bar{3}$ ④ 5 ; ① $\bar{2}$ $\bar{3}$ ④ $\bar{5}$;
1 2 3 $\bar{4}$ ⑤ ; $\bar{1}$ 2 $\bar{3}$ 4 ⑤ ; 1 ② $\bar{3}$ $\bar{4}$ 5 ; $\bar{1}$ ② $\bar{3}$ $\bar{4}$ ⑤ ;
1 2 3 ④ $\bar{5}$; $\bar{1}$ ②③④ $\bar{5}$; ① 2 $\bar{3}$ ④ 5 ;

(4) Obvious, since the other components are in an indeterminate state.

(5) Consider the non-failed components of a pseudo-critical opetating state. As the intersection of the necessarily non-failed components is equal to $\bar{P}_i$, it follows that the union of the indeterminate components is equal to the complement of $\bar{P}_i$.

In the case where a pseudo-critical operating state only involves one critical operating subset, $A = 1$. Otherwise, consider the system formed by the components whose state is indeterminate and having as path the paths of the system with the components of $\bar{P}_j$ excluded.

A is equal to the probability of the system operating, i.e. its availability.

(6) If there were a first-order cut set in this subsystem, we could increase $\bar{P}_j$ by one component. For the subsystem to have failed, it is therefore necessary that at least two components have failed. Consequently, if the components are of good quality, $A \simeq 1$.

(7) Let PS_1 and PS_2 be two pseudo-critical operating states. From question 2, the necessarily

failed components of PS_1 are not included in those of $PS_2^{\bullet}$. In order to find a common component it is necessary to select, if possible, the missing components from the components whose state is indeterminate. The reasoning is similar. The common critical operating states, if they exist, therefore belong to a set whose probability has been disregarded by taking $A = 1$. We obtain a good approximation for $\tilde{\Lambda}(t)$ by taking:

$$\tilde{\Lambda}'(t) = \frac{\sum \lambda_r \prod_{l \in P_j} q_l \prod_{m \in C_j - \{r\}} \bar{q}_m}{\text{Availability}}$$

Application: We find 10 pseudo-critical operating states for the bridge system.

① $\bar{2}$ A (4; 3 → 5)

④ $\bar{5}$ A (1; 2 → 3)

① $\bar{3}$ 4 $\bar{5}$

1 $\bar{2}$ $\bar{3}$ ④

$\bar{1}$ ② A (5; 3 → 4)

$\bar{4}$ ⑤ A (2; 1 → 3)

$\bar{1}$ 2 $\bar{3}$ ⑤

② $\bar{3}$ $\bar{4}$ 5

1 $\bar{2}$ ③ $\bar{4}$ 5

$\bar{1}$ 2 ③ 4 $\bar{5}$

(8) $\tilde{\Lambda} = 1.73 \times 10^{-2}\,\text{h}^{-1}$
$\tilde{\Lambda}' = 1.74 \times 10^{-2}\,\text{h}^{-1}$
$\Lambda = 1.68 \times 10^{-2}\,\text{h}^{-1}$

Exercise 4

(1)

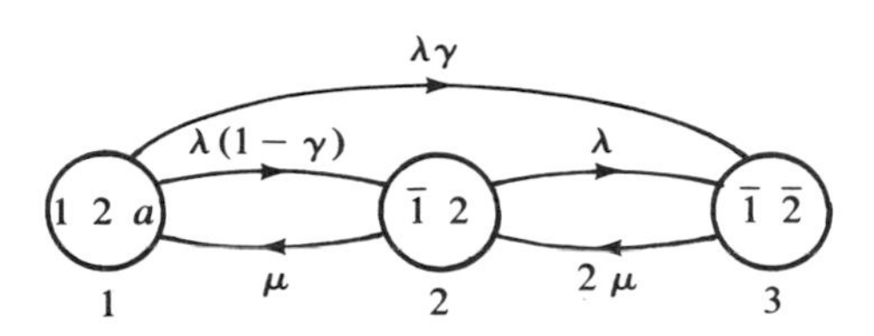

Markov diagram for computing the availability

$$A = \begin{vmatrix} -\lambda & \lambda(1-\gamma) & \lambda\gamma \\ \mu & -(\lambda+\mu) & \lambda \\ 0 & 2\mu & -2\mu \end{vmatrix}$$

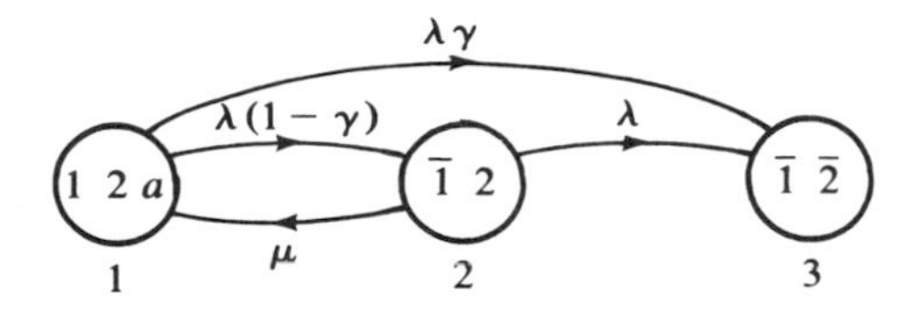

Markov diagram for computing the reliability

$$A_l = \begin{vmatrix} -\lambda & \lambda(1-\gamma) \\ \mu & -(\lambda+\mu) \end{vmatrix}.$$

(2)

$$1 - A_\infty = \frac{\begin{vmatrix} \lambda & -\lambda(1-\gamma) & 0 \\ -\mu & \mu+\lambda & 0 \\ 0 & -2\mu & 1 \end{vmatrix}}{\begin{vmatrix} \lambda & -\lambda(1-\gamma) & 1 \\ -\mu & \mu+\lambda & 1 \\ 0 & -2\mu & 1 \end{vmatrix}} = \frac{\lambda(\lambda+\gamma\mu)}{2\mu^2 + 2\lambda\mu + \lambda^2 + \lambda\gamma\mu}$$

and if $\lambda \ll \mu$ $\quad 1 - A_\infty \simeq \dfrac{\lambda(\lambda+\gamma\mu)}{2\mu(\lambda+\mu)}$.

The eigenvalues β_i of A_l are solutions of:

$$\beta^2 + \beta(\mu + 2\lambda) + \lambda^2 + \lambda\gamma\mu = 0$$

hence the smallest eigenvalue in modulus:

$$\beta_1 = \frac{-\mu - 2\lambda + \sqrt{\mu^2 + 4\lambda\mu(1-\gamma)}}{2} = \frac{-(\mu+2\lambda)}{2}\left[1 - \sqrt{1 - 4\lambda\frac{\lambda+\gamma\mu}{(\mu+2\lambda)^2}}\right]$$

in the case where $\lambda \ll \mu$ and γ is small, we obtain:

$$\beta_1 \simeq -\frac{\mu+2\lambda}{2} \times 2\lambda\frac{\lambda+\gamma\mu}{(\mu+2\lambda)^2} = -\lambda\frac{\lambda+\gamma\mu}{\mu+2\lambda}$$

hence $\Lambda_\infty = -\beta_1 \simeq \dfrac{\lambda(\lambda+\gamma\mu)}{\mu+2\lambda}$.

(3) There are two paths from state 1 to state 3:

hence $\Lambda_\infty \simeq \lambda\gamma + \dfrac{\lambda(1-\gamma)\lambda}{\lambda+\mu} = \dfrac{\lambda(\lambda+\gamma\mu)}{\lambda+\mu}$

and $1 - A_\infty \simeq \frac{\Lambda_\infty}{2\mu} \simeq \frac{\lambda(\lambda + \gamma\mu)}{2\mu(\lambda + \mu)}$.

(4)

$$\mathrm{MTTF}_1 = \frac{\begin{vmatrix} 0 & 1 & 0 \\ 1 & -\lambda & \lambda(1-\gamma) \\ 1 & \mu & -(\lambda+\mu) \end{vmatrix}}{\begin{vmatrix} -\lambda & \lambda(1-\gamma) \\ \mu & -(\lambda+\mu) \end{vmatrix}} = \frac{\mu + 2\lambda - \lambda\gamma}{\lambda(\lambda + \gamma\mu)}$$

We can check that Λ_∞ is a good approximation to $1/\mathrm{MTTF}_1$

$$\mathrm{MTTF}_2 = \frac{\begin{vmatrix} 0 & 0 & 1 \\ 1 & -\lambda & \lambda(1-\gamma) \\ 1 & \mu & -(\lambda+\mu) \end{vmatrix}}{\begin{vmatrix} -\lambda & \lambda(1-\gamma) \\ \mu & -(\lambda+\mu) \end{vmatrix}} = \frac{\lambda + \mu}{\lambda(\lambda + \gamma\mu)}$$

(5) Only state 2 communicates with the failed states.
Therefore $\tilde{\Lambda}_\infty = \frac{1}{\mathrm{MTTF}_2} = \frac{\lambda(\lambda + \gamma\mu)}{\lambda + \mu}$.
We can verify, moreover, that

$$\tilde{\Lambda}_\infty = \frac{\lambda\gamma P_1 + \lambda P_2}{P_1 + P_2} = \frac{\begin{vmatrix} -\lambda & +\lambda(1-\gamma) & \lambda\gamma \\ +\mu & -(\lambda+\mu) & \lambda \\ 0 & 2\mu & 0 \end{vmatrix}}{\begin{vmatrix} -\lambda & \lambda(1-\gamma) & 1 \\ \mu & -(\lambda+\mu) & 1 \\ 0 & 2\mu & 0 \end{vmatrix}} = \frac{\lambda(\lambda + \gamma\mu)}{\lambda + \mu}$$

(6) Availability computation

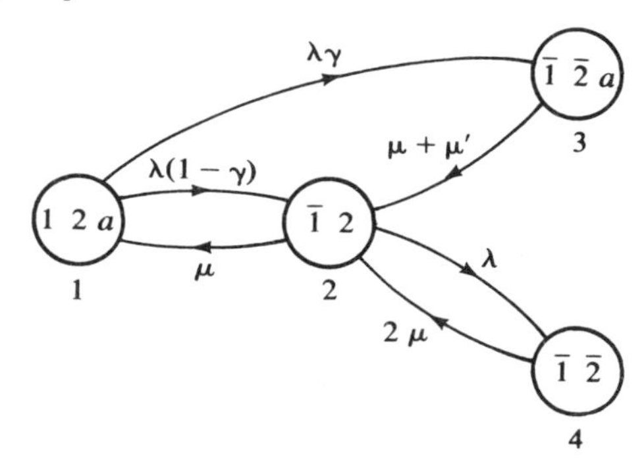

$$A = \begin{vmatrix} -\lambda\gamma & \lambda(1-\gamma) & \lambda\gamma & 0 \\ \mu & -(\lambda+\mu) & 0 & \lambda \\ 0 & \mu + \mu' & -(\mu+\mu') & 0 \\ 0 & 2\mu & 0 & -2\mu \end{vmatrix}$$

$$1 - A_\infty = \frac{\begin{vmatrix} -\lambda & \lambda(1-\gamma) & \lambda\gamma & 0 \\ \mu & -(\lambda+\mu) & 0 & 0 \\ 0 & \mu+\mu' & -(\mu+\mu') & 0 \\ 0 & 2\mu & 0 & 1 \end{vmatrix}}{\begin{vmatrix} -\lambda & \lambda(1-\gamma) & \lambda\gamma & 1 \\ \mu & -(\lambda+\mu) & 0 & 1 \\ 0 & \mu+\mu' & -(\mu+\mu') & 1 \\ 0 & 2\mu & 0 & 1 \end{vmatrix}} = \frac{\lambda^2(1-\gamma)}{2\mu(\gamma+\mu)} + \frac{\lambda\gamma}{\mu+\mu'} + \frac{\lambda\gamma\lambda(\mu'-\mu)}{2\mu(\lambda+\mu)(\mu+\mu')}$$

The graph for the availability computation is unaltered. The matrix A_l is therefore unchanged; therefore Λ_∞, MTTF_1 and MTTF_2 retain their values. The same applies to $\tilde{\Lambda}_\infty$.

Approximation by the method in section 2.2.2.

$$\Lambda_\infty \simeq \lambda\gamma + \frac{\lambda(1-\gamma)\lambda}{\lambda+\mu}$$

$$1 - A_\infty \simeq \frac{\lambda\gamma}{\mu+\mu'} + \frac{\lambda^2(1-\gamma)}{(\lambda+\mu)2\mu}$$

Exercise 5

(1) $$\text{MTTF} = \int_0^\infty R(u)\,du = \lim_{t\to\infty} \int_0^t R(u)\,du = \lim_{t\to\infty} \int_0^t R(u)\,du + \frac{R(t)}{\Lambda_\infty}$$

since $R(t) \to 0$ when $t \to \infty$

$$\text{MTTF} - \rho_1 = \int_{t_0}^\infty R(t)\,dt - \frac{R(t_0)}{\Lambda_\infty} = R(t_0)\left(\int_{t_0}^\infty \frac{R(t)}{R(t_0)}\,dt - \frac{1}{\Lambda_\infty}\right)$$

hence, according to the hypothesis

$$\int_{t_0}^\infty e^{-\Lambda_\infty(1+\varepsilon)(t-t_0)}\,dt - \frac{1}{\Lambda_\infty} \le \text{MTTF} - \rho_1 \le \int_{t_0}^\infty e^{-\Lambda_\infty(1-\varepsilon)(t-t_0)}\,dt - \frac{1}{\Lambda_\infty}$$

$$\frac{1}{\Lambda_\infty(1+\varepsilon)} - \frac{1}{\Lambda_\infty} \le \text{MTTF} - \rho_1 \le \frac{1}{\Lambda_\infty(1-\varepsilon)} - \frac{1}{\Lambda_\infty}$$

$$|\text{MTTF} - \rho_1| \le \frac{\varepsilon}{\Lambda_\infty(1-\varepsilon)}$$

$\int_0^{t_0} R(t)\,dt \leqslant t_0$; consequently, if t_0 is small compared with $1/\Lambda_\infty$, P_1 is approximately $R(t_0)/\Lambda_\infty$. As this system is of good quality, it follows that 1/MTTF is a good approximation for Λ_∞ if, in the initial state, no component has failed.

(2) $$E[T^2] = \int_0^\infty t^2 R(t)\Lambda(t)\,dt = 2\int_0^\infty tR(t)\,dt$$

giving the limit required since $R(t) \to 0$ when $t \to \infty$

$$E[T^2] - \rho_2 = 2\int_{t_0}^\infty tR(t)\,dt - \frac{2\,t_0\,R(t_0)}{\Lambda_\infty} - \frac{2\,R(t_0)}{\Lambda_\infty^2}$$

hence:

$$E[T^2] - \rho_2 \le 2\,R(t_0)\int_{t_0}^\infty t\,e^{-\Lambda_\infty(1-\varepsilon)(t-t_0)}\,dt - \frac{2\,t_0\,R(t_0)}{\Lambda_\infty} - \frac{2\,R(t_0)}{\Lambda_\infty^2}$$

putting $t - t_0 = u$

$$E[T^2] - \rho_2 \leq 2\,R(t_0) \int_0^\infty (u + t_0)\,e^{-\Lambda_\infty(1-\varepsilon)u}\,du - \frac{2\,t_0\,R(t_0)}{\Lambda_\infty} - \frac{2\,R(t_0)}{\Lambda_\infty^2}$$

$$\leq \frac{2\,R(t_0)}{\Lambda_\infty^2(1-\varepsilon)^2} - \frac{2\,R(t_0)}{\Lambda_\infty^2} + \frac{2\,t_0\,R(t_0)}{\Lambda_\infty(1-\varepsilon)} - \frac{2\,t_0\,R(t_0)}{\Lambda_\infty}$$

$$\leq \frac{2\,R(t_0)\,\varepsilon(2-\varepsilon)}{\Lambda_\infty^2(1-\varepsilon)^2} + \frac{2\,t_0\,R(t_0)\,\varepsilon}{\Lambda_\infty(1-\varepsilon)}$$

$$E[T^2] - \rho_2 \leq \frac{2\,\varepsilon\,R(t_0)}{\Lambda_\infty(1-\varepsilon)}\left(t_0 + \frac{2}{\Lambda_\infty(1-\varepsilon)}\right)$$

analysis of the lower bound gives the desired result.

(3) We see that MTTF $\simeq 1/\Lambda_\infty$, consequently the relative error on the MTTF is of the order of ε. We can therefore calculate ρ_1 by integrating the reliability, as convergence is obtained for t_0, which is of the order of a few times the largest repair time of the system components.

Similarly, we see that we can use as an approximation for the variance

$$\sigma^2 = \rho_2 - \rho_1^2.$$

In practice, we obtain a value very near to $1/\Lambda_\infty^2$ (the distribution approximates to an exponential distribution with parameter Λ_∞).

REFERENCES

BARLOW R. E. and PROSCHAN F. (1975): Importance of system components and fault tree events; *Stochastic Processes and their Applications*, **3**, pp. 153–173.

BATTS J. R., BEADLES R. L. and NELSON A. C. (1970): A computer program for approximating system reliability: *IEEE Transactions on Reliability*, **R19**, no. 2, pp. 61–90.

BERGE C. (1968): *Principles de combinatoire*; Dunod, Paris.

BIRNBAUM Z. W. (1969): On the importance of different components and a multicomponent system; In: Multivariate Analysis II, P. R. KRISHNAIAH (Ed.); Academic Press, New York.

CALDAROLA L. and WICKENHAUSER A. (1977): The Karlsruhe computer program for the evaluation of the availability and reliability of complex repairable systems; *Nuclear Engineering and Design*, **43**, pp. 463–470.

GONDRAN M. (1975): Fiabilité des grands systèmes réparables; *Bulletin de la Direction des Etudes et Recherches, d'Electricité de France*, Series C, no. 2.

GONDRAN M. and LALEUF J. C. (1981): Indisponibilité des systèmes à taux de défaillance et à temps de réparation constants; *Bulletin de la Direction des Etudes et Recherches, d'Electricité de France*, Series C, no. 1, pp. 19–26.

LALEUF J. C. (1983): Calcul de la fiabilité des systèmes; Cours de *l'Ecole Nationale de la Statistique et de l'Administration Economique*, Paris.

LAMBERT H. E. (1975): *Fault Trees for Decision Making in Systems Analysis*; PhD Thesis, Lawrence Livermore Laboratory.

MULET MARQUIS D. and DUBREUIL-CHAMBARDEL A. (1984): Une ensemble de logiciels pour l'evaluation qualitative et quantitative de la fiabilité d'un système; Note EDF HI 4737-02.

MURCHLAND J. (1973): *Fundamental Probability Relations for Repairable Items*; NATO Advanced Study Institute on Generic Techniques in Systems Reliability Assessment, University of Liverpool.

PAGES A. (1976): *Calcul de la fiabilité et de la disponibilité des systèmes réparables dont les pannes de faible durée sont admissibles*; EDF Note, no. HI 2203/02.

PAGES A. (1977): *Calcul de la fiabilité des systèmes non-markoviens réparables*; EDF Note, no. HI 2391/02.

PARODI M. (1959): *La localisation des valeurs caracteristiques des matrices et ses applications*; Gauthier-Villars, Paris.

SINGH C. (1972): *Reliability Modelling and Evaluation in Electric Power Systems*; PhD Thesis, University of Saskatchewan, Saskatchewan, Canada.

VARGA R. S. (1962): *Matrix Iterative Analysis*; Prentice-Hall, U.S.A.

VESELY W. E. (1970): A time dependant methodology for fault tree evaluation; *Nuclear Engineering and Design*, **13**, pp. 337–360.

CHAPTER 7

STATISTICAL TREATMENT OF RELIABILITY DATA

The computational methods discussed in the previous chapters are only useful insofar as it is possible to obtain values for the parameters involved (failure rate, repair rate, etc.). Although there are areas for which a considerable amount of good-quality data is available (e.g. electronic components), in most cases data are scarce or even non-existent. This being so, it is necessary to perform reliability trials or to observe identical equipment in the field in order to gather data.

In the first section, we shall examine the problem of estimating parameters from a sample. This section basically consists of a review of statistical methods. For a more detailed analysis, the reader is invited to consult a statistics textbook (e.g. [Fourgeaud and Fuchs, 1972; Hoel, 1971]).

In the second section, we shall see how to obtain data from trials and subsequently, in section 3, from in-service results.

Sections 4 and 5 are devoted to statistical processing of raw data (classical estimation of the parameters of the main distributions used in reliability analysis, and hypothesis testing).

Section 6 introduces Bayesian methods of processing data.

Whatever method is employed for evaluating a parameter, our knowledge of it is not perfect; in section 7, we shall examine the propagation of this uncertainty in the course of reliability computations.

1. THE PROBLEM OF ESTIMATION

1.1. Point estimation

When performing trials in order to estimate the parameters $\theta = (\theta_1, \theta_2, \ldots, \theta_p)$ of a probability distribution (or of a function of these parameters), we wish to determine a value $\hat{\theta}$ which is reasonably close to the unknown value θ. In order to produce the *estimate* $\hat{\theta}$, we use an *estimator* (or *statistic*) which is a random variable S, an *a priori* function of the values observed during the trial.

For a given trial, we obtain a particular value $\hat{\theta}$ assumed by this function.

In order to simplify the notation, let us assume that we are attempting to evaluate the single parameter θ of a probability distribution $f(t, \theta)$; θ could be, for example, the failure rate λ of an exponential distribution $f(t, \lambda) = \lambda e^{-\lambda t}$. The choice of estimator must be based on its particular properties. We shall now examine some useful estimator properties.

1.2. Estimator bias

An estimator S is termed *unbiased* if

$$E(S) = \theta. \tag{1}$$

The mean is taken over all possible samples; the difference $E(S) - \theta$ is called the bias.

Consider, for example, the estimator M known as the sample mean and defined by:

$$M = \sum_{i=1}^{n} \frac{t_i}{n} \quad (n \text{ being the sample size}).$$

Clearly, $E(M)$ is equal to the unknown value μ of the mean.

It can be shown that an unbiased estimator of the sample variance is:

$$\sigma^2 = \frac{\sum_{i=1}^{n} (t_i - M)^2}{n - 1} \tag{2}$$

(see Exercise 1).

1.3. Estimator consistency

An estimator is termed *consistent* if it converges in probability on the parameter to be estimated when the sample size increases indefinitely.

i.e.:

$$\forall \varepsilon > 0 \qquad \mathcal{P}[| S - \theta | < \varepsilon] \to 1 \qquad \text{when } n \to \infty.$$

In this case, the larger the sample, the closer the estimated value $\hat{\theta}$ will approximate in probability to the true value.

The mean of a consistent estimator therefore tends to the true value of the parameter as the sample size increases indefinitely. A biased consistent estimator is therefore asymptotically unbiased.

Example: Empirical mean of a sample from a distribution whose variance σ^2 is finite.

The central limit theorem (cf. Ch. 1, section 4.2.3) indicates that the empirical mean tends to a normal distribution $N(\mu, \sigma/\sqrt{n})$ as the size n of the sample increases indefinitely. In this case, the empirical mean is therefore a consistent estimator of the mean.

1.4. Estimator efficiency

An estimator is said to be *efficient* if it is consistent, unbiased and of *minimum variance*.

Let us assume that the range of variation of t does not depend on the parameter θ. We can then demonstrate the following inequality, known as the Frechet-Darmois inequality

$$V(S) \geqslant \frac{1}{I(\theta)} \tag{3}$$

with

$$I(\theta) = -nE\left(\frac{\partial^2 \log f(t, \theta)}{\partial \theta^2}\right). \tag{4}$$

In the case of the exponential distribution $\lambda e^{-\lambda t}$ we obtain:

$$\log f(t, \lambda) = \log \lambda - \lambda t$$

$$\frac{\partial^2 \log f(t, \lambda)}{\partial^2 \lambda} = \frac{-1}{\lambda^2}$$

$$I(\lambda) = \frac{n}{\lambda^2}$$

and

$$V(S) \geqslant \frac{\lambda^2}{n}.$$

This inequality shows that the estimation always involves a random error, whatever the sample size.

This inequality can be extended to the general case where r parameters $\theta_1, \theta_2, \ldots, \theta_r$ are to be evaluated. We shall call S_i their respective estimators and I the *information matrix* defined by:

$$I_{ij} = -nE\left(\frac{\partial^2 \log (f(t, \theta_1, \theta_2, \ldots, \theta_r))}{\partial \theta_i \, \partial \theta_j}\right) \quad \begin{array}{l} i = 1, \ldots, r \\ j = 1, \ldots, r. \end{array}$$

If the estimators S_i are unbiased and the range of variation of t does not depend on the parameters θ_i, denoting the positive semi-definite matrix of the variances-covariances by V, Eqn (3) can therefore be written as:

$$V - I^{-1} \text{ is positive semi-definite.} \tag{5}$$

Let us return to the case where a single parameter is to be estimated.

Given two unbiased estimators S_1 and S_2 of the same parameter, the ratio $\sigma^2_{S_2}/\sigma^2_{S_1}$ of the variances of the two estimators is called the *relative efficiency* of estimator S_1 with respect to estimator S_2.

1.5. Estimator sufficiency

An estimator S is termed *sufficient* if, given any estimator S', the conditional distribution of S', given S, does not depend on θ. All sampling information on the parameter θ is contained in the value of the estimator S. There is no sufficient estimator for all the probability distributions.

Estimators possessing all these properties rarely exist. Generally speaking, we have to make do with estimators obtained by general methods and possessing only some of these properties. Computational simplicity also plays a large part in the choice of estimator.

We shall now examine two general methods of determining estimators, the maximum likelihood method and the method of moments, the latter only being used when the computations arising from the first method are too complicated.

1.6. Maximum likelihood method

1.6.1. *Definitions*

The *likelihood* of a sample is the probability density of that sample.

For example, in the case of a sample of n independent observations t_i whose probability density is $f(t, \theta)(\theta \in \Omega)$, the likelihood $\mathscr{L}$ is written as:

$$\mathscr{L}(t_1, t_2, \ldots, t_n, \theta) = \prod_{i=1}^{n} f(t_i, \theta) \tag{6}$$

with $\theta = (\theta_1, \theta_2, \ldots, \theta_r)$.

The *maximum likelihood* method consists of taking $\hat{\theta}$ as an estimate such that

$$\mathscr{L}(t, \hat{\theta}) \geqslant \mathscr{L}(t, \theta) \qquad \forall \theta \in \Omega .$$

If there is a unique optimum value over Ω and if $\mathscr{L}$ is continuously differentiable, it follows that at the optimum

$$\frac{\partial \mathscr{L}}{\partial \theta_j}(t, \hat{\theta}) = 0 \qquad \forall j = 1, r. \tag{7}$$

When the function $\mathscr{L}$ is convex, this necessary condition is also sufficient.

Note that Eqns (7) are equivalent to:

$$\frac{\partial \log \mathscr{L}(t, \hat{\theta})}{\partial \theta_j} = 0 \qquad \forall j = 1, \ldots, r. \tag{8}$$

In some cases, it is possible to solve the system of equations (7) analytically. In others, a non-linear programming method is called for.

1.6.2. *Example*

Consider a sample of n independent components whose uptime distribution is exponential with parameter λ. Let $t_1, t_2, \ldots, t_n$ be the uptime before failure of each of the n components.

The likelihood of the sample is written as:

$$\mathscr{L}(t_1, t_2, \ldots, t_n, \lambda) = \prod_{i=1}^{n} (\lambda \, e^{-\lambda t_i}) = \lambda^n \, e^{-\lambda T}$$

with $T = \sum_{i=1}^{n} t_i$.

Hence:

$$\log \mathcal{L} = n \log \lambda - \lambda T \qquad \lambda > 0$$

$$\frac{\partial \log \mathcal{L}}{\partial \lambda} = \frac{n}{\lambda} - T$$

$$\frac{\partial^2 \log \mathcal{L}}{\partial \lambda^2} = \frac{-n}{\lambda^2} < 0\,.$$

Consequently, $\hat{\lambda} = n/T$ is the estimator of λ by the maximum likelihood method, since the sign of the second derivative indicates that the extremum is a maximum.

1.6.3. *Properties of estimators obtained using the maximum likelihood method*

Subject to some very commonly satisfied hypotheses, we can show that the estimator obtained in this way has the following properties:

—it is consistent (but may be biased),
—if a sufficient estimator S of θ exists, the maximum likelihood equation has a unique solution $\hat{\theta}$,
—it is *asymptotically* normal and efficient.

The asymptotic variance-covariance matrix is equal to the inverse of the information matrix.
Equation (4) can be written:

$$I(\theta) = -E\left(\frac{\partial^2 \log \mathcal{L}(t, \theta)}{\partial \theta^2}\right).$$

When the sample becomes sufficiently large, we can take the following matrix as an *approximation* for $I(\theta)$:

$$\bar{I}(\theta) = -E\left(\frac{\partial^2 \log \mathcal{L}(t, \hat{\theta})}{\partial \theta^2}\right) \qquad (9)$$

Therefore, using the asymptotic properties, we obtain an approximation for the variance-covariance matrix V by taking $\bar{I}^{-1}(\theta)$.

Example: In the case of the exponential distribution, the variance-covariance matrix V reduces to a scalar quantity $V(S)$ and, as we have seen, $V(S)$ is greater than or equal to λ^2/n. For comparatively large n, we can take:

$$V(S) = \frac{\lambda^2}{n}\,.$$

If h is a function possessing a reciprocal over Ω, $h(S)$ is a maximum likelihood estimator for the parameter $h(\theta)$.

1.7. Method of moments

This method generally produces estimators with fewer properties than those obtained using the maximum likelihood method. It is therefore only used when the other involves very complex calculations.

This method simply consists of finding the number of equations necessary for determining the parameters by equating the sampling moments with the moments which are functions of the unknown parameters.

Gamma distribution example (cf. Ch. 1, section 4.2)

Assume that $f(t, \theta_1, \theta_2) = \dfrac{\lambda^{\beta} t^{\beta-1} e^{-\lambda t}}{\Gamma(\beta)}$ $\quad (\theta_1 = \lambda \quad \theta_2 = \beta) \quad \begin{matrix} \lambda > 0 \\ \beta > 0 \\ t \geqslant 0 \end{matrix}$

Remembering that
$$E[t] = \frac{\beta}{\lambda}$$
$$\sigma^2[t] = \frac{\beta}{\lambda^2}.$$

Applying the method of moments we can therefore write:

$$\bar{t} = \frac{\sum_{i=1}^{n} t_i}{n} = \frac{\hat{\beta}}{\hat{\lambda}}$$

$$\sigma^2 = \sum_{i=1}^{n} \frac{(t_i - \bar{t})^2}{n} = \frac{\hat{\beta}}{\hat{\lambda}^2}$$

hence
$$\hat{\lambda} = \frac{\bar{t}}{\sigma^2}, \quad \hat{\beta} = \frac{\bar{t}^2}{\sigma^2}.$$

1.8. Estimation by confidence interval

The point estimate $\hat{\theta}$ of θ (in the set Ω) may differ appreciably from the true value, especially when the sample size is small. In this case, it is preferable to determine a random region of Ω such that there is a given *a priori* probability α of θ belonging to this region.

In the case of a unique parameter, this region will be a random interval $[\theta_-, \theta_+]$ such that θ_- and θ_+ are the order α_1 and $1 - \alpha_2$ quantiles respectively of the distribution of S for given θ ($\alpha_1 + \alpha_2 = \alpha$). If the respective values of α_1 and α_2 are not specified, it is advisable to select $\alpha_1 = \alpha_2 = \alpha/2$.

θ_+ is the upper confidence limit and θ_- the lower confidence limit; the interval $[\theta_-, \theta_+]$ is called the confidence interval at $100(1 - \alpha_1 - \alpha_2)\%$.

After determining such an interval, if we state that θ belongs to this interval, there is a probability α of being wrong. On the other hand, it would be inappropriate to say that θ belongs to this interval with probability $1 - \alpha$; in fact, as θ is not random, it either belongs or does not belong to the interval computed.

Example: Confidence interval of the mean m of a normal distribution of known standard deviation.

The estimator of m is the empirical mean M. The distribution of this estimator is a normal distribution $N(m, \sigma/\sqrt{n})$.

Consequently:

$$m_{+} = m + u_{1-\alpha_2}\frac{\sigma_0}{\sqrt{n}}$$

$$m_{-} = m + u_{\alpha_1}\frac{\sigma_0}{\sqrt{n}} = m - u_{1-\alpha_1}\frac{\sigma_0}{\sqrt{n}}$$

where u_α is the order α quantile of the reduced normal distribution which is tabulated (cf. Appendix 1).

A general method of determining confidence intervals is given in Exercise 2. We can usually obtain a good approximation for a confidence interval by using an asymptotic property of the information matrix.

Let us assume, in fact, that θ belongs to R^k. It can be shown (cf. [Fourgeaud and Fuchs, 1972]) that the region $\mathscr{D}_\alpha$ defined by

$$\mathscr{D}_\alpha = \{ u \in R^k : u' \bar{I}(\theta)\, u \leq \chi^2_\alpha(k) \}$$

is a confidence region for θ at the $1 - \alpha$ level, asymptotically. ($\bar{I}(\theta)$ is given by Eqn (9).)

For $k = 2$, an ellipse is obtained.

2. RELIABILITY TRIALS

2.1. Trial designs

Parameter estimation methods naturally require samples. Reliability trials are used to provide these samples. Of the large number of possible trials, we shall confine ourselves to describing those most frequently employed. The following notation will be used:

n: number of components or different equipments under test.
M: trial designs in which the failed components are not replaced.
V: trial designs in which the failed components are immediately replaced by new components.
r: number of failures which we wish to observe *a priori* during the trial. These designs, in which the criterion for terminating the observations is a number of failures, are often known as type II.
T: *a priori* duration of the trial. These designs, in which the termination criterion is a trial duration, are often called type I.

The simplest designs can then be denoted as follows:

$[n, M]$: uncensured trials, continued until n components have failed. These designs provide so-called complete samples.
$[n, M, T]$ and $[n, V, T]$: censured type I trial designs.
$[n, M, r]$ and $[n, V, r]$: censured type II trial designs.
$[n, M, (r, T)]$ and $[n, V, (r, T)]$: composite designs whereby observations continue until the rth failure if the latter takes place at time $t_r < T$, or until time T otherwise.
There are also progressive designs whereby the decision to terminate observations

depends on the results already obtained; these designs are mainly used to check the value of parameters obtained elsewhere (cf. [Peyrache and Schwab, 1969]).

In the following section, we shall also use multicensured or progressively censured designs. Full details of trial designs used in reliability analysis are given in [Peyrache and Schwab, 1969; Beliav et al., 1972].

If the reliability parameters depend on external parameters (temperature, vibration, etc.), the operating conditions must be identical for all components. By repeating the trial to take account of various stresses, it is possible to determine the effect of environmental influences.

2.2. Determining the likelihood of samples obtained from trial designs

We shall assume that the items under test are independent.

In this case, if we call $F(t, \theta)$ the distribution function of the density $f(t, \theta)$ whose parameters we wish to evaluate, the following results are obtained:

Design type	Likelihood $\mathscr{L}(t_i, \theta)$ proportional to	
$[n, M]$	$\prod_{i=1}^{n} f(t_i, \theta)$	(10)
$[n, M, T]$	r = number of failures observed $\left(\prod_{i=1}^{r} f(t_i, \theta)\right)(1 - F(T, \theta))^{n-r}$	(11)
$[n, V, T]$	$\prod_{i=1}^{n}\left(\prod_{j=1}^{n_i} f(t_i^j - t_i^{j-1}, \theta)(1 - F(T - t_i^{n_i}, \theta))\right)$ component i has failed at times $t_i^j, j = 1, \ldots, n_i\ (t_i^0 = 0)$	(12)
$[n, M, r]$	$\left(\prod_{i=1}^{r} f(t_i, \theta)\right)(1 - F(t_r, \theta))^{n-r}$ t_r = time of last failure	(13)
$[n, V, r]$	$\prod_{i=1}^{n}\left(\prod_{j=1}^{n_i} f(t_i^j - t_i^{j-1}, \theta)(1 - F(T_i - t_i^{n_i}, \theta))\right)$ component i has failed at times $t_i^j, j = 1, \ldots, n_i (t_i^0 = 0)$ and t_r is the time of the rth failure.	(14)

The likelihood of the composite designs is equal to that of the corresponding type T or r designs, depending on the nature of the termination.

3. USING IN-SERVICE EQUIPMENT STATISTICS

3.1. Objectives

For technical or financial reasons, it is not always possible to carry out reliability trials. This being so, we can try to work out statistics characterizing the operation of the components whose reliability parameters we wish to determine. However, in this case the equipments are not always exactly identical and the operating conditions may differ markedly. It is therefore necessary to take special precautions if we suspect that the required parameters are very sensitive to certain external causes and to carry out a sensitivity analysis if necessary and if the number of observations permits.

On the other hand, the operating times or the number of stresses are generally different for each piece of equipment, with the result that the designs envisaged in the previous sections cannot be used. However, in most cases we can equate the operation of the equipment to a *multicensured* (or *progressively censured*) trial design.

3.2. Multicensured trial designs

There are two types of multicensured trials: type I, for which the censure times T_i of each of the components are determined in advance, and type II in which termination of the trial is determined by the number of failures. The failed components may or may not be replaced.

In the case of repairable systems, we generally use type I multicensured designs.

In the rest of this chapter, we shall simply call these plans 'multicensured' or '*MC*' to simplify the notation. We shall denote by T_i the time after which component i is censured.

The likelihood of the samples obtained can therefore be written as:

Design type	Likelihood proportional to	
$[n, M, MC]$	$\prod_{i=1}^{r} f(t_i, \theta) \prod_{i=r+1}^{n} (1 - F(T_i, \theta))$ the first r components failed at times $t_i < T_i$	(15)
$[n, V, MC]$	$\prod_{i=1}^{n} \left(\prod_{j=1}^{n_i} f(t_i^j - t_i^{j-1}, \theta)(1 - F(t_r - t^{n_i}, \theta))\right)$ component i failed at times $t_i^j, j = 1, \ldots, n_i (t_i^0 = 0)$	(16)

3.3. Equating in-service operation to a multicensured trial

The censure times T_i are merely equated with the effective operating times at the moment of sampling.

In the case of repairable components, we can only equate in-service operation to a trial design in which the failed components are replaced by new ones. In this case, we often have to assume that the repair is equivalent to a renewal, an assumption which is not always valid. Of course, it would be possible not to take account of components after they have been repaired. However, this would deprive us of a fairly large portion of the information. Note also that certain potential failures may be detected during inspection or preventive maintenance, thereby invalidating the statistics.

On the other hand, the equipment is observed under actual operating conditions, which is not always the case during trials. If the environmental conditions vary too greatly for the different equipments monitored in service, it may be necessary to categorize the components, with each component in the same category subject to similar stresses. Provided the number of components monitored is sufficiently high, it will then be possible to examine the variation of the reliability parameters as a function of the environment using *data analysis techniques* (cf., for example [Ligeron, 1972]).

Another advantage of using in-service statistics is that the precise unavailability time due to a failure can be observed. In fact, this time is made up of:

— the failure detection time,
— the repair delay time,
— the fault correction time proper,
— the restoration time.

These various parameters are often difficult to evaluate from trials.

4. CLASSICAL METHODS OF EVALUATING THE PARAMETERS OF COMMON DISTRIBUTIONS

The adjective 'classical' is used here in contradistinction to Bayesian methods which will be discussed in Ch. 6.

We shall examine the following distributions:

— exponential,
— Weibull,
— normal,
— lognormal,
— gamma,
— binomial.

4.1. Distribution selection

Detailed analysis of the physical phenomena surrounding a failure generally enable us to select an *a priori* distribution. When this is impossible, selection can be based on the following characteristics:

— the coefficient of variation v which is the quotient of the standard deviation σ divided by the mean $m: v = \sigma/m$. (This is one in the case of an exponential distribution.)

— the Fisher coefficients which are independent of the origin and have no units:

$$\text{(a)}\ \beta_1 = \frac{\mu_3^2}{\mu_2^3} \quad \text{and} \quad \text{(b)}\ \beta_2 = \frac{\mu_4}{\mu_2^2}. \tag{17}$$

$\sqrt{\beta_1}$ is called the coefficient of skewness and $\beta_2 - 3$ is known as the coefficient of kurtosis.

The skewness and kurtosis coefficients of the normal distribution are zero. The ratio of the skewness coefficient to the coefficient of variation of the lognormal distribution is equal to $3 + v^2$. Consequently, when the coefficient v is small, this ratio is approximately 3. This ratio is exactly 2 in the case of the gamma distribution.

4.2. Estimation of the exponential distribution parameter

The exponential distribution plays an important part in reliability analysis, both in terms of theory and applications. Lack of data often prevents us from being able to make an effective choice from among the various distributions and it is then advisable to consider the exponential. Consequently, the following section devoted to the exponential is particularly detailed.

In most texts, the parameter estimated is the mean time to failure θ such that:

$$f(t, \theta) = \frac{1}{\theta} \exp\left(-\frac{t}{\theta}\right). \tag{18}$$

In order to bring in the failure rate λ, we prefer to write the probability density in the form:

$$f(t, \lambda) = \lambda \, e^{-\lambda t}. \tag{19}$$

4.2.1. [*n, V, T*] *type designs*

Applying expression (12) enables us to compute the sample likelihood:

$$\mathcal{L}(t, \lambda) = \prod_{i=1}^{n} \lambda^{n_i} \, e^{-\lambda T}.$$

Or, denoting the number of failures observed over the interval $[0, T]$ by $r(T) = \Sigma n_i$:

$$\mathcal{L}(t, \lambda) = \lambda^{r(T)} e^{-\lambda n T}.$$

Whence immediately follows the maximum likelihood estimate $\hat{\lambda}$.

$$\hat{\lambda} = \frac{r(T)}{nT}. \tag{20}$$

As the components are independent and renewed, the failure process is Poissonian with parameter $n\lambda$, hence:

$$\mathcal{P}(r(T) = r) = \frac{(\lambda n T)^r}{r\,!} e^{-\lambda n T}. \tag{21}$$

Consequently:

$$E[\hat{\lambda}] = \sum_{r=0}^{\infty} \frac{r}{nT} \frac{(\lambda n T)^r}{r\,!} e^{-\lambda n T} = \lambda. \tag{22}$$

Estimator (20) is therefore unbiased.

Equation (21) also allows it to be shown that:

$$\sigma^2(\hat{\lambda}) = \frac{\lambda}{nT}. \tag{23}$$

It can also be shown that the estimator is efficient.

Note that the sufficient statistic is simply the number of failures $r(T)$, the failure times yielding no information.

Equation (21) also enables a confidence interval $[\lambda_-, \lambda_+]$ to be determined.

In fact:

$$\mathcal{P}(r(T) \leq r) = \sum_{k=0}^{r} \frac{1}{k!} (\lambda n T)^k \, e^{-\lambda n T} = \int_{\lambda n T}^{\infty} \frac{z^r}{r!} e^{-z} \, dz .$$

From the relationship existing between the incomplete gamma function and the chi-squared distribution [Johnson and Kotz, 1970], it follows that:

$$\mathcal{P}(r(T) \leq r) = \mathcal{P}(\chi^2(2\,r + 2) > 2\,\lambda n T) .$$

Similarly:

$$\mathcal{P}(r(T) > r) = \mathcal{P}(\chi^2(2\,r) < 2\,\lambda n T) .$$

This therefore gives the bounds of the confidence interval ($\alpha_1 = \alpha_2 = \alpha/2$):

$$\left.\begin{aligned} \lambda_- &= \frac{\chi^2_{1-\alpha/2}(2\,r)}{2\,nT} && (a) \\ \lambda_+ &= \frac{\chi^2_{\alpha/2}(2\,r + 2)}{2\,nT} && (b) \end{aligned}\right\} \qquad (24)$$

Note that λ_+ is definite even when $r(T) = 0$.

The values of $\chi^2_\alpha(v)$ will be found in Appendix 2.

Example: Consider the design $N = 100$, $T = 100$ h. The lifetimes of components with a failure rate λ of 0.001 h^{-1} were drawn at random. We obtained $r = 9$ failures, therefore:

$$\hat{\lambda} = 0.9\,10^{-3}\,h^{-1}$$

$$\lambda_+ = \frac{\chi^2_{0.05}(20)}{2 \cdot 10\,000} = 1.57\,10^{-3}\,h^{-1}$$

$$\lambda_- = \frac{\chi^2_{0.95}(18)}{2 \cdot 10\,000} = 0.47\,10^{-3}\,h^{-1}$$

4.2.2. [*n*, *V*, *r*] *type designs*

Expression (14) shows that the sample likelihood is proportional to:

$$\lambda^r \, e^{-\lambda n t_r} .$$

Hence the maximum likelihood estimate $\hat{\lambda}$ is:

$$\hat{\lambda} = \frac{r}{n t_r} . \qquad (25)$$

Computation of the mathematical expectation of $\hat{\lambda}$ shows that:

$$E[\hat{\lambda}] = \frac{r\lambda}{r - 1} \quad \text{for} \quad r > 1 . \qquad (26)$$

The resulting estimate is therefore biased. We can eliminate this bias by taking as estimate:

$$\tilde{\lambda} = \frac{r-1}{nt_r} \quad r > 1 . \tag{27}$$

The variance of this estimator is then written as:

$$\sigma^2(\tilde{\lambda}) = \frac{\lambda^2}{r-2} \quad r > 2 . \tag{28}$$

It is only finite for $r > 2$.

It can be shown (cf. Exercise 3) that the confidence interval $[\lambda_-, \lambda_+]$ is defined by:

$$\left.\begin{aligned} \lambda_- &= \frac{\chi^2_{1-\alpha/2}(2r)}{2\,nt_r} \quad (a) \\ \lambda_+ &= \frac{\chi^2_{\alpha/2}(2r)}{2\,nt_r} \quad (b) \end{aligned}\right\} \tag{29}$$

Example: Consider the design $n = 100, r = 10$. The lifetimes of components whose failure rate λ is 0.001 h^{-1} were sampled at random. We obtained $t_r = 123.715$ h

hence $\hat{\lambda} = 0.73\;10^{-3}\,h^{-1}$

$$\lambda_+ = \frac{\chi^2_{0.05}(20)}{2 \times 12371.5} = 1.3\;10^{-3}\,h^{-1}$$

$$\lambda_- = \frac{\chi^2_{0.95}(20)}{2 \times 12371.5} = 0.44\;10^{-3}\,h^{-1}.$$

4.2.3. [*n, M, r*] *type designs*

Expression (13) shows that the sample likelihood is proportional to:

$$\lambda^r \exp\left(-\lambda \sum_{i=1}^{r} t_i\right) e^{-\lambda(n-r)t_r} . \tag{30}$$

Therefore, calling the total operating time S_r, the maximum likelihood estimate is:

$$\hat{\lambda} = \frac{r}{S_r} = \frac{r}{\sum_{i=1}^{r} t_i + (n-r)\,t_r} . \tag{31}$$

The sufficient statistic is constituted by the total operating time S_r.

Note that S_r can also be written:

$$S_r = \sum_{i=1}^{r} t_i + (n-r)\,t_r = nt_1 + (n-1)(t_2 - t_1) + \dots + (n-r+1)(t_r - t_{r-1}) . \tag{32}$$

On the other hand, the likelihood is proportional to $\lambda^r e^{-\lambda S_r}$, the proportionality coefficient being equal to the number of arrangements of n components taken r to r, i.e. $n(n-1)\dots(n-r+1)$. Hence, $\mathcal{L}(t, \lambda) = n(n-1)\dots(n-r+1)\lambda^r e^{-\lambda S_r}$.

Putting $t'_1 = t_1$

$t'_k = t_k - t_{k-1} \quad k = 2, \dots, r .$

and taking account of Eqn (42) and the fact that the Jacobian of the transformation $t_i \to t'_i$ is 1, we obtain:

$$\mathscr{L}(t, \lambda) = n\lambda\, e^{-\lambda n t'_1}(n-1)\lambda\, e^{-\lambda(n-1)t'_2} \ldots (n-r+1)\lambda\, e^{-\lambda(n-r+1)t'_r}.$$

Consequently, the variables t'_i are mutually independent and the density of t'_k is equal to $(n-k+1)\lambda\, e^{-(n-k+1)\lambda t}$ and that of $(n-k+1)t'_k$ is equal to $\lambda\, e^{-\lambda t}$.

S_r is therefore the sum of r independent variables distributed according to the same exponential law. Consequently, S_r is Erlangian with parameters λ and r, and its probability density can be written as:

$$S_r(t) = \lambda^r \frac{t^{r-1}}{(r-1)!} e^{-\lambda t}. \tag{33}$$

This expression enables us to calculate the mean $E[\hat{\lambda}]$

$$E[\hat{\lambda}] = \int_0^\infty \frac{r}{t} \lambda^r \frac{t^{r-1}}{(r-1)!} e^{-\lambda t}\, dt = \frac{r}{r-1}\lambda \qquad r > 1.$$

Once again the estimate is biased. We can take as an unbiased estimate the value $\tilde{\lambda}$:

$$\tilde{\lambda} = \frac{r-1}{S_r} \qquad r > 1 \tag{34}$$

whose variance can be easily computed using Eqn (33):

$$\sigma^2(\tilde{\lambda}) = \frac{\lambda^2}{r-2} \qquad r > 2.$$

Note that, in this case, the estimate $\hat{\theta}$ of $\theta = 1/\lambda$ defined by:

$$\hat{\theta} = \frac{S_r}{r} \text{ is unbiased.}$$

Equation (33) implies that the confidence interval $[\lambda_-, \lambda_+]$ is defined by:

$$\left.\begin{aligned} \lambda_- &= \frac{\chi^2_{1-\alpha/2}(2r)}{2S_r} \qquad (a) \\ \lambda_+ &= \frac{\chi^2_{\alpha/2}(2r)}{2S_r} \qquad (b) \end{aligned}\right\} \tag{35}$$

Example: Consider the design $n = 100, r = 11$. The lifetimes of components whose failure rate λ is $0.001\,h^{-1}$ were sampled at random. We obtained $t_r = 96.23$ h and $S_r = 9093.42$ h.

Therefore $\tilde{\lambda} = 1.1\;10^{-3}\,h^{-1}$

$$\lambda_+ = \frac{\chi^2_{0.05}(22)}{2 \cdot 9093.42} = 1.87\;10^{-3}\,h^{-1}$$

$$\lambda_- = \frac{\chi^2_{0.95}(22)}{2 \cdot 9093.42} = 0.68\;10^{-3}\,h^{-1}.$$

4.2.4. *[n, M, T] type designs*

Expression (11) shows that the sample likelihood is proportional to:

$$\lambda_r \exp\left(-\lambda \sum_{i=1}^{r} t_i\right) e^{-\lambda(n-r)T}.$$

Denoting the total operating time by S_T, the maximum likelihood estimator $\tilde{\lambda}$ is therefore:

$$\tilde{\lambda} = \frac{r}{S_T} = \frac{r}{\sum_{i=1}^{r} t_i + (n-r)\,T}. \tag{36}$$

This estimator is biased. The same applies to the estimator $\hat{\theta}$ of the mean uptime. Computing the bias is complicated and will not be attempted here.

We can obtain a confidence interval on λ by determining a confidence interval on $R(T)$. The probability of obtaining r failures during the interval T can be written:

$$\mathcal{P}(k = r) = C_n^r\, R(T)^{n-r}\,(1 - R(T))^r.$$

The problem is then one of determining a confidence interval on a proportion $\gamma = R(T)$ (see section 4.7.2, determination of a confidence interval on the parameter of a binomial distribution).

Example: Consider the design $n = 100$, $T = 90$ h. The lifetimes of components whose failure rate λ is 0.001 h^{-1} were sampled at random. We obtained $r = 9$.

The confidence interval on the parameter γ of a binomial distribution is given by chart 1 (section 4.7.2). For $n = 100$, $r/n = 0.09$,

we obtain:

$$\gamma_+ = 0.155$$

$$\gamma_- = 0.045.$$

The equations giving λ_- and λ_+ can be written respectively as:

$$1 - e^{-\lambda_- T} = \gamma_-$$

$$1 - e^{-\lambda_+ T} = \gamma_+$$

hence

$$\lambda_+ = 1.87\,10^{-3}\,h^{-1}$$

$$\lambda_- = 0.5\,10^{-3}\,h^{-1}.$$

The point estimate of γ is $r/n = 0.09$.

Bartholomew describes how the confidence interval can be determined directly [Bartholomew, 1963].

4.2.5. *Multicensured designs*

It is clear from expressions (15) and (16) that the estimator $\hat{\lambda}$ is still equal to the quotient of the number of failures observed divided by the total operating time S of the items tested. Cohen indicates that confidence interval (35) is a good approximation for the actual confidence interval [Cohen, 1963].

4.3. Estimation of the parameters of a Weibull distribution

4.3.1. *Point estimation*

The Weibull is a three-parameter distribution which may provide a relatively good fit for a number of experimental results. The distribution function can be written:

$$F(t) = 1 - \exp\left(-\left(\frac{t-\gamma}{\eta}\right)^{\beta}\right) \quad \text{if} \quad t \geq \gamma \quad \eta > 0 \quad \beta > 0$$
$$= 0 \qquad \text{if} \quad t < \gamma .$$

The lifetimes being positive, we assume that $\gamma = 0$. Wingo examines the case where γ is unknown [Wingo, 1973]. We shall consider directly the type I multicensured designs without replacement. From expression (15), the likelihood is proportional to:

$$\prod_{i=1}^{r} \frac{\beta}{\eta}\left(\frac{t_i}{\eta}\right)^{\beta-1} \exp\left(-\left(\frac{t_i}{\eta}\right)^{\beta}\right) \prod_{i=r+1}^{n} \exp\left(-\left(\frac{t_i}{\eta}\right)^{\beta}\right) \tag{37}$$

with t_i = failure time of component $i, i = 1, \ldots, r$
or t_i = censure time T_i of non-failed component $i, i = r+1, \ldots, n$;
hence

$$\log \mathscr{L}(t, \beta, \eta) = r \log \beta - \beta r \log \eta + (\beta - 1) \sum_{i=1}^{r} \log t_i - \sum_{i=1}^{n} \left(\frac{t_i}{\eta}\right)^{\beta} + \text{constant.}$$

It is then a matter of determining $\hat{\beta}$ and $\hat{\eta}$ such that:

$$\log \mathscr{L}(t, \hat{\beta}, \hat{\eta}) \geq \log \mathscr{L}(t, \beta, \eta) \quad \forall \beta, \eta$$
$$\hat{\beta} > 0$$
$$\hat{\eta} > 0 .$$

It can be shown that there is a unique solution $(\hat{\beta}, \hat{\eta})$ which satisfies:

$$\frac{\partial \log \mathscr{L}}{\partial \beta}(\hat{\beta}, \hat{\eta}) = \frac{r}{\hat{\beta}} - r \log \hat{\eta} + \sum_{i=1}^{r} \log t_i - \sum_{i=1}^{n} \left(\frac{t_i}{\hat{\eta}}\right)^{\hat{\beta}} \log\left(\frac{t_i}{\hat{\eta}}\right) = 0$$

$$\frac{\partial \log \mathscr{L}}{\partial \eta}(\hat{\beta}, \hat{\eta}) = -\frac{\hat{\beta} r}{\hat{\eta}} + \frac{\hat{\beta}}{\hat{\eta}} \sum_{i=1}^{n} \left(\frac{t_i}{\hat{\eta}}\right)^{\hat{\beta}} = 0 .$$

For all positive fixed values of β_0, the function $\dfrac{\partial \log \mathscr{L}}{\partial \eta}$ is maximum for

$$\eta = \left(\frac{1}{r} \sum_{i=1}^{n} (t_i)^{\beta_0}\right)^{1/\beta_0} .$$

It can therefore be shown that $\partial \log \mathscr{L} / \partial \beta$ has one and only one positive root, such that $g(\beta) = 0$ with:

$$g(\beta) = \frac{r}{\beta} - r \frac{\sum_{i=1}^{n} (t_i)^{\hat{\beta}} \log t_i}{\sum_{i=1}^{n} (t_i)^{\hat{\beta}}} + \sum_{i=1}^{r} \log t_i . \tag{38}$$

In the case of $[n, M, r]$ or $[n, M, T]$ designs, we can determine β by putting (t_s = duration of trial):

$$f(\beta) = \cfrac{1}{\cfrac{\sum_{i=1}^{r} t_i^{\beta} \log t_i + (n-r) t_s^{\beta} \log t_s}{\sum_{i=1}^{r} t_i^{\beta} + (n-r) t_s^{\beta}} - \cfrac{1}{r} \sum_{i=1}^{r} \log t_i} .$$

We therefore select an initial value of β (e.g. 1) and operate iteratively, replacing β by $(f(\beta) + \beta)/2$. This method therefore converges rapidly. However, in the case of multicensured designs, it may produce negative values of β. It is therefore preferable to use the following method whereby we solve $g(\beta) = 0$ using Newton's method.

We select an initial value β_0 of β (e.g. $\frac{1}{\beta_0} = \log(\max_i t_i) - \frac{1}{r} \sum_{i=1}^{r} t_i$).

We compute $\beta' = \beta - \frac{g(\beta)}{g'(\beta)}$ and repeat.

We can also use a non-linear programming method for solving:

$$\max \log \mathscr{L}(t_i, \beta, \eta)$$

under the constraints

$$\beta \geqslant \varepsilon$$
$$\eta \geqslant \varepsilon$$

ε being a very small value (e.g. 10^{-5}). The reduced gradient method provides good solutions [Wolfe, 1962], as does the Hooke and Jeeves method [Hooke and Jeeves, 1961].

4.3.2. *Estimation of confidence interval*

It is difficult to compute the confidence intervals on β and γ. Tables for determining them can be found in [Meeker and Nelson, 1976]. Of course, if β is known, the confidence interval on γ can be easily obtained by putting $y = t^\beta$ and reverting to the exponential distribution.

When the sample is not excessively small, we can also use the asymptotic properties of the maximum likelihood estimates (see sections 1.4 and 1.6).

A discussion on the validity of this approximation can be found in [Wingo, 1973].

4.4. Estimation of the normal distribution parameters

4.4.1. *Point estimation*

The normal distribution is a two-parameter distribution with mean m and standard deviation σ. Its probability density is written as:

$$f(t, m, \sigma) = \frac{1}{\sigma\sqrt{2\pi}} \exp\left(-\frac{(t-m)^2}{2\sigma^2}\right).$$

In the case of multicensured designs without replacement, we therefore obtain, from expression (15):

$$\mathscr{L}(t, m, \sigma) \propto \prod_{i=1}^{r} \frac{\exp\left(-\frac{(t_i - m)^2}{2\sigma^2}\right)}{\sigma\sqrt{2\pi}} \prod_{i=r+1}^{n} \left(1 - \int_{-\infty}^{t_i} \frac{\exp\left(-\frac{(x-m)^2}{2\sigma^2}\right)}{\sigma\sqrt{2\pi}} \mathrm{d}x\right) \quad (39)$$

therefore, putting $x_i = \frac{t_i - m}{\sigma}$, $\varphi(u) = \frac{1}{\sqrt{2\pi}} \exp\left(-\frac{u^2}{2}\right)$

$$\mathscr{L}(x_i, m, \sigma) \propto \prod_{i=1}^{r} \frac{1}{\sigma} \varphi(x_i) \prod_{i=r+1}^{n} \left(1 - \int_{-\infty}^{x_i} \varphi(u)\,\mathrm{d}u\right).$$

The maximum likelihood equations can therefore be written as:

$$\left.\begin{aligned}
\frac{\partial \log \mathscr{L}}{\partial m} &= \frac{1}{\hat{\sigma}} \sum_{i=1}^{r} \hat{x}_i + \frac{1}{\hat{\sigma}} \sum_{i=r+1}^{n} z_i = 0 \\
\frac{\partial \log \mathscr{L}}{\partial \sigma} &= = \frac{-r}{\hat{\sigma}} + \frac{1}{\hat{\sigma}} \sum_{i=1}^{r} \hat{x}_i^2 + \frac{1}{\hat{\sigma}} \sum_{i=r+1}^{n} \hat{x}_i z_i = 0 \\
\text{with } z_i &= \frac{\varphi(\hat{x}_i)}{1 - \int_0^{\hat{x}_i} \varphi(u)\,du}, \quad \hat{x}_i = \frac{t_i - m}{\sigma}
\end{aligned}\right\} \qquad (40)$$

A method of solving Eqns (40) using Newton's method is described in [Cohen, 1963].

For simpler designs, these equations can be solved more easily. Estimation is also greatly facilitated when either of the two parameters is known.

For example, in the case of an n sample, i.e. for a type $[n, M]$ design, we obtain the well-known estimates of the mean and the variance:

$$\hat{m} = \frac{1}{n} \sum_{i=1}^{n} t_i$$

$$\hat{\sigma}^2 = \frac{1}{n} \sum_{i=1}^{n} (t_i - \hat{m})^2.$$

4.4.2. *Determination of a confidence region using the variance-covariance matrix*

The information matrix is given by Eqn (4). We can equate the means to the values obtained by replacing the parameters by their estimates (cf. [Cohen, 1963] and section 1.6).

$$\begin{bmatrix} v(\hat{m}) & \operatorname{Cov}(\hat{m}, \hat{\sigma}) \\ \operatorname{Cov}(\hat{m}, \hat{\sigma}) & v(\hat{\sigma}) \end{bmatrix} = \begin{bmatrix} -E\dfrac{\partial^2 \mathscr{L}(\hat{m}, \hat{\sigma})}{\partial^2 m} & -E\dfrac{\partial^2 \mathscr{L}(\hat{m}, \hat{\sigma})}{\partial m\, \partial \sigma} \\ -E\dfrac{\partial^2 \mathscr{L}(\hat{m}, \hat{\sigma})}{\partial m\, \partial \sigma} & -E\dfrac{\partial^2 \mathscr{L}(\hat{m}, \hat{\sigma})}{\partial \sigma^2} \end{bmatrix}$$

Example: Consider a complete sample and determine the confidence region for $(m, v = \sigma^2)$.

Using expression (39) with $r = n$, we obtain:

$$\log \mathscr{L} = -\frac{n}{2} \log v - \frac{1}{2} \sum_{i=1}^{n} \left(\frac{t_i - m}{v}\right)^2 + \text{constant}$$

hence:

$$\frac{\partial^2 \log \mathscr{L}}{\partial^2 m}(\hat{m}, \hat{v}) = -\frac{n}{\hat{v}}$$

$$\frac{\partial^2 \log \mathscr{L}}{\partial m\, \partial v}(\hat{m}, \hat{v}) = 0$$

$$\frac{\partial^2 \log \mathscr{L}}{\partial^2 v}(\hat{m}, \hat{v}) = -\frac{n}{2\hat{v}^2}$$

and the matrix:

$$\bar{I}(\theta) = \begin{bmatrix} \frac{n}{\hat{\sigma}^2} & 0 \\ 0 & \frac{n}{2\hat{\sigma}^4} \end{bmatrix}$$

The confidence region is therefore defined by $(\hat{m}, \hat{\sigma}^2) + \frac{\mathcal{D}_\alpha}{\sqrt{n}}$ where

$$\mathcal{D}_\alpha = \left\{ u = (u_1, u_2) : \frac{u_1^2}{\hat{\sigma}^2} + \frac{u_2^2}{2\hat{\sigma}^4} \leq \chi^2_\alpha(2) \right\}.$$

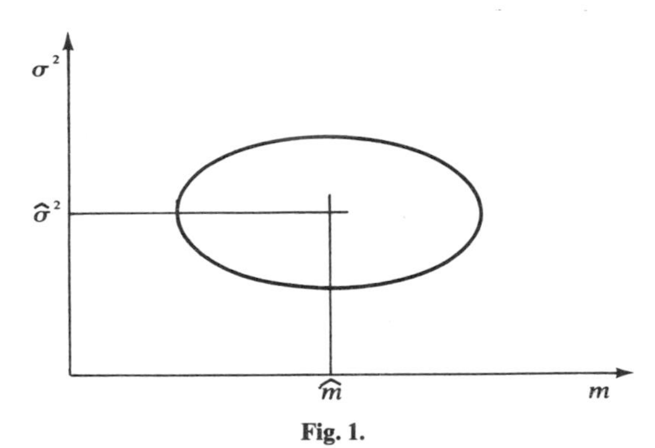

Fig. 1.

4.5. Estimation of the lognormal distribution parameters

The lognormal is a three-parameter distribution whose probability density is written as:

$$f(t, a, b) = \frac{1}{b\sqrt{2\pi}(t-\gamma)} \exp\left(-\frac{1}{2}\left(\frac{\log(t-\gamma)-a}{b}\right)^2\right) \quad \text{if } t \geq \gamma$$

$$f(t, a, b) = 0 \quad \text{if } t < \gamma .$$

In the lifetime distribution case, t is positive. And so we generally take γ as being equal to zero. This will be assumed here. A complete analysis for the unknown case $\gamma \neq 0$ can be found in [Cohen, 1976].

By putting $y_i = \log t_i$, we therefore arrive at the case of the normal distribution. We can also write the likelihood equations and solve them using an iterative method (cf. [Cohen, 1976]).

4.6. Estimation of the gamma distribution parameters

The gamma is a three-parameter distribution whose probability density can be written as:

$$f(t, \beta, \lambda, \gamma) = \frac{(t-\gamma)^{\beta-1}\lambda^\beta}{\Gamma(\beta)} e^{-\lambda(t-\gamma)} \quad \text{if } t \geq \gamma$$

$$f(t, \beta, \lambda, \gamma) = 0 \quad \text{if } t < \gamma .$$

For multicensured samples without replacement, the likelihood logarithm can be written, using expression (15), as:

$$\log \mathcal{L} = -r \log \Gamma(\beta) + r\beta \log \lambda - \lambda \sum_{i=1}^{r} (t_i - \gamma) +$$

$$+ (\beta - 1) \sum_{i=1}^{r} \log (t_i - \gamma) + \sum_{i=r+1}^{n} \log \left(1 - \int_{\gamma}^{t_i} f(t)\, \mathrm{d}t \right) + \text{constant} \tag{41}$$

with t_i = failure time of component $i, i = 1, \ldots, r$;
or t_i = censure time T_i of non-failed component $i, i = r + 1, \ldots, n$.

In the case where $\beta = 1$, we find the exponential distribution.

In the case where $\beta > 1$, we can easily solve the likelihood equations by an iterative method (cf. [Cohen and Norgaard, 1977]).

On the other hand, when β is less than 1, the sample likelihood becomes infinite as γ tends to the smallest observation t_1. However, we can get round this difficulty by assuming $\gamma \leqslant t_1 - \eta/2$, where η represents the accuracy of the observations. We then show that the optimum is obtained for $\gamma = t_1 - \eta/2$. The two equations

$$\frac{\partial \log \mathcal{L}}{\partial \beta} = 0$$

$$\frac{\partial \log \mathcal{L}}{\partial \lambda} = 0$$

then enable us to determine β and λ (cf. [Cohen and Norgaard, 1977]).

4.7. Estimation of the binomial distribution parameter

4.7.1. *Point estimation*

This involves determining the parameter p, such that the probability of obtaining x favourable cases over n attempts is equal to:

$$f(x, p) = C_n^x p^x (1 - p)^{n-x} \quad x = 0, 1, 2, \ldots, n$$

therefore, the likelihood of the sample of size r is:

$$\mathcal{L}(x_i, p) \propto \prod_{i=1}^{r} p^{x_i} (1 - p)^{n - x_i} \tag{42}$$

hence $\log \mathcal{L} = \sum_{i=1}^{r} x_i \log p + \left(rn - \sum_{i=1}^{r} x_i \right) \log (1 - p) + \text{constant}$

$$\frac{\partial \log \mathcal{L}}{\partial p} = \frac{\sum_{i=1}^{r} x_i}{p} - \frac{rn - \sum_{i=1}^{r} x_i}{1 - p} = \frac{\sum_{i=1}^{r} x_i - rnp}{p(1 - p)}.$$

Therefore the estimate $\hat{p}$ by the maximum likelihood method is given by:

$$\hat{p} = \frac{\sum_{i=1}^{r} x_i}{rn} = \text{empirical frequency.} \tag{43}$$

Noting that $\sum_{i=1}^{r} x_i$ is binomial (rn, p), we can easily compute:

$$E(\hat{p}) = p$$

$$\sigma^2(\hat{p}) = \frac{p(1-p)}{nr}$$

the estimator is therefore unbiased and consistent. It can be shown that it has minimum variance and is sufficient.

4.7.2. *Estimation of the confidence interval*

In order to determine the confidence interval $[p_-, p_+]$, we merely have to solve the following equations, putting $x = \sum_{i=1}^{r} x_i$

$$\left.\begin{aligned} \sum_{i=x}^{n} C_n^i p_-^i (1-p_-)^{n-i} &= \alpha_2 \qquad (a) \\ \sum_{i=0}^{x} C_n^i p_+^i (1-p_+)^{n-i} &= \alpha_1 \qquad (b) \end{aligned}\right\} \tag{47}$$

These equations can be solved by Newton's method (with a few precautions, as it is possible that this method may go outside the interval [0, 1]; in this case, it is necessary to replace the unwanted iteration by a dichotomy). In the case where p is small, the form b is the more useful. Equation (47(a)) can therefore be written as:

$$\sum_{i=0}^{x-1} C_n^i p_-^i (1-p\ )^{n-i} = 1 - \alpha_2 \,. \tag{47a}$$

We therefore arrive at a type (b) form.

Chart 1, which is valid for small values of $p = x/n$, was compiled using this method.

We can also determine a confidence interval by using the approximation for the normal distribution if $\hat{p}$ is not close to zero when $x(1 - \hat{p})$ is greater than 18.

The relationship between the binomial distribution and the Fisher distribution F_{ν_1,ν_2}

$$\sum_{i=0}^{x} C_n^i p^i (1-p)^{n-i} = 1 - \mathcal{P}\left(F_{\nu_1,\nu_2} < \frac{n-x}{x+1}\,\frac{p}{1-p}\right) \tag{48}$$

$$\nu_1 = 2(x+1)$$

$$\nu_2 = 2(n-x)$$

enables us to determine a confidence interval using the F-distribution tables (cf. [Morice, 1977]).

The relationship between the binomial distribution and the beta distribution (cf. [Johnson and Kotz, 1970])

$$\sum_{i=n-x}^{n} C_n^i p^i (1-p)^{n-i} = \frac{\Gamma(n+1)}{\Gamma(n-x)\,\Gamma(x+1)} \int_0^p u^{n-x-1} (1-u)^x \, \mathrm{d}u \tag{49}$$

allows us to determine a confidence interval using the beta disttibution tables; conversely, this relationship enables us to determine the quantiles of the beta distribution using the quantiles of the binomial distribution (when the parameters are whole numbers).

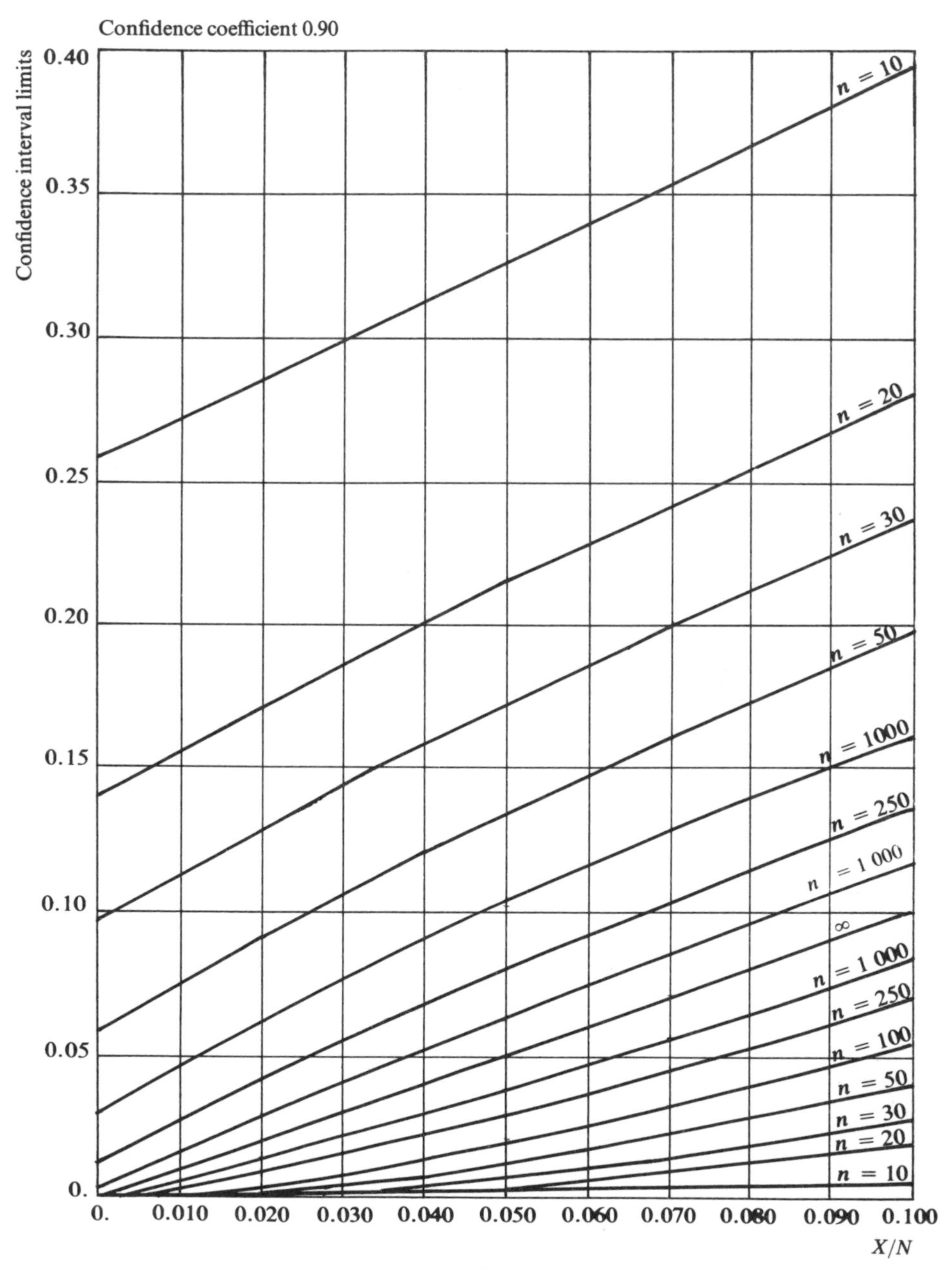
Confidence coefficient 0.90
Confidence interval limits
0.40
0.35
0.30
0.25
0.20
0.15
0.10
0.05
0.
n = 10
n = 20
n = 30
n = 50
n = 1000
n = 250
n = 1 000
∞
n = 1 000
n = 250
n = 100
n = 50
n = 30
n = 20
n = 10
0.
0.010
0.020
0.030
0.040
0.050
0.060
0.070
0.080
0.090
0.100
X/N

Chart 1.

5. HYPOTHESIS TESTING

5.1. Objectives

Having estimated the parameters of a distribution, can we tell if the random variable is actually distributed in this way? More generally, the following types of question arise:

—are two or more samples produced from the same distribution?
—assuming that the distribution (with distribution function F_0) is known and with unknown parameters $\theta_1, \ldots, \theta_r$, do the parameters have values $\hat{\theta}_1, \hat{\theta}_2, \ldots, \hat{\theta}_r$?
—does this sample correspond to a given type of *a priori* distribution?
—etc.. . . .

These questions amount to a hypothesis; let us therefore assume that when the hypothesis is true, we can define a probability density φ of a certain function ϕ of the sample. We therefore compute the values m_- and m_+ such that:

$$\int_{m_+}^{\infty} \varphi(u)\,\mathrm{d}u = \alpha_2$$

$$\int_{-\infty}^{m_-} \varphi(u)\,\mathrm{d}u = \alpha_1$$

The interval $[m_-, m_+]$ is called the acceptance interval of the hypothesis at the $(\alpha_1 + \alpha_2)$ level of error. In fact, if we take a sample and compute the value m of the function ϕ for this sample, the probability of m not belonging to the interval $[m_-, m_+]$ is equal to $(\alpha_1 + \alpha_2)$.

When m belongs to the interval $[m_-, m_+]$, the test is said to be non-significant. However, this is not an infallible conclusion, as we accept that we are running the risk $\alpha = \alpha_1 + \alpha_2$; α is called a type I risk (that of rejecting the hypothesis when it is true).

However, it is possible that the hypothesis is false and that another is true. The risk β of accepting this false hypothesis is called a type II risk. The strength of a test is defined by $1 - \beta$. We shall examine the two following tests in turn:

Tests	Function ϕ of sample x_j
Chi-squared test	$\sum_{i=1}^{k} \frac{(n_i - n_{i_0})^2}{n_{i_0}}$ n_i = number of $x_j \in i$th interval n_{i_0} = theoretical number of $x_j \in i$th interval
Kolmogorov-Smirnov test	$\mathrm{Sup}_{x_j} \mid F_n(x_j) - F_0(x_j) \mid$ F_n = experimental distribution function

5.2. Chi-squared test

The hypothesis to be tested is as follows: the random variable X is distributed with theoretical distribution function F_0. There is a known experimental distribution function F.

The interval of variation of X is partitioned into k classes C_i of respective frequencies n_i. Pearson has shown that the random variable

$$X = \frac{\sum_{i=1}^{k} (n_i - n_{i_0})^2}{n_{i_0}}$$

is asymptotically χ_v^2 distributed (n_{i_0} is the theoretical frequency of class C_i).

If the distribution is entirely determinate, the degree of freedom v of the χ^2 is equal to $k - 1$. If it is necessary to determine l parameters from the sample, $v = k - 1 - l$.

The χ^2 test is not very sensitive. In general, it is necessary to have a minimal sample size $n = 50$ and the frequencies of each class must not be too low.

5.3. Kolmogorov-Smirnov test

The hypothesis to be tested is the following: the random variable X is distributed with a completely specified, continuous theoretical distribution function F_0 (i.e. one with *known parameters*). An experimental distribution function F_n and the parameters of the F_0 distribution are known.

Kolmogorov has shown that the probability distribution of the random variable $D_n = \sup_x |F_n(x) - F_0(x)|$ is distributed independently of F_0. More precisely:

$$\lim_{n \to \infty} \mathcal{P}(\sqrt{n}\, D_n < x) = K(x)$$

with

$$\begin{aligned} K(x) &= 0 & \text{if } x \leq 0 \\ &= \sum_{j=-\infty}^{+\infty} (-1)^j \, e^{-2j^2x^2} & \text{if } x > 0\,. \end{aligned}$$

This function has been tabulated (cf., for example [Peyrache and Schwab, 1969, p. 200]). The test ceases to be effective when F_0 is not completely specified, i.e. when the parameters of the distribution have been determined from the sample.

6. BAYESIAN ESTIMATION OF PARAMETERS

6.1. Bayes' theorem

The methods of evaluating the parameters of a probability distribution described in section 4 enable us to determine a confidence region from samples.

However, it is impossible to take account of additional information of a subjective nature such as confidence in the manufacturer or confidence in well-known older equipment from which the equipment under test has been derived. If we can incorporate this subjective information by assigning an *a priori* distribution to the parameters to be evaluated, it will be possible, using Bayes' theorem, to combine the subjective information with the objective data (contained in a sample).

Bayes' formula was examined in Ch. 1 (section 2.5) in the discrete case. In the continuous case, taking into account two random variables x and y and using the following notation:

$f'(\theta)$ the *a priori* density of θ, θ belonging to the domain Ω,
$g(x|\theta)$ the conditional density of x with respect to θ,
$f''(\theta|x)$ the conditional density of θ with respect to x,

it can be written as:

$$f''(\theta \mid x) = \frac{g(x \mid \theta) f'(\theta)}{\int_\Omega g(x \mid \theta) f'(\theta) \, d\theta} . \tag{50}$$

As a rule, we use $'$ to indicate *a priori* densities and $''$ to indicate *a posteriori* densities. Whatever densities are used, applying formula (50) results in very complicated computations. The formula is therefore generally used with pairs (f', g) which possess useful properties. We shall attempt to define these properties and then determine the pairs (f', g) exhibiting them. First, a few definitions.

6.2. Definitions

Given an *a priori* distribution $f'(\theta)$ of parameter θ, the likelihood of a sample $\mathscr{L}(x, \theta)$ is a conditional density which we shall denote by $l(x|\theta)$. We saw several examples of this in section 4. We shall therefore call the *marginal likelihood* of this sample, given the *a priori* distribution $f'(\theta)$, the expression defined by:

$$l^*(x \mid f') = \int_\Omega l(x \mid \theta) f'(\theta) \, d\theta .$$

The sample x is said to belong to the spectrum of f' if $l^*(x|f')$ is positive. In the lifetime distribution case, l is positive and so is l^*. Bayes' theorem (50) can then be applied to find an *a posteriori* density of θ:

$$f''(\theta \mid x) = f'(\theta) \, l(x \mid \theta) \, . \, N(x) \tag{51}$$

with
$$N(x) = \frac{1}{l^*(x \mid f')} .$$

Any function $k(\theta)$ such that:

$$f(\theta) = \frac{k(\theta)}{\int_\Omega k(\theta) \, d\theta} .$$

is called the *kernel of a density* $f(\theta)$.

This relation is denoted by $\propto$ (proportional to):

$$f(\theta) \propto k(\theta) .$$

Any function $k(x|\theta)$ such that

$$l(x \mid \theta) \propto k(x \mid \theta)\, \rho(x)$$

is called the *kernel of a conditional density* $l(x|\theta)$.
$\rho(x)$ is known as the residual.

Equation (51) can therefore be written as:

$$f''(\theta \mid x) = k'(\theta)\, k(x \mid \theta) \frac{\rho(x)\, N(x)}{\int_{\Omega} k'(\theta)\, \mathrm{d}\theta}$$

hence
$$f''(\theta \mid x) \propto k'(\theta)\, k(x \mid \theta)\,. \tag{52}$$

The proportionality coefficient is determined by the equation:

$$\int_{\Omega} f''(\theta \mid x)\, \mathrm{d}\theta = 1\,.$$

Therefore, the kernel of the a posteriori *density is equal to the product of the kernel of the* a priori *density and the kernel of the conditional density* $l(x|\theta)$.

Example: θ is the intensity λ of a Poisson distribution and x is equal to the number r of observed events at time t. Therefore

$$l(r \mid \lambda) = \frac{\mathrm{e}^{-\lambda t}(\lambda t)^r}{r\,!}\,.$$

Assume that the *a priori* distribution of λ is gamma (t', r'):

$$f'(\lambda) = \frac{\mathrm{e}^{-\lambda t'}(\lambda t')^{r'-1}}{(r'-1)\,!}\, t'\,.$$

The respective kernels can then be written as:

$$k(r \mid \lambda) = \mathrm{e}^{-\lambda t}\, \lambda^{r}$$
$$k'(\lambda) = \mathrm{e}^{-\lambda t'}\, \lambda^{r'-1}$$

hence:
$$f''(\lambda \mid r) \propto \mathrm{e}^{-\lambda(t+t')}\, \lambda^{r+r'-1} = \mathrm{e}^{-\lambda t''}\, \lambda^{r''-1}$$

and
$$\int_0^{\infty} \mathrm{e}^{-\lambda t''}\, \lambda^{r''-1}\, \mathrm{d}\lambda = \frac{(r''-1)\,!}{t''^{r''}}\,.$$

6.3. Desirable properties of an *a priori* F-distribution

In general, our subjective knowledge is somewhat vague and it is impossible to specify an *a priori* distribution very precisely. This means that we have a large latitude in the selection of an *a priori* distribution. We can therefore choose from distributions possessing some of the following properties.

— F must be easy to use with respect to the three following points:

(a) the *a posteriori* distribution must be simple to calculate from the *a priori* distribution and the sample;
(b) it must be easy to compute the means of useful functions associated with F;

(c) the *a posteriori* distribution must be of the same type as the *a priori* distribution in order to permit an iterative computation.

—F must be comparatively complex in order to be able to represent a large number of situations.

—F must be parameterizable, and the parameters must admit of physical interpretation.

6.4. Case where a fixed dimension sufficient statistic exists

Let us assume, therefore, that we have independent samples of a distribution possessing a sufficient statistic y of dimension s, $y = (y_1, y_2, \dots, y_s)$. These assumptions are expressed by the following relations:

—given two samples of size r and $(n - r)$:

$$l(x_1, x_2, \dots, x_r \mid \theta) \times l(x_{r+1}, \dots, x_n \mid \theta) = l(x_1, x_2, \dots, x_n \mid \theta)$$

—there exists a function k such that:

$$l(x_1, x_2, \dots, x_n \mid \theta) \propto k(y \mid \theta) .$$

We can therefore prove the following theorem: (cf. [Raiffa and Schlaifer, 1961]):

Theorem.—*If* $y^{(1)} = y(x_1, x_2, \dots, x_r)$
and $y^{(2)} = y(x_{r+1}, \dots, x_n)$

then there exists a binary operation $*$ *such that*:

$$y^* = y^{(1)} * y^{(2)}$$

and possessing the following properties:

$$\begin{aligned} l(x_1, x_2, \dots, x_n \mid \theta) &\propto k(y^* \mid \theta) \qquad (a) \\ k(y^* \mid \theta) &\propto k(y^{(1)} \mid \theta) \times k(y^{(2)} \mid \theta) \qquad (b) \end{aligned} \tag{53}$$

Example of the binomial distribution with parameter p

$$l(x_1, x_2, \dots, x_n \mid p) = p^r (1 - p)^{n-r}$$

with $\quad r = \sum_{i=1}^{n} x_i$

the statistic $y = (r, n)$ is sufficient and of dimension 2.

Given two samples, it can be easily verified that:

$$(r^*, n^*) = (r_1, n_1) * (r_2, n_2) = (r_1 + r_2, n_1 + n_2) . \tag{54}$$

The binary operation $*$ is therefore very simple in this case.

6.5. Natural conjugates

The kernel k is a function $k(\cdot \mid \theta)$ with parameter θ. However, we can also regard it as a function $k(y \mid \cdot)$ with parameter y. Consider, therefore, a function $f(\theta \mid y)$ defined by:

$$f(\theta \mid y) = N(y)\, k(y \mid \theta) . \tag{55}$$

If $N(y)$ is a positive function and if $\int_\Omega k(y(\theta))\,d\theta$ exists, $f(\theta|y)$ is a probability density. It is called the *natural conjugate* with parameter y of kernel k.

Computing the *a posteriori* distribution is very simple when the *a priori* distribution is the natural conjugate of the kernel of the sample likelihood, i.e. when $f'(\theta) \propto k(y'|\theta)$.

In fact, let $k(y|\theta)$ be the kernel of the likelihood of sample x, given θ, y being a sufficient statistic of fixed dimension.

We know that:

$$f''(\theta \mid y) \propto f'(\theta)\, k(y \mid \theta)$$

$$k(y^{(1)} \mid \theta) \,.\, k(y^{(2)} \mid \theta) = k(y^{(1)} * y^{(2)} \mid \theta)\,.$$

Therefore if $f'(\theta)$ is the natural conjugate of k with parameter y'

$$f'(\theta) \propto k(y' \mid \theta)$$

therefore

$$f''(\theta \mid y) \propto k(y' * y \mid \theta)\,. \tag{56}$$

Consequently, the way in which the kernel of the a priori *density combines with the kernel of the sample is the same as the way in which two samples combine. Using natural conjugates therefore provides an easy means of merging objective and subjective data.*

Example 1: Case of the parameter p of the binomial distribution.

Let r be the number of favourable cases over a sample of size n. The sufficient statistic y is therefore the pair (r, n).

The kernel of the sample likelihood is:

$$k(y \mid p) = p^r (1-p)^{n-r}$$

consequently (cf. Eqn (55)):

$$N(y) = \frac{1}{\int_0^1 k(y \mid p)\,dp} = \frac{1}{B(r+1, n-r+1)}\,.$$

$B(p, q)$ represents the complete beta function equal to $\dfrac{\Gamma(p)\,\Gamma(q)}{\Gamma(p+q)}$

hence

$$f'(p \mid y') = \frac{p^{r'-1}(1-p)^{n'-r'-1}}{B(r', n'-r')} \quad \text{with} \quad 0 < r' < n'\,. \tag{57}$$

The *a posteriori* density is then defined by the operation:

$$(r, n) * (r', n') = (r + r', n + n') = y^*$$

hence

$$f''(\theta \mid y^*) = \frac{p^{r+r'-1}(1-p)^{n+n'-r-r'-1}}{B(r+r', n+n'-r-r')}\,. \tag{58}$$

Example 2: The exponential distribution parameter λ.

The sufficient statistic is the number r of failures observed during the total operating time t.

The kernel of the sample likelihood is:

$$k(y \mid \lambda) = \mathrm{e}^{-\lambda t} \lambda^{r}$$

consequently (cf. Eqn (55)):

$$N(y) = \frac{1}{\int_0^\infty k(y \mid \lambda)\, \mathrm{d}\lambda} = \frac{t^{r+1}}{r\,!}$$

hence

$$f'(\lambda \mid y') = \frac{\mathrm{e}^{-\lambda t'} \lambda^{r'-1} t'^{r'}}{(r'-1)} \quad \text{with } t' > 0, \; r' > 0 \tag{59}$$

the *a posteriori* density is therefore defined by the operation

$$(r, t) * (r', t') = (r + r', t + t') = y^*$$

therefore:

$$f''(\lambda \mid y^*) = \frac{\mathrm{e}^{-\lambda(t+t')} \lambda^{r+r'-1} (t + t')^{r+r'}}{(r + r' - 1)}. \tag{60}$$

6.6. Selection and interpretation of an *a priori* distribution

If we have ample information on the parameter θ, we can determine the *a priori* distribution best fitting this information, the method of fit generally having no effect on the result.

However, in most cases we only have 'some ideas' about this parameter. It is therefore necessary to transform these *a priori* ideas into qualitative data for determining the distribution parameters. However, although we may be perfectly satisfied with an *a priori* distribution determined in this way, we may be in disagreement with the *a posteriori* distribution obtained after combination with a sample. We can therefore determine the parameters of the *a priori* distribution by imposing both an *a priori* and an *a posteriori* condition relative to a certain sample (cf. [Raiffa and Schlaifer, 1961, p. 59]).

Is there a relationship between the value of the a priori *distribution parameters and the quantity of* a priori *data available*?

In fact, it would be desirable that some parameter domains should have a very low *a priori* information content, with sample knowledge subsequently predominating.

First, *a positive response*.

Take the example of a normal distribution of known standard deviation σ and unknown mean μ.

The sufficient statistic of a sample of size n is composed of the pair $\left(m = \frac{\sum_{i=1}^{n} x_i}{n}, n \right)$. Consequently, the natural conjugate has a kernel:

$$f(\mu \mid m', \sigma n') \propto \exp\left(-\frac{n'}{2\sigma^2} (m' - \mu)^2 \right) \quad n' > 0. \tag{61}$$

It can be easily shown that the parameters of the *a posteriori* distribution are defined by:

$$(m'', n'') = (m', n') * (m, n) = \left(\frac{m'n' + mn}{n + n'}, n + n' \right). \tag{62}$$

n may be interpreted as a value characterizing the 'quantity of information' contained in the sample. The parameter m therefore becomes an estimate of μ based on n 'information units'. We can make the same interpretation of the pair (n', m'). As n' tends to zero, m'' and n'' tend respectively to m and n. On the other hand, the variance on the estimator $\tilde{\mu}$ of μ, worked out from the *a priori* information, tends to infinity and $\tilde{\mu}$ tends to uniformity in the sense that the quotient of the probability of two intervals of the same length tends to 1.

In addition, from Eqn (62), m'' and n'' tend respectively to m and n, i.e. the information provided by the sample is predominant.

Therefore, in this example, very vague *a priori* information amounts to regarding the *a priori* distribution as 'uniform' and favours the sample information.

Now, *a negative response.*

Consider the parameter p of a binomial distribution. The sufficient statistic of a sample of size n is represented by the pair $(r, n) = \left(\sum_{i=1}^{n} x_i, n\right)$.

We know that the natural conjugate of the likelihood of this sample is a β distribution:

$$f'(p \mid r', n') \propto p^{r'-1}(1-p)^{n'-r'-1} \quad n' > r' > 0$$

The *a posteriori* distribution is therefore defined by the parameters (r'', n'')

$$(r'', n'') = (r + r', n + n').$$

In this case, as n' tends to zero, the distribution of $\tilde{\mu}$ tends to a discrete distribution defined by:

$$\mathcal{P}(p = 0) = 1 - m'$$
$$\mathcal{P}(p = 1) = m'.$$

Under no circumstances can this distribution be regarded as vague. On the contrary, it is extremely precise.

Similarly, in the case of the normal distribution with known variance, the distribution of the parameter $\eta = 1/\mu$ tends to a Dirac distribution as n' tends to zero.

Further discussion on this important point can be found in (Raiffa and Schlaifer, 1961]. To say that a large-variance distribution is the expression of a vague opinion is not always justified.

Another important point is to what extent the results are influenced by the selection of an *a priori* distribution [Raiffa and Schlaifer, 1961; Kapur and Lamberson, 1977].

In practical terms, a method of determining the parameters of an *a priori* gamma distribution on the basis of subjective data is presented in [Johnson et al., 1977].

6.7. Bayesian confidence intervals

6.7.1. *Definition*

The Bayesian estimate $\hat{\theta}$ of a parameter θ is obtained by taking the *a posteriori* mean (which minimizes the variance of the estimate):

$$\hat{\theta} = E[f''(\theta \mid y)] = \int_{-\infty}^{+\infty} \theta f''(\theta \mid y)\, d\theta. \tag{63}$$

Similarly, it is easy to define a confidence interval using the following equations:

$$\left.\begin{array}{ll} \theta_{-} = q(y) \text{ such that } \displaystyle\int_{-\infty}^{q(y)} f''(\theta \mid y)\, d\theta = \alpha_1 & a) \\ \theta_{+} = q'(y) \text{ such that } \displaystyle\int_{q'(y)}^{\infty} f''(\theta \mid y)\, d\theta = \alpha_2 & b) \end{array}\right\} \tag{64}$$

6.7.2. *The exponential distribution parameter*

Computing a failure rate involves computing the quantiles of the gamma distribution.

We try to find $g(\alpha)$ such that

$$\int_0^{g(\alpha)} \frac{x^{\beta-1}\lambda^{\beta}e^{-\lambda x}}{\Gamma(\beta)}\,dx = \alpha . \tag{65}$$

By changing variables, we arrive at the computation of the quantiles of a chi-squared distribution, but with a non-integer degree of freedom:

$$g(\alpha) = \frac{\chi^2_{1-\alpha}(2\beta)}{2\lambda} .$$

The method proposed in [Goldstein, 1973] gives good results for comparatively large β, more precisely for $\beta \geqslant 1 + 2|u_\alpha|$. For small values of β we can operate in the following indirect manner if we have programs for computing the distribution function of the gamma distribution and u_α (cf. Ch. 1, section 4.2.3). We can then solve Eqn (65) using Newton's method, which gives good results for these values of β.

By applying these two methods, we obtain Table 1 giving the values of $g(\alpha)$ for $\lambda = 1$.

6.7.3. *The binomial distribution parameter*

Computing a proportion involves computing the quantiles of the beta distribution. For example, we can arrive at the computation of the quantiles of the binomial distribution using Eqn (49). However, in a reliability context, this proportion is often small and the simplest method is to revert to the previous case, noting that the binomial distribution in this case approximates to a Poisson distribution.

Generally speaking, the confidence intervals thus defined are narrower than the intervals obtained by classical methods. Morlat shows that the two methods do not provide different solutions to the same problem but rather solutions to different problems [Morlat, 1978].

Typical application

Suppose we have to estimate the parameter λ of an exponential distribution. This exponential distribution is the lifetime distribution of a new type of equipment. Given the precautions taken, we think that this failure rate will be of the order of $\lambda_0 = 10^{-3}\,h^{-1}$, λ_0 being the failure rate of the previous-generation equipment. We also think that the standard deviation is of the order of $\lambda_0/2$. More precisely, we think that:

$$E(\lambda) = 10^{-3}$$
$$v(\lambda) = 25 \,.\, 10^{-8} .$$

Therefore, using the value of the mean and the variance of the gamma distribution given by Eqn (59):

$$\frac{r'}{t'} = 10^{-3} \; ; \quad \frac{r'}{t'^2} = \frac{1}{4}\,10^{-6}$$

we find $r' = 4$, $t' = 4000$.

BETA	$\alpha = 0.95$	$\alpha = 0.90$	$\alpha = 0.80$	$\alpha = 0.20$	$\alpha = 0.10$	$\alpha = 0.05$
0.10	.5805	.2662	.694E-01	.622E-07	.607E-10	.593E-13
0.20	1.031	.6049	.2635	.209E-03	.653E-05	.204E-06
0.30	1.372	.8848	.4601	.327E-02	.324E-03	.321E-04
0.40	1.662	1.130	.6456	.134E-01	.235E-02	.415E-03
0.50	1.921	1.353	.8212	.321E-01	.790E-02	.197E-02
0.60	2.159	1.561	.9890	.588E-01	.181E-01	.564E-02
0.70	2.383	1.757	1.151	.923E-01	.331E-01	.122E-01
0.80	2.595	1.945	1.307	.1315	.530E-01	.219E-01
0.90	2.799	2.127	1.460	.1754	.772E-01	.350E-01
1.00	2.996	2.303	1.609	.2231	.1054	.513E-01
1.10	3.187	2.474	1.756	.2742	.1371	.708E-01
1.20	3.373	2.641	1.900	.3280	.1719	.931E-01
1.30	3.554	2.806	2.042	.3842	.2096	.1183
1.40	3.733	2.967	2.182	.4425	.2497	.1459
1.50	3.907	3.126	2.321	.5026	.2922	.1759
1.60	4.079	3.282	2.458	.5643	.3367	.2081
1.70	4.249	3.437	2.594	.6275	.3831	.2422
1.80	4.416	3.589	2.728	.6920	.4311	.2783
1.90	4.581	3.740	2.862	.7577	.4808	.3160
2.00	4.744	3.890	2.994	.8244	.5318	.3554
2.10	4.905	4.038	3.126	.8921	.5842	.3962
2.20	5.065	4.185	3.257	.9607	.6378	.4385
2.30	5.223	4.330	3.387	1.030	.6925	.4820
2.40	5.380	4.475	3.516	1.100	.7484	.5268
2.50	5.535	4.618	3.645	1.171	.8052	.5727
2.60	5.690	4.761	3.773	1.243	.8629	.6198
2.70	5.843	4.902	3.900	1.315	.9215	.6678
2.80	5.995	5.043	4.027	1.388	.9809	.7169
2.90	6.146	5.183	4.153	1.461	1.041	.7668
3.00	6.296	5.322	4.279	1.535	1.102	.8177
3.10	6.445	5.461	4.404	1.609	1.164	.8694
3.20	6.593	5.599	4.529	1.684	1.226	.9218
3.30	6.741	5.736	4.654	1.759	1.289	.9751
3.40	6.888	5.873	4.778	1.835	1.352	1.029
3.50	7.034	6.009	4.902	1.911	1.417	1.084
3.60	7.179	6.144	5.025	1.988	1.481	1.139
3.70	7.324	6.279	5.148	2.064	1.546	1.195
3.80	7.467	6.413	5.271	2.142	1.612	1.251
3.90	7.611	6.547	5.393	2.219	1.678	1.309
4.00	7.754	6.681	5.515	2.297	1.745	1.366
4.10	7.896	6.814	5.637	2.375	1.812	1.425
4.20	8.038	6.946	5.758	2.453	1.879	1.483
4.30	8.179	7.079	5.879	2.532	1.947	1.543
4.40	8.319	7.210	6.000	2.611	2.015	1.602
4.50	8.459	7.342	6.121	2.690	2.084	1.663
4.60	8.599	7.473	6.241	2.769	2.153	1.723
4.70	8.738	7.604	6.362	2.849	2.222	1.784
4.80	8.877	7.734	6.482	2.929	2.292	1.846
4.90	9.015	7.864	6.601	3.009	2.362	1.908
5.00	9.153	7.994	6.721	3.090	2.433	1.970

Quantiles of the standardized gamma distribution: $\int_0^{g_\alpha} \frac{x^{\beta-1} e^{-x}}{\Gamma(\beta)} \, dx = \alpha$.

Table 1.

Let us now assume that six components are observed for 1000 hours and that only one failure is reported; consequently, the likelihood of this sample is proportional to $e^{-6000\lambda}$, i.e. $r = 1$ and $t = 6000$.

Applying Eqn (60) shows that the *a posteriori* distribution of λ is a gamma distribution with parameters $(5, 10^4)$, hence:

$$E(\lambda) = 5 \cdot 10^{-4} \qquad \lambda_+ = 9.15 \cdot 10^{-4} \quad \lambda_- = 1.97 \cdot 10^{-4}$$

$$v(\lambda) = 5 \cdot 10^{-8} \qquad (\alpha_1 = \alpha_2 = 0.05)\,.$$

7. SOME RESULTS: RELIABILITY DATA BASES

In-service equipment monitoring has provided a large amount of raw data which has been processed by the methods described in this chapter and incorporated in reliability data bases or data banks. Building up these data banks is essential, as all the methods of computation would be useless if we did not have at our disposal numerical values for the various parameters (failure rate, mean time to repair, etc.).

Suffice it to say that, in most cases, we are more concerned with explaining the variations in failure rates as a function of various operating conditions than with determining confidence intervals (example: CNET data bank for electronic components).

8. PROPAGATION OF UNCERTAINTIES IN RELIABILITY COMPUTATIONS

We have seen that for various reasons (small sample resulting in a large confidence interval, considerable influence of certain external parameters, etc.) the data available are unreliable. These uncertainties mean that the result of a reliability or availability computation is itself unreliable [Dorleans and Grange, 1969].

Two solutions are therefore possible for determining a confidence interval for the result: combining an analytical method with Monte Carlo simulation or using a purely analytical method.

8.1. Combination of an analytical method with Monte Carlo simulation

We can adopt the following procedure: we equate the parameters to random variables and attempt to determine the moments of the result which is itself a random variable. In general we confine our attention to the mean and the variance. When a *high-speed analytical program* is available, a very simple solution consists of carrying out Monte Carlo simulation organized on the following lines:

(a) We sample at random all the parameter values in accordance with a predetermined distribution.

(b) We then compute the value of the function f (e.g. failure rate or unavailability) with these particular parameter values using the analytical program.

We then repeat the above process a certain number of times n.

The empirical mean $\bar{f}$ can therefore be written as $\bar{f} = \frac{1}{n} \sum_{i=1}^{n} f_i$.

The empirical variance $\bar{\sigma}^2$ is written as $\bar{\sigma}^2 = \frac{1}{n-1} \sum_{i=1}^{n} (f_i - \bar{f})^2$.

The n random variables f_i have the same distribution. If we assume that the mean $E[f]$ and the variance σ^2 of this distribution are finite, the central limit theorem states that the variable $\bar{f}$ is asymptotically normally distributed with mean $E[f]$ and variance σ^2/n. This enables us to determine the parameter n by laying down an *a priori* accuracy for the computation of $\bar{f}$ (cf. Ch. 5). This method also makes it possible to determine all the quantiles of the distribution of f, and therefore to determine a confidence interval.

8.2. Entirely analytical method

If the above method cannot be used, we have to attempt to compute an approximation for the moments of f by an analytical method. The reader will recall the relations giving $\tilde{\Lambda}_\infty$ and A_∞ (using the notations in Ch. 6):

$$\tilde{\Lambda}_\infty = \frac{\sum_{i \in M_c} \Lambda_i P_i(\infty)}{A(\infty)} \simeq \sum_{i \in M_c} \Lambda_i P_i(\infty)$$

$$A_\infty = \sum_{i \in M} P_i(\infty)$$

with

$$P_i(\infty) = \prod_{j \in \mathcal{P}_i} q_j(\infty) \prod_{j \in \mathcal{M}_i} (1 - q_j(\infty)), \quad \tau_j = \frac{1}{\mu_j}.$$

Let us consider the very common case in which the system is made up of components of good availability, i.e. $\lambda_j \tau_j \ll 1$.

In this case:

$$P_i(\infty) \simeq \prod_{j \in \mathcal{P}_i} \lambda_j \tau_j \prod_{j \in \mathcal{M}_i} (1 - \lambda_j \tau_j) \simeq \prod_{j \in \mathcal{P}_i} \lambda_j \tau_j.$$

In the high-reliability area, we can also write

$$1 - R(t) \simeq e^{-\Lambda_\infty t} \simeq 1 - \Lambda_\infty t.$$

In short, we can therefore assume that the functions f which we are computing are polynomials in λ_i and τ_i.

Example 1: Two components (λ_1, τ_1) and (λ_1, τ_2) in an active redundancy configuration.

$$ID_\infty = 1 - A_\infty \simeq \lambda_1 \lambda_2 \tau_1 \tau_2$$

$$\tilde{\Lambda}_\infty = \lambda_1 \lambda_2 (\tau_1 + \tau_2).$$

Example 2: Two components (λ_1, τ_1) and (λ_2, τ_2) in series.

$$ID_\infty = 1 - A_\infty \simeq \lambda_1 \tau_1 + \lambda_2 \tau_2$$
$$\tilde{\Lambda}_\infty = \lambda_1 + \lambda_2 .$$

As the function f is a polynomial in λ_i and τ_i, we therefore have to examine the sum and the product of random variables. Consider two random variables X and Y with respective finite means $\bar{x}$ and $\bar{y}$ and with respective finite variances σ_x^2 and σ_y^2.

Let us recall the results obtained in Ch. 1:

$$E(X + Y) = \bar{x} + \bar{y} \tag{65}$$

and if X and Y are independent:

$$\text{Var}(X + Y) = \sigma_x^2 + \sigma_y^2 \tag{66}$$
$$E(XY) = \bar{x}\bar{y} \tag{67}$$
$$\text{Var}(XY) = \sigma_x^2 \sigma_y^2 + \bar{x}^2 \sigma_y^2 + \bar{y}^2 \sigma_x^2 . \tag{68}$$

We denote the median of random variable X by med (X).

In the case of a *lognormal distribution* for X and Y, we additionally have the following properties:

$$\text{med}(XY) = \text{med}(X)\,\text{med}(Y) \tag{69}$$
$$\text{med}(X^2) = (\text{med}(X))^2 . \tag{70}$$

These latter expressions explain why the median is used instead of the mean for the propagation of uncertainties [U.S.N.R.C., 1975].

In fact, Eqn (69) follows from Eqn (67) but Eqn (70) is much simpler to use than:

$$E(X^2) = \bar{x}^2 + \sigma_x^2 . \tag{71}$$

However, in the following we shall use the mean in preference to the median, as in general we have:

$$\text{med}(Y_1 + Y_2) \neq \text{med}(Y_1) + \text{med}(Y_2) . \tag{72}$$

Example 1 (continued)

Obviously we have:

$$E(ID_\infty) = \overline{ID}_\infty = \bar{\lambda}_1 \bar{\lambda}_2 \bar{\tau}_1 \bar{\tau}_2 .$$

If, in addition, the λ_i and τ_i are lognormally distributed, we obtain:

$$\text{med}(ID_\infty) = \text{med}(\lambda_1)\,\text{med}(\lambda_2)\,\text{med}(\tau_1)\,\text{med}(\tau_2) .$$

In order to analyse $\tilde{\Lambda}_\infty$ we have to formulate it as $\lambda_1\lambda_2(\tau_1 + \tau_2)$ and not as $\lambda_1\lambda_2\tau_1 + \lambda_1\lambda_2\tau_2$ as, in that case, $\lambda_1\lambda_2\tau_1$ and $\lambda_1\lambda_2\tau_2$ are not independent.

It follows therefore that:

$$E(\tilde{\Lambda}_\infty) = \bar{\Lambda}_\infty = \bar{\lambda}_1 \bar{\lambda}_2 (\bar{\tau}_1 + \bar{\tau}_2) .$$

We put $\sigma_i^2 = \text{var}(\lambda_i)$, $r_i^2 = \text{var}(\tau_i)$.

To calculate the variance of ID_∞ and $\tilde{\Lambda}_\infty$ we first compute:

$$E(\lambda_1 \lambda_2) = \bar{\lambda}_1 \bar{\lambda}_2, \text{Var}(\lambda_1 \lambda_2) = \sigma_1^2 \sigma_2^2 + \sigma_2^2 \bar{\lambda}_1^2 + \sigma_1^2 \bar{\lambda}_2^2$$

$$E(\tau_1 \tau_2) = \bar{\tau}_1 \bar{\tau}_2, \text{Var}(\tau_1 \tau_2) = r_1^2 r_2^2 + r_2^2 \bar{\tau}_1^2 + r_1^2 \bar{\tau}_2^2$$

$$E(\tau_1 + \tau_2) = \bar{\tau}_1 + \bar{\tau}_2, \text{Var}(\tau_1 + \tau_2) = r_1^2 + r_2^2$$

hence

$$\text{Var}(ID_\infty) = \text{Var}(\lambda_1 \lambda_2) \text{Var}(\tau_1 \tau_2) + \text{Var}(\lambda_1 \lambda_2) [E(\tau_1 \tau_2)]^2 + \\ + \text{Var}(\tau_1 \tau_2) [E(\lambda_1 \lambda_2)]^2$$

$$\text{Var}(\tilde{\Lambda}_\infty) = \text{Var}(\lambda_1 \lambda_2) \text{Var}(\tau_1 + \tau_2) + \text{Var}(\lambda_1 \lambda_2) [E(\tau_1 + \tau_2)]^2 + \\ + \text{Var}(\tau_1 + \tau_2) [E(\lambda_1 \lambda_2)]^2 .$$

Note, therefore, that thanks to factorizations, we usually only need to consider the sums and the products of independent random variables, and can therefore use Eqns (65)–(68).

It is consequently necessary to attempt to factorize the polynomials ID_∞ and $\tilde{\Lambda}_\infty$ before applying Eqns (65)–(68). However, factorization is not always possible, even in the case of statistically independent components. There may also exist statistical dependencies between the components (e.g. components from the same batch and also possible of poor quality). Exercise 4 contains an example of these various situations.

These methods, as well as methods of estimating confidence intervals from the first two moments, are discussed in [Apostalakis and Lee, 1976].

9. SUMMARY

Reliability data are determined by the *parameters of the F-distributions* of successful operation or repair. These parameters are determined experimentally using samples. These samples can be obtained from *reliability trials*. However, the latter are generally costly and it is generally preferable to use the *in-service* results of the equipment.

The parameters are evaluated from samples by means of *estimators*, which are random variables enabling us to obtain a *point value* of the parameters but also, and this is preferable, to determine a confidence interval covering the true value with a specified *confidence level.*

As far as the parameter λ of the exponential distribution — the most frequently used distribution in reliability analysis — is concerned, the estimator is a function of the number of observed failures r and the total operating times S of the equipment constituting the sample. For a multicensured design, which provides the best model of a sample reflecting in-service results, the failure rate estimator $\hat{\lambda}$ is:

$$\hat{\lambda} = \frac{r}{S}.$$

In addition, the symmetrical confidence interval at confidence level α is given, to a good approximation, by:

$$\lambda_- = \frac{\chi^2_{1-\alpha/2}(2\,r)}{2\,S}$$

$$\lambda_+ = \frac{\chi^2_{\alpha/2}(2\,r)}{2\,S}.$$

Hypothesis tests subsequently enable us to check that the uptime or time to repair is actually F-distributed with the estimated parameters. The chi-squared test is somewhat insensitive, and so the Kolmogorov-Smirnov test is generally used.

When we have subjective information concerning a parameter of an F-distribution, *Bayesian estimation* enables us to add it to the objective information contained in a sample. For this purpose, we regard the parameter θ as a random variable of the given distribution ϕ. Knowledge of additional information simply modifies the distribution ϕ of the parameter. In order to simplify this modification, it is necessary that the *a priori* distribution ϕ of the parameter and the F-distribution can be simply combined. The simplest case is where the introduction of additional information does not modify the type of the distribution ϕ but simply alters the parameters of that distribution. ϕ is then said to be a *natural conjugate* of the F-distribution. The natural conjugate of an exponential distribution is a gamma distribution and the natural conjugate of a binomial distribution is a beta distribution. The Bayesian estimator of parameter θ is equal to the mean of the *a posteriori* distribution, as this mean minimizes the variance of the estimate. Determining a confidence interval then reduces to evaluating the quantiles of the *a posteriori* distribution.

The fact that the data are not known precisely but with a degree of uncertainty, poses the problem of the propagation of this uncertainty in reliability computations.

If we have a rapid method of computing the availability or the reliability, we can use it in conjunction with a Monte Carlo simulation method to analyse the distribution of the result. If not, we must find an analytical approximation for the mean and the variance of the result.

EXERCISES

Exercise 1 — Unbiased estimator of the variance of a sample of unknown mean

Consider a sample $(x_1, x_2, \ldots, x_n)$ of a random variable X of unknown mean m and unknown standard deviation σ. The x_i are assumed to be independent.

(1) Compute the variance σ_m^2 of the estimator $\hat{m} = \frac{1}{n}\sum_{i=1}^{n} x_i$ of the mean.

(2) Consider the estimator $S = \frac{1}{n}\sum(x_i - \hat{m})^2$ of the variance σ^2. Show that $E(S) = \sigma^2 - \sigma_m^2$.

(3) Deduce an unbiased estimator $\hat{\sigma}^2$ of the variance.

Exercise 2 — General method of determining a confidence interval in the case of a parameter having an unbiased estimator

(1) *Acceptance interval of a random variable.*
Consider a real random variable with a known continuous probability distribution.
Let there be a set of intervals I of R such that:

$$\mathcal{P}(X \in I) = 1 - \alpha \qquad (0 < \alpha < 1).$$

If we state that X is always located on I, we are putting forward a proposition which has a probability $1-\alpha$ of being true and a probability α of being false. We shall call I the acceptance interval of the variable X at risk level α.

We are interested only in the acceptance intervals consisting of a single interval. We shall denote this interval by $[x', x'']$.

Put $\alpha_1 = \mathcal{P}(X \leqslant x')$

$\alpha_2 = \mathcal{P}(X > x'')$

$\alpha = \alpha_1 + \alpha_2$.

Assume that the cost of an overprediction is equal to C_1 and the cost of an underprediction is C_2. Compute α_1 and α_2 for which the average under- and overprediction costs are equal.

If we have no information on C_1 and C_2, we take $C_1 = C_2$. We then have a two-sided interval with symmetrical tails.

Show that in this case the distribution of X is symmetric and the length of the interval I is minimal (for given α).

(2) *Confidence interval of an estimator.*

Consider a continuous distribution dependent on a parameter θ. Assume that we know an unbiased estimator S of θ. ($E(S) = \theta$).

S is a random variable and we assume that for given θ its probability distribution is known. For all θ, we can define the acceptance interval at level $\alpha(\alpha_1, \alpha_2)$ of estimator S.

Let $S'(\theta)$ and $S''(\theta)$ be the bounds of this interval. These are the α_1 and $1-\alpha_2$ order quantiles of the distribution of S for given θ.

Show that in the trial design (θ, S) the curves $S'(\theta)$ and $S''(\theta)$ are located on either side of the first bisectrix for small α. The curves $S'(\theta)$ and $S''(\theta)$ encountered in practice are monotonically increasing and with constant sign concavity: $S'(\theta)$ rotates its concavity towards positive S and $S''(\theta)$ towards negative S. We shall assume for the rest of the exercise that this is so.

(3) *Confidence interval of a parameter to be estimated.*

Two numerical values S_0 and θ_0 are said to be in correspondence if S_0 belongs to the acceptance interval of S when $\theta = \theta_0$. Given S_0, consider the set Ω of the values of θ such that S_0 is in the acceptance interval of θ.

(3.1) Demonstrate a graphical method of constructing this set.

(3.2) Deduce from hypotheses made on the curves $S'(\theta)$ and $S''(\theta)$ that Ω is an interval in itself and that $\theta_0 = S_0 \in \Omega$.

(3.3) Putting $\Omega = [\theta', \theta'']$,
show that $S''(\theta') = S(\theta'')$

$$\mathcal{P}(S > S_0 \mid \theta = \theta') = \alpha_2$$

$$\mathcal{P}(S < S_0 \mid \theta = \theta') = \alpha_1 .$$

S_0 is both the upper bound of the acceptance interval of S when $\theta = \theta'$ and the lower bound of the acceptance interval of S when $\theta = \theta''$.

(3.4) Show that Ω covers the true value of θ with probability $1-\alpha$. Deduce that Ω is the confidence interval of θ at $100(1-\alpha_1-\alpha_2)\%$.

Exercise 3—Estimation of the exponential distribution parameter in an $[n, V, r]$ design

(1) Determine the probability density of the random variable I = time of occurrence of the rth failure.

(2) Determine the mean of the maximum likelihood estimator. Deduce from it an unbiased estimator.

(3) Compute the variance of this unbiased estimator.

(4) Compute the symmetrical confidence interval bounds at the $100(1-\alpha)\%$ level.

Exercise 4—Propagation of uncertainties in reliability computations

(1) *Compute the mean and the variance* of ID_∞ and Λ_∞ in Example 2, section 8.2.

(2) *Factorization impossible.*
Consider three components $(\lambda_1, \tau_1), (\lambda_2, \tau_2), (\lambda_3, \tau_3)$ in an active redundancy setup.

(2.1) Compute ID_∞ and $\tilde{\Lambda}_\infty$.

(2.2) Compute $E(ID_\infty)$ and $\mathrm{Var}(ID_\infty)$.

(2.3) Compute $E(\tilde{\Lambda}_\infty)$.

(2.4) Compute $\mathrm{Var}(\tau_1\tau_2 + \tau_2\tau_3 + \tau_3\tau_1)$.

(2.5) Deduce $\mathrm{Var}(\tilde{\Lambda}_\infty)$.

(3) *Components from the same batch.*
Consider a system comprising two components in an active redundancy setup. Both components come from the same batch. Their failure and repair rates λ and μ are *equal* but known with a degree of uncertainty.

(3.1) Compute ID_α and $\tilde{\Lambda}_\infty$.

(3.2) Assume that τ is constant and that λ is lognormally distributed with parameters a and b such that:

$$\bar{\lambda} = E(\lambda) = \exp\left(a + \frac{b^2}{2}\right)$$

$$\sigma^2 = \mathrm{Var}(\lambda) = e^{2a+b^2}(e^{b^2} - 1)\,.$$

Compute $E(ID_\infty)$, $E(\tilde{\Lambda}_\infty)$, $\mathrm{Var}(ID_\infty)$ and $\mathrm{Var}(\tilde{\Lambda}_\infty)$.

(3.3) Given a random variable X, compute the variance of X^2 as a function of moments of order less than or equal to 4.
Application to the special case of a normal distribution.

(3.4) Compute the variance of $\tilde{\Lambda}_\infty$ when λ is normally distributed.

(4) Consider a system comprising a single component whose failure rate λ and mean time to repair τ are lognormally distributed. Assume that λ and τ are independent.

Let the limit unavailability $q_\infty = \dfrac{\lambda\tau}{1 + \lambda\tau}$, which is therefore a random variable.

Show that $\dfrac{1 - q_\infty}{q_\infty}$ is lognormally distributed.
Deduce a confidence interval on q_∞.

SOLUTIONS TO EXERCISES

Exercise 1

(1) $\sigma_m^2 = E(\hat{m} - m)^2 = E\left(\frac{1}{n}\sum_{i=1}^{n} x_i - m\right)^2 = \frac{1}{n}E(x_i)^2\,.$

As the variables x_i are independent, consequently:

$$E(x_i\, x_j) = E(x_i)\, E(x_j)$$

therefore

$$\sigma_m^2 = \frac{1}{n}E(x_i)^2 + \frac{n-1}{n}[E(x_i)]^2 - m^2 = \frac{E(x_i^2) - [E(x_i)]^2}{n} = \frac{\sigma^2}{n}\,.$$

$$(2)\quad E(S) = E\left(\frac{\sum_{i=1}^{m}(x_i - \hat{m})^2}{n}\right) = E\left(\frac{\sum_{i=1}^{n} x_i^2 - \hat{m}^2}{n}\right) = E\left(\frac{\sum_{i=1}^{n} x_i^2}{n}\right) - E(\hat{m}^2)$$

$$= \sigma^2 + m^2 - E(\hat{m}^2) = \sigma^2 - \sigma_m^2\,.$$

(3) $E(S) = \sigma^2 - \sigma_m^2 = \sigma^2 - \dfrac{\sigma^2}{n} = \dfrac{n-1}{n}\sigma^2.$

It follows that $\hat{\sigma}^2 = \dfrac{1}{n-1}\sum_{i=1}^{n}(x_i - \hat{m})^2$ is an unbiased estimator of σ^2.

Exercise 2

(1) Average cost of overprediction: $\alpha_1 C_1$.
Average cost of underprediction: $\alpha_2 C_2$

$$\text{therefore } \alpha_1 = \frac{\alpha C_2}{C_1 + C_2} \qquad \alpha_2 = \frac{\alpha C_1}{C_1 + C_2}$$

if $C_1 = C_2$, $\alpha_1 = \alpha_2 = \dfrac{\alpha}{2}$.

Let there be an interval $[x', x''[$ such that $\mathcal{P}(X \in [x', x''[) = 1 - \alpha$.

Let there be another interval $[x' + dx', x'' + dx''[$ also containing the probability $1 - \alpha$. Consequently:

$$f(x')\,dx' = f(x'')\,dx''$$

the length increase is therefore $dx'' - dx'$.

If the distribution of X is symmetrical and $\alpha_1 = \alpha_2$, it follows that $f(x') = f(x'')$, therefore $dx'' = dx'$ and the length increase is zero. As the derivatives of $f(x)$ in x' and x'' are of opposite sign, a minimum is involved.

(2) Since $\alpha = \alpha_1 + \alpha_2$ is small, the mathematical expectation of S lies within the acceptance interval

$$S'(\theta) < E(S) < S''(\theta)$$

the estimator being unbiased, it follows that:

$$S'(\theta) < \theta < S''(\theta)$$

the first bisectrix $S = \theta$ is therefore between the curves $S'(\theta)$ and $S''(\theta)$.

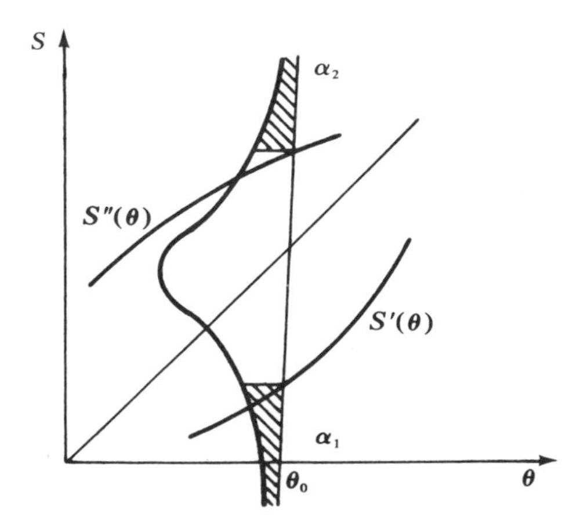

(3.1) For S_0 to belong to the acceptance interval of S for $\theta = \theta_0$, it is necessary and sufficient that the two lines $S = S_0$ and $\theta = \theta_0$ should intersect between the lines $S'(\theta)$ and $S''(\theta)$. Ω is therefore obtained by plotting the line $S = S_0$. This line intersects the curves $S'(\theta)$ and $S''(\theta)$ respectively at points θ' and θ'' on the abscissa.

(3.2) Consequently, $\Omega = [\theta', \theta'']$ and $\theta_0 \in \Omega$.

(3.3) In graphical terms, the following situation is obtained:

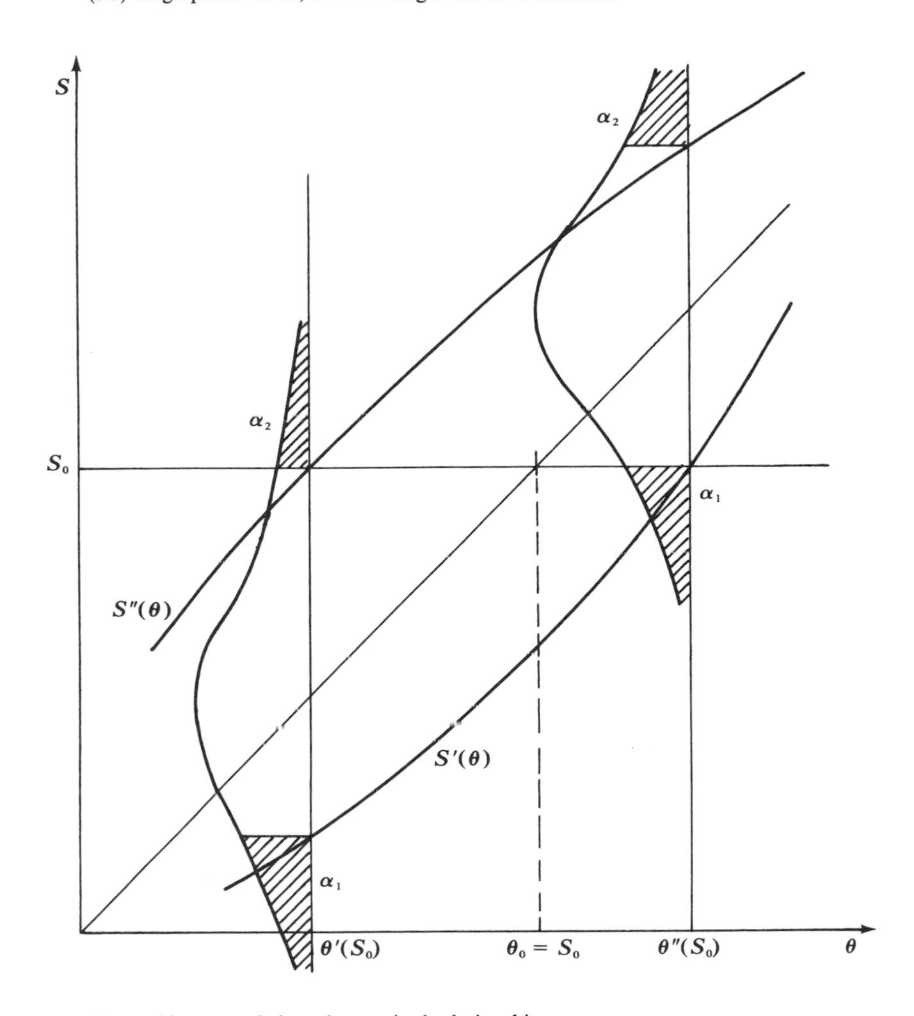

From this we can deduce the required relationships.

(3.4) Assume that the true value of θ is θ^*. Let S^* be the estimate obtained from a sample. Therefore:

$\theta^* \in [\theta'(S^*), \theta''(S^*)] \Leftrightarrow$ the coordinate point (θ^*, S^*) is between the curves $S'(\theta)$, $S''(\theta)$

$S^* \in$ acceptance interval of S for $\theta = \theta^*$.

This means that there is a probability $1 - \alpha$ of Ω covering the true value. This is therefore a confidence interval at $100(1 - \alpha)\%$.

Exercise 3

(1) As failed components are replaced, the failure density is constant and equals $n\lambda$. Denoting by T_r the random variable of the occurrence of the rth failure, we therefore obtain:

$$f(t_r)\,\mathrm{d}t_r = \mathcal{P}(t_r \leq T_r \leq t_r + \mathrm{d}t_r) = \mathcal{P}\begin{pmatrix} r-1 \text{ failures} \\ \text{during } [0, t_r] \end{pmatrix} \times n\lambda\,\mathrm{d}t_r.$$

As the failure density is constant, the number of failures over $[0, t_r]$ is Poisson-distributed with parameter $n\lambda$. Hence

$$f(t_r) = n\lambda \frac{(n\lambda t_r)^{r-1}\, e^{-n\lambda t_r}}{(r-1)!}.$$

We of course obtain the density of an Erlangian distribution.

(2) Let $\hat{\lambda} = r/nt_r$ be the maximum likelihood estimator

$$E[\hat{\lambda}] = \int_0^\infty \frac{r}{nt_r} \frac{(n\lambda)^r\, t_r^{r-1}\, e^{-n\lambda t_r}}{(r-1)!} \mathrm{d}t_r = \frac{r}{r-1}\lambda \text{ for } r > 1$$

if $r \leqslant 1$, $E[\hat{\lambda}]$ does not exist.

$\hat{\lambda}$ is therefore a biased estimator. As an unbiased estimator we take $\tilde{\lambda} = \dfrac{r-1}{nt_r}$.

(3)
$$E[\tilde{\lambda}^2] = \int_0^\infty \frac{(r-1)^2}{n^2 t_r^2} f(t_r)\,\mathrm{d}t_r = \frac{\lambda^2(r-1)}{r-2} \qquad r > 2$$

therefore
$$\sigma^2 = \frac{\lambda^2(r-1)}{r-2} - \lambda^2 = \frac{\lambda^2}{r-2}.$$

(4) Compute the density of the random variable $u = n\lambda t_r$. Using the result of the first question, we obtain:

$$f(u) = \frac{u^{r-1}\, e^{-u}}{(r-1)!} \text{ which does not depend on the unknown } \lambda.$$

Let X therefore be such that:

$$\int_0^X \frac{u^{r-1}\, e^{-u}}{(r-1)!} \mathrm{d}u = \frac{\alpha}{2} = \mathcal{P}(u \leq X) = \mathcal{P}\left(\lambda \leq \frac{X}{nt_r}\right)$$

putting $u = t/2$ and $2r = v$, we obtain:

$$\int_0^{2X} \frac{t^{\nu/2-1} \exp\left(-\frac{t}{2}\right)}{2^{\nu/2}\, \Gamma\left(\frac{\nu}{2}\right)} \mathrm{d}t = \frac{\alpha}{2}$$

hence $X = \frac{1}{2}\chi^2_{1-\alpha/2}(2r)$.

Consequently

$$\lambda_- = \frac{\chi^2_{1-\alpha/2}(2r)}{2\,nt_r}.$$

similarly

$$\lambda_+ = \frac{\chi^2_{\alpha/2}(2r)}{2\,nt_r}.$$

Exercise 4

We put $E(\lambda_i) = \bar{\lambda}_i \qquad E(\tau_i) = \bar{\tau}_i$

$\mathrm{Var}(\lambda_i) = \sigma_i^2 \qquad \mathrm{Var}(\tau_i) = r_i^2$

(1) $E(ID_\infty) = \bar{\lambda}_1 \bar{\tau}_1 + \bar{\lambda}_2 \bar{\tau}_2$

$E(\tilde{\Lambda}_\infty) = \bar{\lambda}_1 + \bar{\lambda}_2$

$\mathrm{Var}(ID_\infty) = \sigma_1^2 r_1^2 + \sigma_1^2 \bar{\tau}_1^2 + \bar{\lambda}_1^2 r_1^2 + \sigma_2^2 r_2^2 + \sigma_2^2 \bar{\tau}_2^2 + \bar{\lambda}_2^2 r_2^2$

$\mathrm{Var}(\tilde{\Lambda}_\infty) = \sigma_1^2 + \sigma_2^2.$

(2) (2.1) $ID_\infty = \lambda_1 \lambda_2 \lambda_3 \tau_1 \tau_2 \tau_3$

$\tilde{\Lambda}_\infty = \lambda_1 \lambda_2 \lambda_3 (\tau_1 \tau_2 + \tau_2 \tau_3 + \tau_3 \tau_1)$ (cf. ch.6).

(2.2) $E(ID_\infty) = \bar{\lambda}_1 \bar{\lambda}_2 \bar{\lambda}_3 \bar{\tau}_1 \bar{\tau}_2 \bar{\tau}_3$

$$\mathrm{Var}(\lambda_1 \lambda_2 \lambda_3) = \sigma_1^2 \sigma_2^2 \sigma_3^2 + \bar{\lambda}_1^2 \sigma_2^2 \sigma_3^2 + \bar{\lambda}_2^2 \sigma_1^2 \sigma_3^2 + \bar{\lambda}_3^2 \sigma_1^2 \sigma_2^2 + \bar{\lambda}_1^2 \bar{\lambda}_2^2 \sigma_3^2 + \bar{\lambda}_1^2 \bar{\lambda}_3^2 \sigma_2^2 + \bar{\lambda}_2^2 \bar{\lambda}_3^2 \sigma_1^2$$

$$\mathrm{Var}(\tau_1 \tau_2 \tau_3) = r_1^2 r_2^2 r_3^2 + \bar{\tau}_1^2 r_2^2 r_3^2 + \bar{\tau}_2^2 r_1^2 r_3^2 + \bar{\tau}_3^2 r_1^2 r_2^2 + \bar{\tau}_1^2 \bar{\tau}_2^2 r_3^2 + \bar{\tau}_1^2 \bar{\tau}_3^2 r_2^2 + \bar{\tau}_2^2 \bar{\tau}_3^2 r_1^2$$

$$E(\lambda_1 \lambda_2 \lambda_3) = \bar{\lambda}_1 \bar{\lambda}_2 \bar{\lambda}_3 \quad E(\tau_1 \tau_2 \tau_3) = \bar{\tau}_1 \bar{\tau}_2 \bar{\tau}_3 .$$

We then apply Eqn (68) to determine $\mathrm{Var}(\lambda_1 \lambda_2 \lambda_3 + \tau_1 \tau_2 \tau_3)$.

(2.3) $E(\tilde{\Lambda}_\infty) = \bar{\lambda}_1 \bar{\lambda}_2 \bar{\lambda}_3(\bar{\tau}_1 \bar{\tau}_2 + \bar{\tau}_2 \bar{\tau}_3 + \bar{\tau}_3 \bar{\tau}_1)$.

(2.4) $T = \tau_1 \tau_2 + \tau_2 \tau_3 + \tau_3 \tau_1$

$$E(T) = \bar{\tau}_1 \bar{\tau}_2 + \bar{\tau}_2 \bar{\tau}_3 + \bar{\tau}_3 \bar{\tau}_1 = \bar{T}$$

$$\mathrm{Var}(T) = r_1^2 r_2^2 + r_2^2 r_3^2 + r_3^2 r_1^2 + r_1^2(\bar{\tau}_2 + \bar{\tau}_3)^2 + r_2^2(\bar{\tau}_1 + \bar{\tau}_3)^2 + r_3^2(\bar{\tau}_1 + \bar{\tau}_2)^2 .$$

(2.5) $\mathrm{Var}(\tilde{\Lambda}_\infty) = \mathrm{Var}(\lambda_1 \lambda_2 \lambda_3)(\mathrm{Var}\, T + \bar{T}^2) + \mathrm{Var}(T)(\bar{\lambda}_1 \bar{\lambda}_2 \bar{\lambda}_3)^2$

(3) (3.1) $ID_\infty \simeq (\lambda t)^2$

$$\tilde{\Lambda}_\infty \simeq 2 \lambda^2 \tau .$$

(3.2) λ^2 is lognormally distributed with parameters $2a$, $2b$. We put $\sigma^2 = \mathrm{Var}(\lambda)$, hence:

$$\mathrm{var}(\lambda^2) = e^{4a+4b^2}(e^{4b^2} - 1) = (\sigma^2 + \bar{\lambda}^2)\left(\left(1 + \frac{\sigma^2}{\bar{\lambda}^2}\right)^4 - 1\right)$$

$$E(ID_\infty) = (\bar{\lambda}^2 + \sigma^2)\tau^2$$

$$E(\tilde{\Lambda}_\infty) = 2(\bar{\lambda}^2 + \sigma^2)\tau$$

$$\mathrm{var}(ID_\infty) = \mathrm{var}(\lambda^2)\tau^2$$

$$\mathrm{var}(\tilde{\Lambda}_\infty) = 4\, \mathrm{var}(\lambda^2)\tau .$$

(3.3) $\mathrm{var}(X^2) = \mu_4 + 4\mu_1' \mu_3 + 4\mu_1'^2 \mu_2 - \mu_2^2$

$$\mu_i' = E(X^i), \; \mu_i = E((X - \mu_1')^i)$$

if X is normal, $\mu_3 = 0$, $\mu_4 = 3\sigma^4$ and $\mathrm{var}(X^2) = 4\bar{\lambda}^2 \sigma^2 \left(1 + \frac{1}{2}\frac{\sigma^2}{\bar{\lambda}^2}\right)$.

(3.4) $\mathrm{var}(\tilde{\Lambda}_\infty) = 4(\mathrm{var}(\lambda^2) r^2 + \mathrm{var}(\lambda^2) E^2 + (\bar{\lambda}^2 + \sigma^2)^2 r^2)$

with $\quad \mathrm{var}(\lambda^2) = 4\sigma^2 \bar{\lambda}^2 \left(1 + \frac{1}{2}\frac{\sigma^2}{\bar{\lambda}^2}\right)$.

(4) Putting $\mathrm{Lg} = -\mathrm{Log}\,\lambda\tau = -\mathrm{Log}\,\lambda - \mathrm{Log}\,\tau$
it follows that Lg is normally distributed.

However, $\mathrm{Lg} = \mathrm{Log}\dfrac{1 - q_\infty}{q_\infty}$. Consequently, $\dfrac{1 - q_\infty}{q_\infty}$ is lognormally distributed, hence the confidence interval at the $1 - \alpha_1 - \alpha_2$ level is:

$$\left[\frac{1}{1 + \exp(-(\lambda + \tau) + u_\alpha \sqrt{\sigma^2 + r^2})}, \frac{1}{1 + \exp(-(\lambda + \tau) - u_\alpha \sqrt{\sigma^2 + r^2})}\right].$$

REFERENCES

APOSTOLAKIS G. E. and LEE Y. T. (1976): *Probability Intervals for the Top Event Unavailability of Fault Trees*; U.C.L.A., ENG 7663.

BARTHOLOMEW D. J. (1963): The sampling distribution of an estimate arising in life testing; *Technometrics*, **5**, pp. 361–374.

BELIAEV Y., GNEDENKO B. and SOLOVIEV A. (1972): *Mathematical Methods in Reliability*; MIR, Moscow.
COHEN A. C. (1963): Progressively censured samples in life testing; *Technometrics*, **5**, no. 3.
COHEN A. C. (1976): Progressively censured sampling in the three parameter lognormal distribution; *Technometrics*, **18**, no. 1.
COHEN A. C. and NORGAARD N. J. (1977): Progressively censured sampling in the three parameter gamma distribution, *Technometrics*, **19**, no. 3.
DORLEANS J. and GRANGE J. M. (1969): Précision des calculs prévisionnels de fiabilité; *L'onde électrique*, **49**, no. 5.
FOURGEAUD C. and FUCHS A. (1972): *Statistique*; Dunod, Paris.
GOLDSTEIN R. B. (1973): Chi-square quantiles. Algorithm no. 451; *Communications of the ACM*, **16**, no. 8.
HOEL P. G. (1971): *Introduction to Mathematical Statistics*; Wiley, New York.
HOOKE R. and JEEVES T. A. (1961): Direct search solution of numerical and statistical problems; *JACM*, **8**, no. 2, p. 212.
JOHNSON M. M., WALLER R. A., WATERMAN M. S. and MARTZ H. F. (1977): *Gamma prior distribution selection for Bayesian analysis of failure rate and reliability*; International Conference on *Nuclear Systems Reliability Engineering and Risk Assessment*, Gatlinburg.
JOHNSON N. L. and KOTZ S. (1970): *Discrete Distributions 2*; Houghton Mifflin, Boston.
KAPUR K. C. and LAMBERSON L. R. (1977): *Reliability in Engineering Design*; Wiley, New York.
LIGERON J. C. (1972): Modélisation multidimensionnelle de taux de défaillance; *Congrès National de Fiabilité*; Perros Guirec.
MANN N. R., SCHAFER R. E. and SINGPURWALLA N. D. (1974): *Methods for Statistical Analysis of Reliability and Life Data*; Wiley, New York.
MEEKER W. Q. and NELSON W. (1976): Weibull percentile estimates and confidence limits from singly censured data by maximum likelihood; *IEEE Transactions on Reliability*, **R25**, no. 1.
MORICE E. (1977): Emploie des tables de la loi de F pour le calcul de l'intervalle de confiance du paramètre p d'une loi binomiale; *Revue de Statistique Appliquée*, **XXV**, no. 2, pp. 33, 38.
MORLAT G. (1978): Sur la comparaison des intervalles de confiance classiques et bayésiens; *Revue de Statistique Appliquée*, no. 2.
PEYRACHE G. and SCHWOB M. (1969): *Traité de fiabilité*; Masson, Paris.
RAIFFA H. and SCHLAIFER R. (1961): *Applied Statistical Decision Theory*; Harvard University, Boston.
U.S.N.R.C. (1975): Nuclear Regulatory Commission, *Reactor Safety Study. An assessment of Accident Risk in U.S. Commercial Nuclear Power Plants*; WASH 1400 (NUREG–75/014). Appendix III. Failure Data. Washington D.C.
WINGO D. R. (1973): Solution of the three parameter Weibull equations by constrained modified quasilinearization (progressively censured samples); *IEEE Transactions on Reliability*, **R22**, no. 2.
WOLFE P. (1962): *The Reduced Gradient Method*; Rand Document.

CHAPTER 8

OPTIMIZATION AND MAINTENANCE

This chapter introduces some optimization and maintenance problems which may arise at various stages in the life of a system.

Section 1 examines a problem frequently encountered at the system design stage: how to optimize redundancy.

Section 2 deals with a fundamental problem: the common mode failures of a system; failures of this kind can actually cancel out the gain in reliability achieved by using redundant components or systems.

Sections 3, 4 and 5 are devoted to much more traditional maintenance problems: stock control of spares, the impact of testing standby components on system availability and the role of preventive maintenance.

1. REDUNDANCY OPTIMIZATION

1.1. Statement of the problem

Systems which have to perform their missions under maintenance-free conditions (aircraft in flight, satellites, rockets, etc.) must be designed to ensure maximum reliability (*) throughout their mission. In addition, certain constraints of a technical nature (overall dimensions, weight, floorspace) or of an economic nature (component costs, budget) have to be taken into account.

Consequently, a specification may list requirements such as "the item shall operate at maximum reliability over a time T, its cost shall not exceed C and its weight shall not exceed P", or "the system reliability shall exceed R_{min} over a period T at minimum cost".

For present purposes, we shall merely consider a series-parallel configuration, i.e. one made up of K stages all of which must be operating for the system to be able to function.

(*) This is a case of non-repairable systems where reliability $\equiv$ availability.

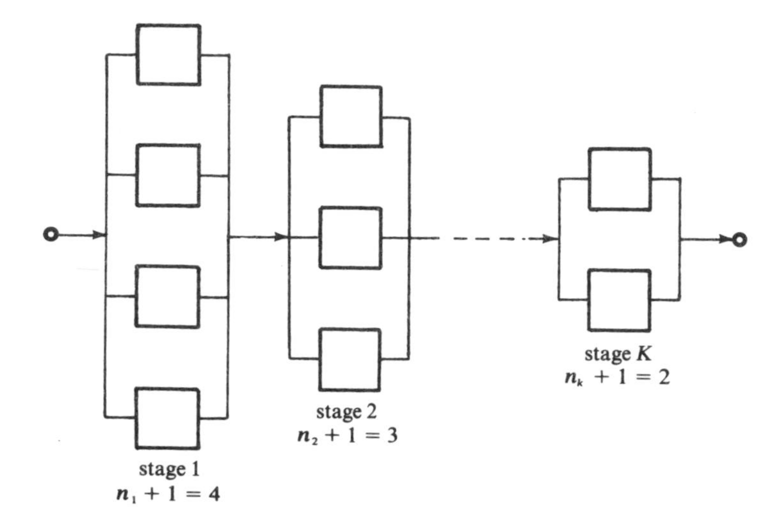

Fig. 1. *Reliability block diagram of a series-parallel configuration.*

The system reliability R over a given period is the product of the reliabilities of the stages: $R = \prod_{i=1,K} R_i$.

Stage i comprises $n_i + 1$ identical components with reliability p_i. We therefore have $R_i = 1 - (1 - p_i)^{n_i+1}$.

If each type i component is assigned costs C_{ij} ($j = 1, 2, \ldots, m$) the technical and economic constraints (cost, weight, volume, etc.) can be written as:

$$\sum_{i=1,K} C_{ij} n_i \leq C_j \qquad \forall j = 1, 2, \ldots, m. \tag{1}$$

The problem of selecting maximum redundancy then becomes one of determining a vector $n = (n_1, n_2, \ldots, n_K)$ with n_i positive integers, solving the problem:

$$\begin{cases} \text{maximize } R(n) = \prod_{i=1,K} [1 - (1 - p_i)^{n_i+1}] \\ \text{subject to constraints } \sum_{i=1,K} C_{ij} n_i \leq C_j \qquad \forall j = 1, 2, \ldots, m \\ \qquad n_i \text{ positive integer} \qquad \forall i = 1, \ldots, K. \end{cases} \tag{2}$$

This problem is generally very difficult to solve, as the function $R(n)$ is *non-linear* and the required solutions n_i are *integer-valued* (the continuous function of problem (2) is not always much help in determining the integer-valued solution).

In section 1.2, we shall describe a simple case in which the optimum solution is easily obtained.

In section 1.3, we extend this case to more general situations, obtaining excellent approximate solutions.

In section 1.4, we discuss how dynamic programming can be used to obtain precise solutions for *small-scale* problems.

Let us consider this problem further. The constraints $C_j (j = 1, \ldots, m)$ such as cost, weight, volume, etc. are not always precisely defined and it is often useful to take them as parameters and to see how the reliability varies as a function of their variations. This sensitivity analysis is performed explicitly below.

Lastly, in certain problems the constraint may be defined in terms of reliability and the function to be optimized is the cost. The problem can therefore be written (cost corresponding to $j = 1$ in Eqn (1)) as:

$$\left\{\begin{array}{lll} \text{minimize} \quad C(n) = \sum_{i=1,K} C_{1i} n_i & & \\ \text{subject to constraints } R(n) \geq R_0 & & \\ \qquad \sum_{i=1,K} C_{ij} n_i \leq C_j & \forall j = 2, \ldots, m & \\ \qquad n_i \text{ positive integer} & \forall i = 1, \ldots, K. & \end{array}\right. \tag{3}$$

Sensitivity analysis of problem (2) then makes it possible to solve problem (3).

1.2. Solution for a simple case

We shall assume that there is only one constraint (e.g. cost) and that the cost of each type of component is the same.

The reliability maximization problem for a specified number of components N can therefore be written as:

$$\left\{\begin{array}{l} \max R(n) = \prod_{i=1,K} [1 - (1 - p_i)^{n_i + 1}] \\ \text{with} \quad \sum_{i=1,K} n_i = N \\ \qquad n_i \text{ integer} \geq 0. \end{array}\right. \tag{4}$$

Assuming that p_i, the reliability of a stage i component, is large, $q_i = 1 - p_i$ is small and $R(n)$ can be written as:

$$R(n) \simeq 1 - \sum_{i=1,K} q_i^{n_i+1}. \tag{5}$$

We shall adopt this form in the following paragraphs, although the reasoning applied remains valid for the general case (problem (4)).

The following algorithm then provides an optimum solution to the problem.

Algorithm 1

1. (given N). For i from 1 to K, $n_i = 0$.

$$R \simeq 1 - \sum_{i=1,K} q_i.$$

2. Select i_0 such that:

$$r = -q_{i_0}^{n_{i_0}+2} + q_{i_0}^{n_{i_0}+1} = \max_{i=1,K} (-q_i^{n_i+2} + q_i^{n_i+1}).$$

3. Make $n_{i_0} \leftarrow n_{i_0} + 1, R \leftarrow R + r$
$N \leftarrow N - 1$

If $N = 0$, END; if not, go to 2.

Theorem 1 — *The solution* n^* *obtained at the end of the algorithm is optimal.*

Proof

Let n be another solution and compute

$$R(n^*) - R(n) = \sum_{i=1,K} (q_i^{n_i+1} - q_i^{n_i^*+1}).$$

Let I_1 be the set of i such that $n_i < n_i^*$ and I_2 the set of i such that $n_i > n_i^*$. We have:

$$R(n^*) - R(n) = \sum_{i \in I_1} (q_i^{n_i+1} - q_i^{n_i^*+1}) - \sum_{i \in I_2} (q_i^{n_i^*+1} - q_i^{n_i+1}).$$

If $\underline{n_i < n_i^*}$,

$$q_i^{n_i+1} - q_i^{n_i^*+1} = q_i^{n_i^*+1} \frac{1 - q_i^{n_i^*-n_i}}{q_i^{n_i^*-n_i}} = q_i^{n_i^*+1}(1 - q_i)\left(\frac{1}{q_i^{n_i^*-n_i}} + \cdots + \frac{1}{q_i}\right)$$

$$q_i^{n_i+1} - q_i^{n_i^*+1} \geq (n_i^* - n_i)(q_i^{n_i^*} - q_i^{n_i^*+1}).$$

If $\underline{n_i > n_i^*}$.

$$q_i^{n_i^*+1} - q_i^{n_i+1} = q_i^{n_i^*+1}(1 - q_i^{n_i-n_i^*}) = q_i^{n_i^*+1}(1 - q_i)(1 + q_i + q_i^2 + \cdots q_i^{n_i-n_i^*-1})$$

$$q_i^{n_i^*+1} - q_i^{n_i+1} \leq (n_i - n_i^*)(q_i^{n_i^*+1} - q_i^{n_i^*+2}).$$

Finally, therefore:

$$R(n^*) - R(n) \geq \sum_{i \in I_1} (n_i^* - n_i)(q_i^{n_i^*} - q_i^{n_i^*+1}) - \sum_{i \in I_2} (n_i - n_i^*)(q_i^{n_i^*+1} - q_i^{n_i^*+2}).$$

If r^* is the last value of r in the algorithm, we have:

$$r^* \leq q_i^{n_i^*} - q_i^{n_i^*+1} \quad \text{et} \quad r^* \geq q_i^{n_i^*+1} - q_i^{n_i^*+2}$$

hence, ultimately:

$$R(n^*) - R(n) \geq \sum_{i \in I_1} (n_i^* - n_i) r^* - \sum_{i \in I_2} (n_i - n_i^*) r^*$$

$$R(n^*) - R(n) \geq \left(\sum_{i=1,K} n_i^* - \sum_{i=1,K} n_i\right) r^* = (N - N) r^* = 0.$$

It is easily shown that in the case where $R(n) = \prod_{i=1,K} (1 - q_i^{n_i+1})$ is taken rather than Eqn (5), we obtain an optimum solution from algorithm 1′:

Algorithm 1′

1. (given N). For i from 1 to K, $n_i = 0$.

$$R = \prod_{i=1,K} (1 - q_i).$$

2. Select i_0 such that

$$r = \frac{1 - q_{i_0}^{n_{i_0}+2}}{1 - q_{i_0}^{n_{i_0}+1}} = \frac{R_{i_0}(n_{i_0}+1)}{R_{i_0}(n_{i_0})} = \max_i \left(\frac{R_i(n_i+1)}{R_i(n_i)}\right).$$

3. Make $n_{i_0} \leftarrow n_{i_0} + 1$, $R \leftarrow R \times r$

$$N \leftarrow N - 1.$$

If $N = 0$, END; if not, go to 2.

1.3. An approximate solution

We shall define two algorithms for problem (2), one valid for the single-constraint case ($m = 1$), the other applicable to the general case (any m).

Algorithm 2 ($m = 1$)

1. (given C_1). For i from 1 to K, $n_i = 0$.

$$\log R = \sum_{i=1,K} \log R_i(0) = \sum_{i=1,K} \log(1 - q_i).$$

2. Select i_0 such that:

$$r = \frac{1}{C_{i_0 1}}[\log R_{i_0}(n_{i_0}+1) - \log R_{i_0}(n_{i_0})] =$$

$$= \max_i \frac{1}{C_{i1}}[\log R_i(n_i+1) - \log R_i(n_i)].$$

3. If $C_{i_0 1} < C_1$, make $n_{i_0} \leftarrow n_{i_0} + 1$

$$\log R \leftarrow \log R + \log R_{i_0}(n_{i_0}+1) - \log R_{i_0}(n_{i_0})$$

$$C_1 \leftarrow C_1 - C_{i_0 1}$$

go to 2.

If $C_{i_0 1} > {}_1$, END.

In the general case we define a set of numbers $a_1, a_2, \ldots, a_m$ satisfying $a_j \geqslant 0$ for j from 1 to m and $\sum_{j=1,K} a_j = 1$.

A method of selecting these coefficients can be found in [Ketelle, 1962].

Algorithm 3 (any m)

1. (C_j, j from 1 to K, are given).
For i from 1 to K, $n_i = 0$.

$$\text{Log } R = \sum_{i=1,K} \log R_i = \sum_{i=1,K} \log(1 - q_i).$$

2. Select i_0 such that

$$r = \frac{1}{\sum_{j=1,m} a_j C_{i_0 j}} [\log R_{i_0}(n_{i_0}+1) - \log R_{i_0}(n_{i_0})] =$$

$$= \max_i \frac{1}{\sum_{j=1,m} a_j C_{ij}} [\log R_i(n_i+1) - \log R_i(n_i)].$$

3. If $C_{i_0 j} < C_j \quad \forall j$ from 1 to m,
   make $n_{i_0} \leftarrow n_{i_0} + 1$

   $\log R \leftarrow \log R + \log R_{i_0}(n_{i_0}+1) - \log R_{i_0}(n_{i_0})$

   for all j from 1 to m, $C_j \leftarrow C_j - C_{i_0 j}$

   go to 2.

   If not, END.

The solutions obtained by algorithms 2 and 3 are not necessarily optimal. However, they are generally very good, as we demonstrate below.

A vector $n^0 = (n_1^0, n_2^0, \ldots, n_K^0)$ is *non-dominated* if $R(n) \geqslant R(n^0)$ implies $C_j(n) \geqslant C_j(n^0)$ for all j, where $C_j(n) = \sum_{i=1,K} C_{ij} n_i$.

Theorem 2 [Ketelle, 1962] — *The solution n^* obtained at the end of algorithms 2 and 3 is non-dominated.*

Proof—This can be performed in an identical manner to theorem 1 by taking into account the concavity of $\log R_i(n_i)$.

1.4. Dynamic programming

Problem (2), which can be written as:

$$\begin{cases} \text{maximize } \log R(n) = \sum_{i=1,K} \log R_i(n_i) \\ \text{with} \quad \sum_{i=1,K} C_{ij} n_i \leq C_j \quad \forall j = 1, 2, \ldots, m \\ n_i \text{ positive integer} \quad \forall i = 1, \ldots, K \end{cases} \tag{6}$$

has separated variables in the function to be maximized and in the constraints, and can therefore theoretically be solved by dynamic programming.

Let us take the case $m = 1$ (the 'any m' case is theoretically possible but yields virtually impossible calculations). In order to simplify the notation, we shall assume that the constraint can be written as:

$$\sum_{i=1,K} c_i n_i \leq C. \tag{7}$$

For all c, $0 \leqslant c \leqslant C$ and for all k, $1 \leqslant k \leqslant K$, let us define the problems $P_k(c)$ by:

$$\begin{cases} \text{maximize} \sum_{i=1,k} \log R_i(n_i) \\ \text{with} \quad \sum_{i=1,k} c_i n_i \leq c \\ \quad n_i \text{ positive integer} \quad \forall i \text{ from 1 to } k. \end{cases} \tag{8}$$

Calling $f_k(c)$ the value of the optimal solution of $P_k(c)$, for k from 1 to $K - 1$ we obviously have:

$$f_{k+1}(c) = \max_{0 \leq n_{k+1} \leq \left\lfloor \frac{c}{c_{k+1}} \right\rfloor} [f_k(c - c_{k+1} n_{k+1}) + \log R_{k+1}(n_{k+1})] \tag{9}$$

where $\lfloor \; \rfloor$ represents the lower integer-valued part,

and
$$f_1(c) = \log R_1\left(\left\lfloor \frac{c}{c_1} \right\rfloor\right). \tag{10}$$

If C and c_i are integers (or multiples of the same value), we can deduce the following dynamic programming algorithm:

Algorithm 4 ($m = 1$)

1. For c from 0 to C, make

$$x_1(c) = \lfloor c/c_1 \rfloor$$
$$f_1(c) = \log R(\lfloor c/c_1 \rfloor)$$
$$k = 1.$$

2. Compute $f_{k+1}(c)$ by (9).
   Let $x_{k+1}(c)$ be the value of n_{k+1} for which the maximum is attained

$$k \leftarrow k + 1$$

3. If $k = K$, go to 4.
   If not, go to 2.

4. ($f_K(C)$ is the maximum reliability. The optimum vector n^* is obtained as follows.)
   As long as $k \neq 0$, make

$$n_k = x_k(C)$$
$$C \leftarrow C - c_k n_k$$
$$k \leftarrow k - 1.$$

In practice, as soon as $m > 1$, dynamic programming becomes impossible.

Therefore, if we want to solve problem (6) exactly, we use a Lagrangian method, (cf., for example [Gondran and Minoux, 1979, Ch. 11, Appendix 2]).

2. COMMON MODE FAILURES AND HUMAN ERRORS

When examining a system in detail, we take account of the fact that the various components comprising it are not statistically independent. They all depend on the environment, design, manufacture, assembly and human factors.

These dependences are often lumped together under the term '*common mode failures*', often concealing a lack of knowledge on our part.

In section 2.1, we shall attempt a classification of the common modes. In section 2.2, we concentrate on human errors. Finally, in section 2.3, we quantify these common modes for certain cases.

2.1. Common mode classification

System data collection and probability analysis will vary according to the types of intercomponent dependence. It is therefore useful to draw up a common mode classification applicable to data collection (cf. Ch. 7) and probability analysis. The three following common mode types are proposed [Gondran, 1978]:

(1) A *type A common mode* is an event or a failure which, by itself, is certain to produce *simultaneous (or cascade) failure* of a well-defined set of components.

The simplest example of this type is a redundant system which fails if the common mode event occurs. In terms of the reliability block diagram, it is a block in *series* with the redundant system.

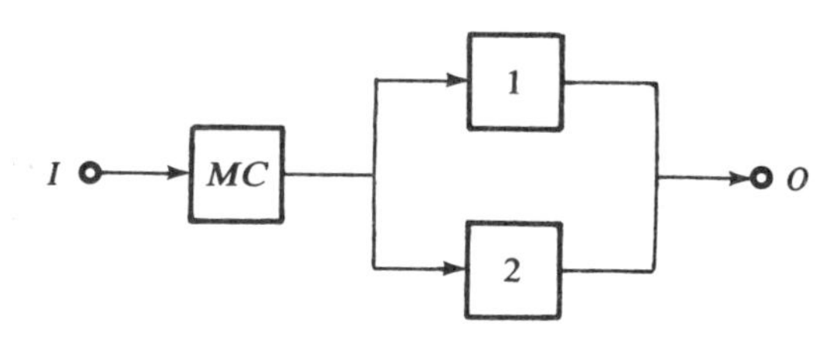

It was this type of common mode which was taken into account for the system logic description in Ch. 2, section 2.2.

If λ and τ are, respectively, the failure rate and the mean time to repair of components 1 and 2, and if λ_{MC} is the common mode occurrence rate, the (approximate) failure rate of the whole is given by:

$$\lambda_{MC} + 2\,\lambda^2\,\tau .$$

There are many phenomena of this type, for example: plane crashes, floods, explosions, fire, earthquakes, common points in a circuit and certain in-service human errors such as operator errors.

(2) A *type B common mode* is an event or failure which, by itself, simultaneously *alters* the *failure probabilities* of a number of components, all in the same direction. Phenomena of this type are extremely numerous.

We note the following:

- dust, dirt, humidity, temperature, vibrations, saline atmosphere, frost, wind, etc.
- manufacturing, assembly and engineering errors, etc.
- human maintenance errors such as routine testing not carried out, maintenance instructions not complied with, calibration or adjustment errors, etc.

Quantitative analysis of some of these common modes associated with the environment (temperature, humidity, etc.) will be given in section 2.3.

(3) A *type C common mode* introduces statistical dependences of any kind.

These could be design or assembly errors and certain human errors during operation and maintenance.

The potential difficulty of taking into account, within a given field, all the possible causes of common mode failures depends to a large extent on the degree of technical expertise and the experience acquired in that field. Analysis charts to ease the task of tracking down common mode failures in nuclear power station systems, as well as ways of reducing their probabilities, can be found in [Carnino and Gachot, 1976; Gachot, 1977] (cf. also Table 1 for an overview).

2.2. **Human errors** [Montmayeul, 1977; Gachot, 1977]

The human element is always present in industrial installations, whether it be in design or operation. We can therefore speak of real man-machine systems in which man is a system component. Reliability analysis cannot therefore provide the reliability of the complete man-machine system unless it takes the human factor into account. The possibility of human failure is obviously associated with the fact that a human being may suddenly find himself incapable of carrying out his/her functions, but more often with the fact that he/she is liable to make mistakes. Objective research into military systems [Meister, 1964], has shown that in some cases more than 40% of all system failures were due to human error.

The table below [Swain, 1976] gives a few examples that are significant, despite their sometimes limited statistics:

Type of system	Type of failure	No. of operations	% due to human error
Missiles	Major system fault	122 faults	35%
Missiles	Equipment fault	1425 faults	20%
Atomic weapons	Manufacturing defect detected by inspection	23 000 faults	82%
Nuclear power stations	Operator errors during reactor operation, tests, repairs	30 incidents	70–80%

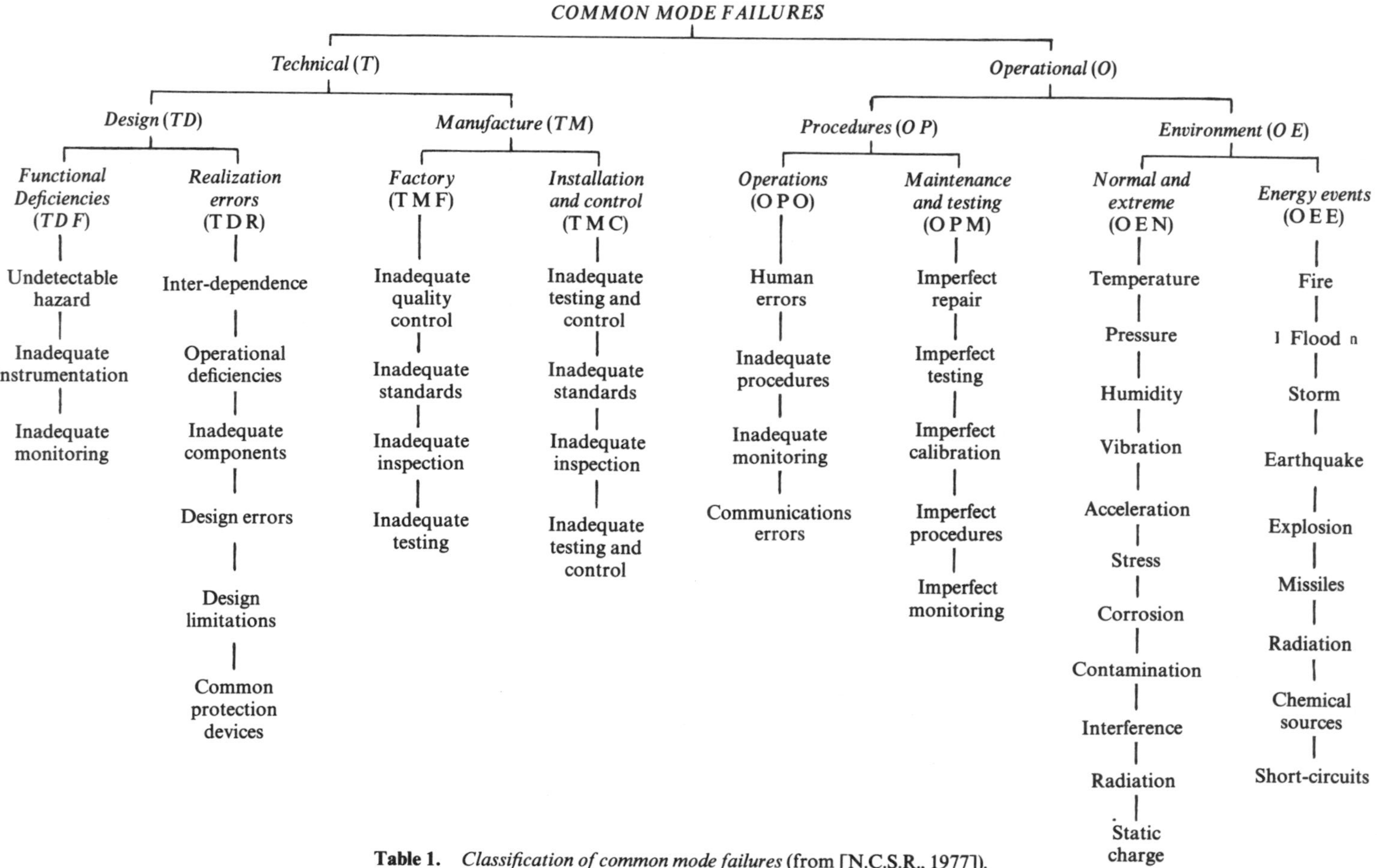

Table 1. *Classification of common mode failures* (from [N.C.S.R., 1977]).

One can only conclude from these results that overall system reliability estimates not taking human reliability into account would, at best, be imprecise and, without doubt, grossly inaccurate in some cases.

Human error is partly responsible for common mode failures due to design, manufacturing or assembly errors. However, rectifying such errors generally involves action at design department, manufacturer or workshop level rather than by staff operating the installation. In addition, the consequences of such errors, which might be described as 'delayed action', mainly come to light at the time of final assembly checks or performance testing and during the first years of plant operation, in the form of equipment failure.

Consequently, when we speak of common mode failures due to human errors, we generally mean human errors during operation in the field or during maintenance.

Given their variety, we cannot claim to give an exhaustive list of common mode failures due to human errors. The list below is, therefore, only an illustration, based on experience in operating nuclear power stations.

Human errors in the field

These may include:

—operator errors due to instructions improperly applied or disregarded, poorly interpreted information, poorly understood processes, etc.;
—alarms that are ignored or misinterpreted as a failure of the alarm system itself;
—faulty transmission of instructions between individuals;
—use of out-of-date or incomplete instructions.

Human errors during routine testing

These may involve:

—periodic tests not carried out or performed incorrectly;
—calibration or adjustment errors;
—use of out-of-date or incomplete testing procedures;
—negligence and oversight resulting in a system being left in a different state from that specified following testing.

Human errors in maintenance operations

These may involve:

—maintenance procedures not adhered to;
—using out-of-date or incomplete maintenance procedures;
—negligence or oversight resulting in a system being left in a state other than that specified following maintenance.

The main methods available for minimizing the frequency of human errors are:

—taking operator behaviour into account at the design stage;
—formulation of operating and maintenance rules which will minimize interpretation and application difficulties;
—in-service quality control: monitoring of operating, testing or maintenance procedures, and checking that they are being properly applied, etc.;
—initial training, updating and subsequent retraining of operators.

Whatever precautions are taken, there will still be a risk of common mode failure of human origin. This risk is very hard to quantify when evaluating system reliability, and the task becomes increasingly difficult when we move away from normal operating practice to examine operator reactions in hypothetical accident situations.

The introduction of ergonomics at the design stage and the use of training simulators helps to prevent potential human errors for which experience-derived data are rare and often biased.

For a more detailed study of these problems, the reader is referred to [Montmayeul, 1977; Gachot, 1977; Swain, 1977].

2.3. Quantitative evaluation of common modes

Most system components operate in a random environment, i.e. the absolute values of the variables defining the environment exhibit random fluctuations in time (temperature, humidity, vibration level and frequency, etc.).

We shall show how placing components in a random environment produces a positive correlation between reliabilities which are assumed to be independent in a constant environment.

Take temperature, for example. We know that, in general, a failure rate varies as a function of the absolute temperature T in accordance with Arrhenius' law [Swain, 1976]:

$$\lambda(T) = \lambda_0 \exp\left(-\frac{E}{T}\right).$$

In the general case, let Ω_c be the set of variations of the stresses C and let $g(c, t)$ be the probability density of stress c at time t (we are assuming here that sudden stress variations cannot occur: $g(c, t)$ is independent of the history of the stresses up to t. For other hypotheses, see [Pollyak, 1963; Roy, 1979; Corazza, 1975, Ch. 2]).

Let $\lambda(t, c)$ be the failure rate of the system subject to stress c at time t. If $R(t)$ is the system reliability at time t, we have:

$$R(t+\mathrm{d}t) = R(t)\left[1 - \int_{\Omega_c} \lambda(t, c)\, g(t, c)\, \mathrm{d}c\right]$$

i.e.:

$$\frac{\frac{\mathrm{d}R}{\mathrm{d}t}(t)}{R(t)} = -\int_{\Omega_c} \lambda(t, c)\, g(t, c)\, \mathrm{d}c$$

from which it follows that:

$$R(t) = \exp\left(-\int_0^t \lambda_m(u)\, \mathrm{d}u\right) \tag{11}$$

hence

$$\lambda_m(t) = \int_{\Omega_c} \lambda(t, c)\, g(t, c)\, \mathrm{d}c \tag{12}$$

is the *mean failure rate* of the system at time t.

Note: Equation (11) is very important as it shows that the mean failure rate of a system is largely dependent on the probability density $g(c, t)$ of the stresses. This explains the very large variations (often by a factor 100) of the failure rate values given in the Rasmussen report [U.S.A.E.C., 1975]; these variations are mainly due to the different operating conditions of the equipment (different stress levels).

2.3.1. *Two components in series*

Let $\lambda_1(t, c)$ and $\lambda_2(t, c)$ be the failure rates of components 1 and 2 at time t subject to stress c. The mean failure rates are:

$$\lambda_1(t) = \int_{\Omega_c} \lambda_1(t, c)\, g(t, c)\, \mathrm{d}t \quad \text{et} \quad \lambda_2(t) = \int_{\Omega_c} \lambda_2(t, c)\, g(t, c)\, \mathrm{d}c$$

and the mean failure rate of the two components in series is:

$$\int_{\Omega_c} [\lambda_1(t, c) + \lambda_2(t, c)]\, g(t, c)\, \mathrm{d}c =$$

$$= \int_{\Omega_c} \lambda_1(t, c)\, g(t, c)\, \mathrm{d}c + \int_{\Omega_c} \lambda_2(t, c)\, g(t, c)\, \mathrm{d}c = \lambda_1(t) + \lambda_2(t)\,.$$

The mean failure rate of the system in series is therefore the sum of the mean failure rates of the components.

In the series case, the stress c may simultaneously alter the failure probabilities of the components in series, but it does not create problems for computing the reliability, as we still have:

$$R(t) = \exp\left(-\int_0^t [\lambda_1(u) + \lambda_2(u)]\, \mathrm{d}u\right) = R_1(t)\, R_2(t)\,. \tag{13}$$

2.3.2. *Two components in parallel*

Here, we shall consider two similar components with failure rate $\lambda(c)$ subject to stress c (therefore independent of time t) arranged in parallel in an environment where the stress c has a probability density $g(c)$ (therefore independent of time t). Let τ be the mean time to repair of one of these components.

Under stress c, we can show (cf. Ch. 6, section 2.3.1) that, when $\lambda(c)\tau \ll 1$, the system failure rate equals $2\lambda^2(c)\tau$.

It therefore follows that the mean system failure rate is equal to:

$$\Lambda_m = \int_{\Omega_c} 2\,\lambda^2(c)\, \tau g(c)\, \mathrm{d}c \tag{14}$$

and the system reliability is $\exp(-\Lambda_m t)$.

As c is a random variable, λ is therefore a random variable with mean

$$\lambda_m = \int_{\Omega_c} \lambda(c)\, g(c)\, dc$$

and variance

$$\sigma_\lambda^2 = \int_{\Omega_c} [\lambda_m - \lambda(c)]^2\, g(c)\, dc\,.$$

We therefore have:

$$\Lambda_m = 2(\lambda_m^2 + \sigma_\lambda^2)\, \tau > 2\, \lambda_m^2\, \tau\,. \tag{15}$$

The system reliability can therefore be written as:

$$R(t) = e^{-\Lambda_m t} = e^{-2\lambda_m^2 \tau t}\, e^{-2\sigma_\lambda^2 \tau t}\,. \tag{16}$$

The random environment therefore amounts to creating an imaginary component in series with the system (the common mode) having a failure rate equal to $2\sigma_\lambda^2\tau$.

2.3.3. *n components in parallel*

The same assumptions concerning the components and the environment will apply as in the case of two components in parallel. The mean system failure rate is therefore (cf. Ch. 6, section 2.3.1):

$$\Lambda_m = \int_{\Omega_c} n\lambda^n(c)\, \tau^{n-1}\, g(c)\, dc$$

$$\Lambda_m = n\tau^{n-1} \int_{\Omega_c} \lambda^n(c)\, g(c)\, dc = n\mu_n'\, \tau^{n-1} \tag{17}$$

where μ_n' is the nth moment of the random variable λ.

The common mode is, here, equal to $n\tau^{n-1}(\mu_n' - \lambda_m^n)$

2.3.4. *Example: the role of temperature*

A component with failure rate $\lambda(T) = \lambda_0\, e^{-E/T}$ in a temperature environment of probability density $g(T)$ will therefore have a mean failure rate:

$$\lambda_m = \lambda_0 \int_{\Omega_T} e^{-E/T}\, g(T)\, dT\,. \tag{18}$$

Let T_0 and σ_T^2 be the mean and the variance, respectively, of the distribution of T. By expanding $\exp(-E/T)$ to second order, we obtain:

$$e^{-E/T} = e^{-E/T_0}\left[1 + \frac{E(T - T_0)}{T_0^2} + \frac{E(E - 2\,T_0)}{2\,T_0^4}(T - T_0)^2 + \cdots\right]$$

which gives, as an expansion of λ_m:

$$\lambda_m \simeq \lambda_0 \, e^{-E/T_0} \left[1 + \frac{E(E - 2\,T_0)}{2\,T_0^4} \, \sigma_T^2 \right]. \tag{19}$$

Now consider the case of two similar components in parallel. The mean failure rate of the system is then given by Eqn (14), i.e.:

$$\Lambda_m = 2\,\lambda_0^2\,\tau \int_{\Omega_T} e^{-2E/T}\, g(T)\, dT \simeq 2\,\lambda_0^2\,\tau\, e^{-2E/T_0} \left[1 + \frac{2\,E(E - T_0)}{T_0^4}\, \sigma_T^2 \right].$$

The temperature distribution amounts to creating an imaginary component in series with the system having a failure rate equal to:

$$\Lambda_m - 2\,\lambda_m^2\,\tau \simeq 2\,\lambda_0^2\,\tau\, e^{2E/T_0}\, \frac{E^2\,\sigma_T^2}{T_0^4}. \tag{20}$$

For further information on this subject, the reader is referred to [Roy, 1979].

3. STOCK CONTROL OF SPARES

The repair time of equipment, and consequently its availability, depends to a large extent on the existence of a stock of spares. In this section, we shall examine some of the most common cases.

3.1. Stock control of large equipment

3.1.1. *Statement of the problem*

Consider the case of a large item of equipment used in a production line, failure of which brings production to a halt (e.g. the primary coolant pumps of a 900 megawatt PWR power plant).

The assumptions concerning the model will be as follows:

—There are n instances of this equipment in service in n production lines (one per line).

—The losses due to a halt in production are proportional to the line downtime; let C_p be the cost per time unit.

—The equipment failure rate is constant and equal to λ.

—When an equipment fails, if a spare part (equipment) is in stock, the cost and the repair (changeout) time are regarded as negligible. Otherwise, a replacement is ordered immediately. The average delivery time is τ. The distribution of this time is assumed to be exponential with parameter $\mu = 1/\tau$.

—The spares in stock are checked periodically such that their 'shelf' failure rates are zero. The storage cost is included in the purchase cost C_a.
—The costs are continuously updated at updating rate i.
—The lifetime of the production lines is T.

The problem is to determine the level k of spares which minimizes the updated operating cost of the entire equipment population.

3.1.2. *Modelling the problem*

The state of the equipment population at each instant will be characterized by the number of items of equipment in operation and the number in stock.

The state transition diagram of the equipment population for an initial stock of k items of equipment will be as follows:

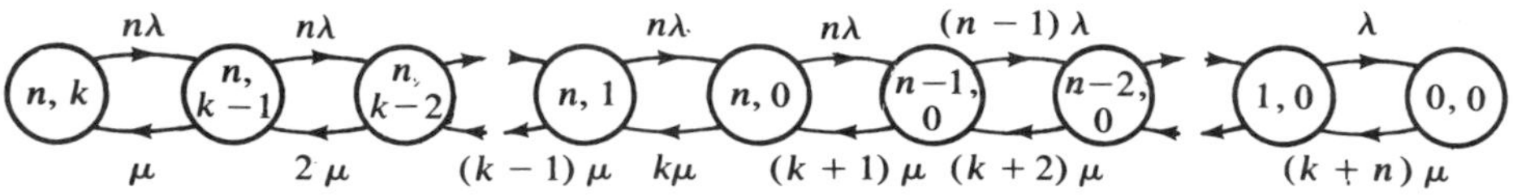

$n\lambda$ $n\lambda$ $n\lambda$ $n\lambda$ $(n-1)\lambda$ λ
(n, k) $(n, k-1)$ $(n, k-2)$ $(n, 1)$ $(n, 0)$ $(n-1, 0)$ $(n-2, 0)$ $(1, 0)$ $(0, 0)$
μ 2μ $(k-1)\mu$ $k\mu$ $(k+1)\mu$ $(k+2)\mu$ $(k+n)\mu$

Let $P_{m,l}(t)$ be the probability of being in state (m, l). We have $P_{n,k}(0) = 1$.
The mean updated cost of being in state $\{n-j, 0\}$, with j from 1 to n, is:

$$jC_p \int_0^T P_{n-j,0}(t)\,\mathrm{e}^{-it}\,\mathrm{d}t$$

therefore, the mean updated operating cost is:

$$C(k) = kC_a + C_p \int_0^T \sum_{j=1,n} jP_{n-j,0}(t)\,\mathrm{e}^{-it}\,\mathrm{d}t\,. \tag{21}$$

The problem therefore becomes one of computing the $P_{m,l}(t)$.

These probabilities are solutions of the set of differential equations (cf. Ch. 4, section 5.2):

$$\begin{cases} \dfrac{\mathrm{d}P}{\mathrm{d}t} = AP \\ P_{n,k}(0) = 1 \\ P_{m,j}(0) = 0 \qquad \text{if } m + j \neq n + k\,. \end{cases} \tag{22}$$

with

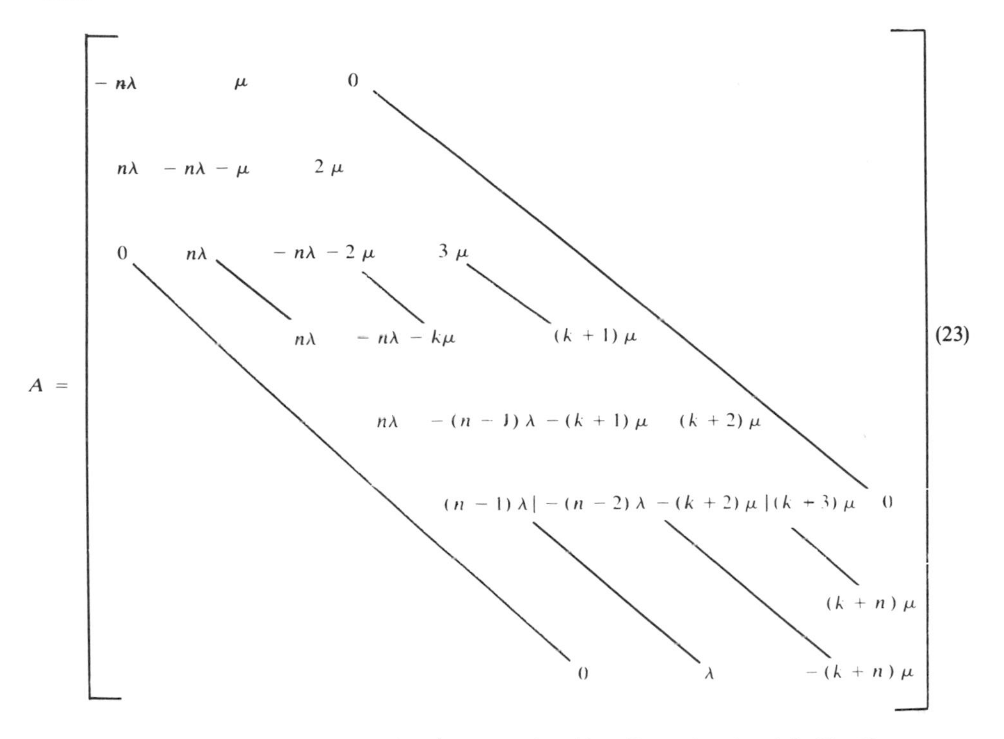

This system of equations can be simply solved by discretization (cf. Ch. 5).

Noting that $C(k)$ is the sum of an increasing function in k, kC_a and of a decreasing function tending to zero when $k \to +\infty$, $C(k)$ is therefore initially decreasing and then increasing.

k^* minimizing $C(k)$ can be easily found by computing $C(1), C(2), C(3), \ldots$, until the function $C(k)$ begins to increase. In practice, we only compute the $C(k)$ and only attempt to find the minimum around the approximate optimum determined by Eqn (29) in section 3.1.3.

3.1.3. *Simplifying the computations*

The $P_{n-j,0}(t)$ tend to a limit $P_{n-j,0}(\infty)$ as T tends to infinity. This limit is reached comparatively quickly, after a few τ, when $\lambda\tau$ is small compared to 1 (cf. Ch. 2). In Eqn (21) we can therefore replace the $P_{n-j,0}(t)$ by the $P_{n-j,0}(\infty)$, giving:

$$C(k) = kC_a + C_p \frac{1 - e^{-iT}}{i} \sum_{j=1,n} jP_{n-j,0}(\infty) . \tag{24}$$

The $P_{m,l}(\infty)$ are therefore solutions of the homogeneous linear set:

$$\begin{cases} AP = 0 \\ \Sigma P_{m,l}(\infty) = 1 . \end{cases} \tag{25}$$

As the matrix A is triangular, the $P_{m,l}(\infty)$ can be computed explicitly (cf. Ch. 4, section 5.2). We therefore have:

$$P_{n,k-l}(\infty) = \frac{\dfrac{(n\lambda\tau)^l}{l\,!}}{1 + \sum\limits_{p=1,k} \dfrac{(n\lambda\tau)^p}{p\,!} + n\,!\,(n\lambda\tau)^k \sum\limits_{j=1,n} \dfrac{(\lambda\tau)^j}{(k+j)\,!\,(n-j)\,!}} \tag{26}$$

$$P_{n-j,0}(\infty) = \frac{\dfrac{n\,!}{(k+j)\,!\,(n-j)\,!}(n\lambda\tau)^k\,(\lambda\tau)^j}{1 + \sum\limits_{p=1,k} \dfrac{(n\lambda\tau)^p}{p\,!} + n\,!\,(n\lambda\tau)^k \sum\limits_{r=1,n} \dfrac{(\lambda\tau)^r}{(k+r)\,!\,(n-r)\,!}}\,. \tag{27}$$

The computation can be further simplified by noting that when $n\lambda\tau$ is small ($n\lambda\tau \ll 1$), the term corresponding to $j = 1$ is predominant in the summation $\sum\limits_{j=1,n} jP_{n-j,0}(\infty)$. Moreover, by the same token, the denominator of Eqns (26) and (27) is equivalent to 1 and we therefore have:

$$P_{n-1,0}(\infty) \simeq \frac{(n\lambda\tau)^{k+1}}{(k+1)\,!}$$

which finally gives:

$$C(k) \simeq kC_a + C_p \frac{1 - e^{-iT}}{i} \frac{(n\lambda\tau)^{k+1}}{(k+1)\,!}\,. \tag{28}$$

It is then obvious how to find k^* minimizing $C(k)$. Note that:

$$C(k+1) - C(k) = C_a - \left(1 - \frac{n\lambda\tau}{k+2}\right) C_p \frac{1 - e^{-iT}}{i} \frac{(n\lambda\tau)^{k+1}}{(k+1)\,!} \simeq$$

$$\simeq C_a - C_p \frac{1 - e^{iT}}{i} \frac{(n\lambda\tau)^{k+1}}{(k+1)\,!}$$

and k^* will be the first k satisfying:

$$\frac{(n\lambda\tau)^{k+1}}{(k+1)\,!} \geq \frac{iC_a}{C_p(1 - e^{-iT})}\,. \tag{29}$$

Numerical application

$\lambda = 0.0768\,y^{-1}$; $\tau = 1.5\,y$; $n = 4$; $C_a = 2.5$ MFF; $C_p = 277.2$ MFF; $i = 0.1\,y^{-1}$.

Equation (28) gives, in turn:

$C(1) = 204$ MFF; $C(2) = 34$ MFF; $C(3) = 10.8$ MFF; $C(4) = 10.3$ MFF; $C(5) = 12.5$ MFF.

The minimum is very flat. We therefore have a choice between $k^* = 3$ and $k^* = 4$. Another selection criterion may discriminate between the two.

3.1.4. *Extensions of the problem*

In many practical cases, there are several large items of equipment in series in each production line. Let p be the number of these components in series.

The failure rate of a production line is then $p\lambda$ and the supply delay time per line remains unchanged.

The above method therefore remains valid if we change λ to $p\lambda$. For other extensions, see [Pagès, 1977, 1979].

3.2. Stock control of small equipment

Section 3.1 dealt with expensive equipment for which the renewal demand was low (few units in operation and low failure rate). We shall now consider mass-produced equipment for which the renewal demand is high (many units in operation and non-negligible failure rate).

It is obvious that a spare is not going to be ordered each time there is a failure. The orders will in fact be lumped together. The problem of stock control of spares is one of predicting the restocking dates and volumes [Desbazeille, 1972; Faure, 1971; Buchan and Koenigsberg, 1965].

3.2.1. *Statement of the problem*

Consider the case of a small item of equipment used in great numbers in a plant. The assumptions concerning the model will be as follows:

— There are n items of equipment in operation.

— The equipment failure rate is constant and equal to λ.

— When equipment fails, if a spare is available, the repair cost and repair (changeout) time will be regarded as negligible. On the other hand, if no spare is stocked, it will be assumed that the cost is proportional to the waiting time before delivery, or C_p the cost per unit time (out-of-stock cost).

— When the stock reaches a level s, known as the *re-order level*, a quantity S' of spares is immediately ordered. The mean delivery (restocking) time is τ. It will be assumed to be constant.

— Let C_l be the *initiation cost*; this represents the administrative expenses of the restocking operation.

— Let C_s be the *storage cost* which is proportional to the quantity stored and the storage time (we can take $C_s = C_a/i$ where C_a is the purchase cost and i the updating rate).

The problem is to determine the re-order level s and the volume S' of an order so as to minimize the cost of managing the entire equipment population.

3.2.2. *Selecting the re-order level s*

During the restocking time τ, the probability of r units failing is:

$$p(r) = \frac{(n\lambda\tau)^r}{r\,!} e^{-n\lambda\tau}.$$

If, during this period τ, the number of failed units is less than the re-order level s, the situation is as shown below:

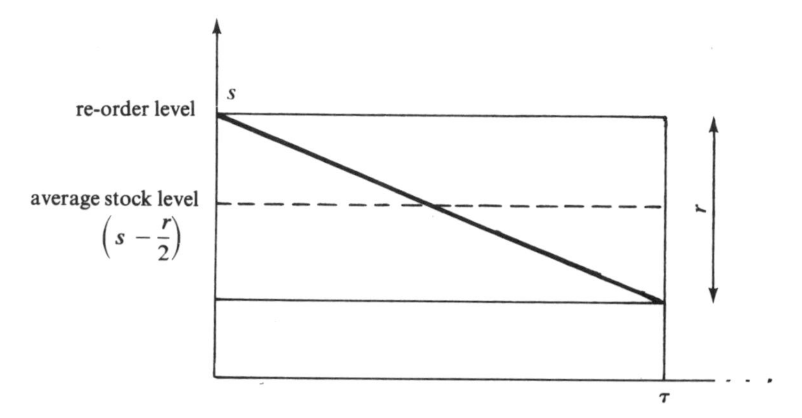

and the cost is:

$$\Gamma_1 = \tau C_s \left(s - \frac{r}{2} \right).$$

If the number of failed units exceeds the re-order level s, the situation is as follows:

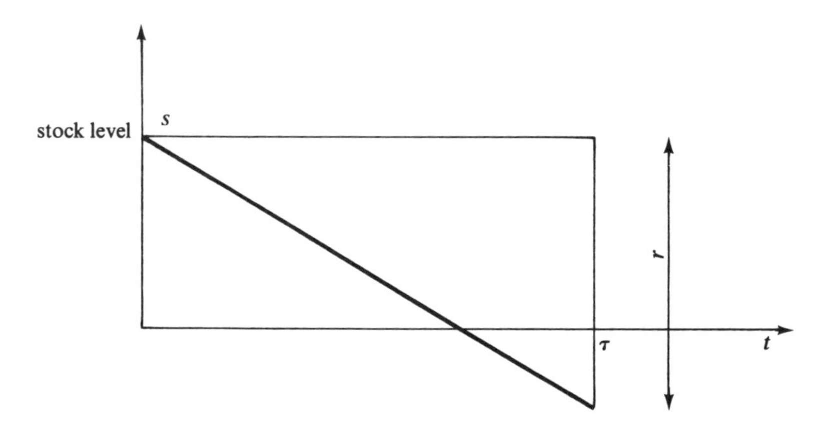

and the cost is therefore:

$$\Gamma_2 = C_s \tau \frac{1}{2} \frac{s^2}{r} + C_p \tau \frac{1}{2} \frac{(r-s)^2}{r}.$$

The mathematical expectation of the management cost is therefore:

$$\Gamma(s) = C_s \tau \sum_{r=0}^{s} \left(s - \frac{r}{2} \right) p(r) + C_s \tau \sum_{r=s+1}^{\infty} \frac{1}{2} \frac{s^2}{r} p(r) + \\ + C_p \tau \sum_{r=s+1}^{\infty} \frac{1}{2} \frac{(r-s)^2}{r} p(r).$$

It can be easily shown that:

$$\Gamma(s+1) = \Gamma(s) + (C_p + C_s)\left[\sum_{r=0}^{s} p(r) + \left(s + \frac{1}{2}\right) \sum_{r=s+1}^{\infty} \frac{p(r)}{r}\right] - C_p .$$

It follows that the minimum of $\Gamma(s)$ will occur for s_0 such that:

$$L(s_0 - 1) < \rho < L(s_0)$$

where $\rho = \dfrac{C_p}{C_p + C_s}$ (often known as the 'out-of-stock rate')

and $L(s) = \sum_{r=0}^{s} p(r) + \left(s + \frac{1}{2}\right) \sum_{r=s+1}^{\infty} \frac{p(r)}{r}.$

This value s_0 will be the re-order level selected.

3.2.3. *Selecting the maximum stock level S*

Defining a re-order level s enabled us to take account of the fact that the demand for spares was random. In order to determine the maximum stock level minimizing the management cost, we shall assume that the demand for spares is constant per unit time and equal to the failure rate $n\lambda$.

Management can therefore be regarded as being exercised at intervals T as shown below:

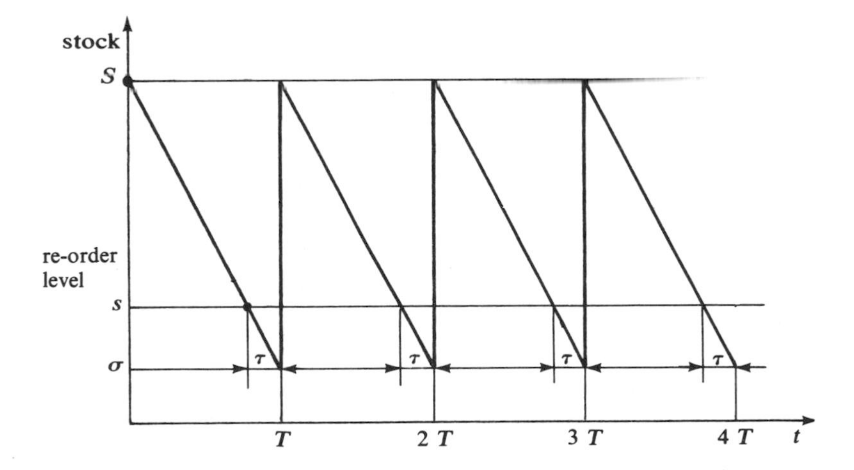

The minimum level σ is therefore:

$$\sigma = s - n\lambda\tau$$

and S can be written as: $S = S' + \sigma$.

The total cost of a volume order is therefore:

$$C_l + \frac{1}{2} C_s S' T$$

with $S' = n\lambda T$.

The cost per unit time is therefore:

$$\frac{C_l}{T} + \frac{1}{2} n\lambda TC_s$$

which is minimum for:

$$T = \sqrt{\frac{2\,C_l}{n\lambda C_s}}\,.$$

It follows that it is necessary to place an order of:

$$S' = \sqrt{2\,n\lambda\frac{C_l}{C_s}} \tag{30}$$

when the re-order level s is reached.

By now, taking into account the random aspect of the demand for spares, we can be sure that the maximum stock level will not exceed $S' + s$.

4. TESTING FREQUENCY FOR STANDBY COMPONENTS

In a large number of systems, some components do not participate directly in the normal operation of the system, but are designed to take over in the event of main component failure. Thus, in PWR nuclear units, some back-up circuits are only activated if the primary circuit goes down.

These components are normally on standby (non-operating). Standby equipment is nevertheless prone to 'standby' failures (due to contamination or corrosion) characterized by a failure rate $\lambda_a(t)$ (standby failure rate or dormant failure rate) which we shall assume to be constant.

The existence of this standby failure rate λ_a and the need to ensure that these standby components have a good availability means that it is necessary to test, periodically, that they are capable of performing their mission.

In section 4.1, we shall determine the test frequency maximizing the availability of one such component.

In section 4.2, we shall assume the existence of failure and testing costs and seek to minimize the cost of the testing policy.

In section 4.3, we shall try to find the best sequence of tests in the case of two identical standby components.

4.1. Maximizing the availability of one standby component

For the standby component, let λ_a be the standby failure rate and γ the probability of non-start-up or of failure during the test period (an item under test being subject to start-up failures due to the additional stresses imposed).

Let λ be the operating failure rate, τ the mean time to repair and T the time interval between two tests. We shall make the following assumptions:

$H1: \lambda_a T \ll 1$,
$H2: \lambda_a \tau \ll 1$,
$H3$: the component's unavailability is low.

Unavailability of a standby component tested at regular intervals is due to two separate causes: failures due to the tests and failures due to the standby status; we shall now evaluate the unavailability associated with each of these causes.

Unavailability due to testing

If T is the interval between two tests and τ the mean time to repair, the unavailability due to testing over a long period $(NT + \gamma N\tau)$ is equal to $\gamma N\tau$.

The mean unavailability ID_t due to testing is therefore equal to:

$$ID_t = \frac{\gamma N\tau}{NT + \gamma N\tau} \simeq \frac{\gamma\tau}{T} \tag{31}$$

assuming $H3$.

Unavailability due to being on standby

Over a long operating period, the mean unavailability ID_a due to being on standby is equal to:

$$ID_a = 1 - \frac{N\int_0^T e^{-\lambda_a t}\,dt}{NT + n\tau}$$

with n = mean number of failures $= N(1 - e^{-\lambda_a T})$ hence:

$$ID_a = 1 - \frac{1 - e^{-\lambda_a T}}{\lambda_a T + \lambda_a \tau(1 - e^{-\lambda_a T})}.$$

Or, based on assumptions $H1$ and $H2$:

$$ID_a \simeq \lambda_a\left(\tau + \frac{T}{2}\right) \tag{32}$$

(cf. Ch. 3, section 5).

The *mean long-term unavailability* ID is therefore equal to:

$$ID = ID_a + ID_t = \lambda_a \tau + \lambda_a \frac{T}{2} + \frac{\gamma\tau}{T}. \tag{33}$$

As maximum availability corresponds to minimum unavailability, if:

$$T = T_0 = \sqrt{\frac{2\gamma\tau}{\lambda_a}}. \tag{34}$$

then the minimum unavailability ID_m is:

$$ID_m = \lambda_a(T_0 + \tau) = \lambda_a \tau + \sqrt{2\lambda_a \gamma\tau}. \tag{35}$$

Effect on ID of variations of T around the optimum T_0

The curve representing Eqn (33) has the following form:

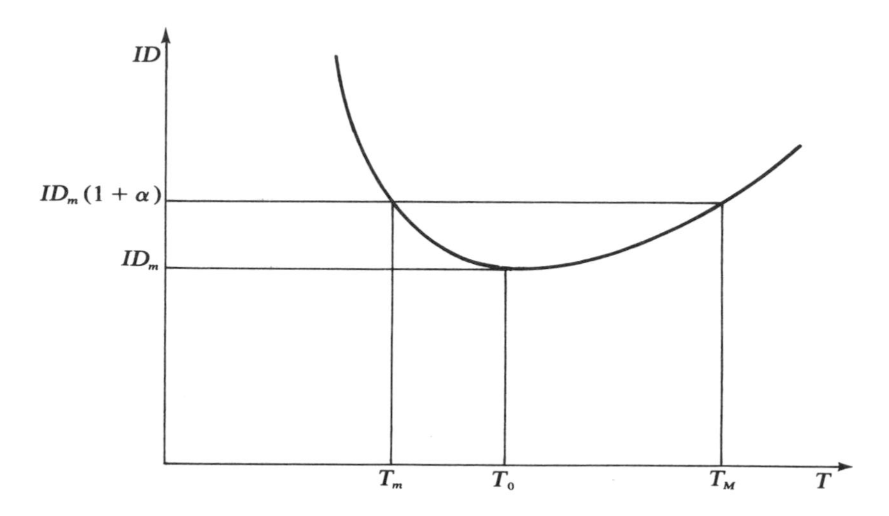

We find the values T_m and T_M of T for which:

$$ID = ID_m(1+\alpha) \qquad (\alpha > 0)\,. \tag{36}$$

We obtain:

$$\begin{cases} T_M = T_0 + \alpha(T_0+\tau)\left(1 + \sqrt{1 + \dfrac{2\,T_0}{\alpha(T_0+\tau)}}\right) \\ T_m = T_0 + \alpha(T_0+\tau)\left(1 - \sqrt{1 + \dfrac{2\,T_0}{\alpha(T_0+\tau)}}\right) \end{cases} \tag{37}$$

Note that $T_M - T_0 > T_0 - T_m$.
If $\alpha \ll 1$ we can take:

$$\frac{T_M - T_0}{T_0} = \sqrt{2\,\frac{T_0+\tau}{T_0}\,\alpha}\,. \tag{38}$$

In applications, only T_M is of interest as, in general, the aim is usually one of reducing the test frequency. Clearly, T_M can vary in relative value by the order of $\alpha^{1/2}$ without ID varying in relative value by more than α. Practical values (day, week, month, etc.) can therefore be taken for T_0.

Note 1: In practice, the standby failure rate λ_a is very inaccurately known. However, we can set a test frequency based on the experience of the manufacturer or user. This is the same as assuming, implicitly, that λ_a is equal to:

$$\lambda_a = \frac{2\,\gamma\tau}{T_0^2}\,. \tag{39}$$

Let us consider a diesel engine used as a standby unit and having the following parameters:

$$\lambda_a T_0 + \gamma = 3 \times 10^{-2} \quad \lambda = 7 \times 10^{-4}\,\mathrm{h}^{-1}$$

$$\tau = 173\,\mathrm{h} \quad T_0 = 168\,\mathrm{h}.$$

Applying Eqns (39) and (35) simultaneously, gives:

$$\gamma = 10^{-2};\ \lambda_a = 1.2 \times 10^{-4}\,\mathrm{h}^{-1};\ ID_m = 4.09 \times 10^{-2}.$$

Taking $\alpha = 0.1$, applying Eqns (37) gives:

$$T_m = 80\,\mathrm{h} \quad T_M = 314\,\mathrm{h}.$$

Note 2: If $\gamma = 0$, Eqn (34) gives $T_0 = 0$, which is not feasible.

In fact, γ will never be precisely zero since, even if the instantaneous probability of not starting may be zero, the probability of failing during the test period τ_t is not. The optimal time between tests is therefore:

$$T_0 = \sqrt{\frac{2\,\lambda \tau \tau_t}{\lambda_a}}. \tag{40}$$

It is sometimes likely to be quite short.

Therefore, if we wish to establish the failure and testing costs, we apply the method described in the next section.

4.2. Minimizing the cost of the testing policy

When $\gamma = 0$, to apply Eqn (34) would involve continuous testing of the standby component. However, economic considerations allow us to determine a test interval T_1 by assigning a cost C_t to each test and a cost C_p per failure of the system as a whole.

Assume that the system S has a failure rate λ_S and a mean time to repair τ_S.

It can then be shown (cf. Ch. 6) that over a long period $t = NT$, a system with a failure rate λ_S having an associated standby component with:

- —a standby failure rate λ_a,
- —an operating failure rate λ,
- —an instantaneous probability γ of not starting,
- —a mean time to repair τ,
- —a mission time t_m,

and tested at a constant interval T, has a reliability of:

$$\exp\left(-\lambda_s t\left[\lambda_a\left(\frac{T}{2}+\tau\right)+\gamma\frac{\tau}{T}+\gamma+\lambda t_m\right]\right).$$

$ID_0 = ID_m + \gamma + \lambda t_m$ is the *operational unavailability* of the component.

Over a long period, the average cost of the tests per unit time is C_t/T and the average cost of the failures per unit time is:

$$C_p \lambda_S \left[\lambda_a\left(\frac{T}{2}+\tau\right)+\gamma\frac{\tau}{T}+\gamma+\lambda t_m\right].$$

Hence the total cost is:

$$C = \frac{C_t + C_p\,\gamma\lambda_S\,\tau}{T} + \frac{C_p\,\lambda_S\,\lambda_a}{2}\,T + C_p\,\lambda_S[\lambda_a\,\tau + \lambda t_m + \gamma]\,. \tag{41}$$

The cost is minimized for:

$$T = T_1 = \sqrt{\frac{2(C_t + \gamma\lambda_S\,\tau C_p)}{\lambda_S\,\lambda_a\,C_p}}\,. \tag{42}$$

It is therefore well determined if $\underline{\gamma = 0}$. In this case

$$T = T_1' = \sqrt{\frac{2\,C_t}{\lambda_S\,\lambda_a\,C_p}}\,. \tag{43}$$

Note that the mean unavailability of the standby component is not a minimum.

If the system S becomes degraded, λ_S will increase and T_1 will have to be reduced in accordance with Eqn (42). Note that the time between tests T_1 must vary as a function of the square root of the mean uptime of the system S. The cost C in fact depends on two terms, one $C_p\lambda_S[\lambda_a\tau + \lambda t_m]$ independent of the test frequency, and the other $\frac{C_t}{T} + \frac{C_p\lambda_S\lambda_a}{2}T$ dependent on it.

It is therefore only worth minimizing the cost if the second term is large compared with the first, i.e. if:

$$\frac{C_t}{T_1} + \frac{C_p\,\lambda_a\,\lambda_S\,T_1}{2}$$

is comparatively large compared with $c_p\lambda_S(\lambda_a\tau + \lambda t_m)$.

4.3. Test sequence for a two-component standby system

This problem is examined for a negligible mean time to repair and a start-up failure rate of zero in [Jacobs and Marriot, 1969]. The tests therefore have to be staggered by $T/2$ between each component in order to minimize the overall unavailability due to the tests, given that each component is tested at intervals of T. We propose to show that these results remain valid with the following assumptions:

*H*4: The unavailabilities $I_1(t)$ and $I_2(t)$ of the two standby components are small compared to 1.

*H*5: While one of the components is being repaired, testing of the other is suspended. Effects due to starting and restarting are disregarded. When one component has been repaired, the other is tested.

*H*6: The probability of the standby system having to perform a mission is exponential with parameter λ_S (failure rate of system S).

Compute the *long-term* probability that the overall system will not perform its mission between 0 and t.

Three types of failure may prevent the overall system from performing its mission:

—The standby system has a mission to perform but both standby components are unavailable or fail to start.

—The standby system has a mission to perform but only one component is available or only one starts and fails before the other has been repaired.

—The standby system has a mission to perform, both components start and both fail.

We shall regard these three types of failures as being independent (they have a very low probability of occurring and are separated in time).

In certain types of operation, the third type of failure does not occur. In any case, the probability of the first two types of failure occurring is independent of the time by which testing of the two standby units is staggered. We therefore only compute the probability $P_1(t)$ of a failure of the first type.

Over a long period NT, the probability of a failure of the first type occurring is:

$$P_1(t) = N \int_0^T \lambda_S [I_1(t) I_2(t) + I_1(t)(1 - I_2(t))\gamma + I_2(t)(1 - I_1(t))\gamma + \\ + (1 - I_1(t))(1 - I_2(t))\gamma^2] \, dt .$$

Or, based on assumption $H4$ and taking $t = NT$

$$P_1(t) = \frac{\lambda_S t}{T} \int_0^T I_1(t) I_2(t) \, dt + \lambda_S t [2 \gamma ID_m + \gamma^2] .$$

It can be shown (cf. Exercise 1) that, if x is the stagger between the two tests:

$$P_1(t) = \lambda_S t \left[\lambda_a^2 \left[\frac{x^2}{2} - \frac{xT}{2} + \frac{T^2}{3} + \tau \left(1 + \frac{\gamma}{\lambda_a T} \right) (\tau + T) \right] + 2 \gamma ID_m + \gamma^2 \right] .$$

$P_1(t)$ is therefore at a minimum when the *tests are staggered by* $T/2$ ($x = T/2$). We then have:

$$\frac{P_1(t)}{\lambda_S t} = \frac{5}{24} \lambda_a^2 T^2 + \lambda_a \tau (\lambda_a T + \gamma) \left(1 + \frac{\tau}{T} \right) . \tag{44}$$

When the stagger is zero, this value must be increased by:

$$\frac{\Delta P_1(t)}{\lambda_S t} = \frac{\lambda_a^2 T^2}{8} . \tag{45}$$

In the case of the strategy outlined in section 4.1 ($T = T_0$), Eqn (44) becomes:

$$\frac{P_1(t_0)}{\lambda_S t} = \frac{17}{24} \lambda_a^2 T_0^2 + \lambda_a^2 \tau \left(\tau + \frac{3 T_0}{2} \right) .$$

Or if $\tau \ll T_0$

$$\frac{\Delta P_1(t)}{P_1(t)} = \frac{3}{17} \tag{46}$$

if $\tau = T_0$

$$\frac{\Delta P_1(t)}{P_1(t)} = \frac{3}{77} . \tag{47}$$

Under the strategy outlined in section 4.2 for the case when $\gamma \simeq 0$ and τ is small compared to T, Eqn (44) becomes:

$$\frac{P_1(t)}{\lambda_S\, t} = \frac{5}{24}\,\lambda_a^2\, T_2^2$$

or

$$\frac{\Delta P_1(t)}{P_1(t)} = \frac{3}{5}\,. \tag{48}$$

This last value is the one obtained in [Jacobs and Marriot, 1969]. *Therefore, the existence of a non-zero start-up failure rate and a considerable mean time to repair significantly reduces the advantage of staggered testing (for two components).*

Other factors reduce this advantage still further:

—The required probability is equal to the sum of three types of failure probabilities denoted by P_1, P_2, P_3. Therefore, the relative advantage is, in fact:

$$\frac{\Delta P_1}{P_1 + P_2 + P_3}$$

and may be negligible if P_1 is small compared to $P_2 + P_3$.

Moreover, $\dfrac{P_2(t)}{\lambda_S\, t} = \gamma^2 + 2\gamma I D_m$ and this term is predominant when γ is non-negligible.

—At system start-up and following each failure, it is necessary to test a standby system after time $T/2$ has elapsed, therefore at a non-optimum interval.

The test sequence therefore has a considerable bearing on the reliability of the general system when $\gamma = 0$. On the other hand, when $\gamma \neq 0$ and τ is comparatively large, the test sequence has a negligible bearing on the reliability of the overall system.

5. PREVENTIVE MAINTENANCE

Let us now consider a system comprising a particularly fragile component whose reliability is $R(t)$. If we wait for this component to fail before replacing it, we shall have to bear the cost of the consequences of an unscheduled shutdown of the system. If, on the other hand, we decide to replace this component routinely (during system maintenance time) as soon as it reaches age θ, there will only be a $1 - R(\theta)$ probability of having to bear the failure cost. What is the best strategy?

When no preventive maintenance is carried out, we pay for:

—replacement of the component at price p.
—the cost of the failure P (lost production, etc.).

Consequently, the average cost per unit time is

$$C_\infty = \frac{p + P}{m} \tag{49}$$

where m is the mean uptime of the component:

$$m = \int_0^\infty R(t)\,\mathrm{d}t .$$

When, on the other hand, the fragile component is replaced at age θ, we pay for:

— replacement of the component at price p,
— the cost of failure p with probability $1 - R(\theta)$, or $[1 - R(\theta)]P$.

This gives an average cost per unit time of:

$$C(\theta) = \frac{p + [1 - R(\theta)]\,P}{m_\theta} \tag{50}$$

where m_θ is the mean uptime of the component replaced at age θ: $m_\theta = \int_0^\theta R(t)\,\mathrm{d}t$, cf. Ch. 1 and Fig. 1.

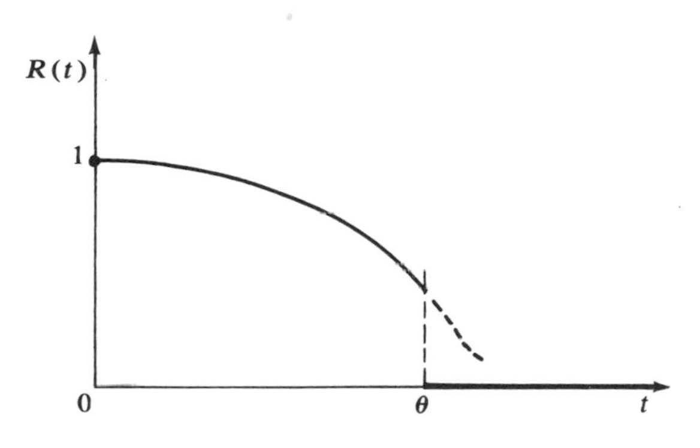

Fig. 1.

Note that the no-maintenance case corresponds to $\theta = +\infty$.

The problem is therefore reduced to analysing the curve $C(\theta)$ and finding out whether it has a smaller minimum than its asymptote C_∞.

When $C(\theta)$ has a minimum, it is obtained by the equation:

$$\frac{\mathrm{d}C}{\mathrm{d}\theta} = \frac{\left\{-\dfrac{\mathrm{d}R}{\mathrm{d}\theta}P\right\}m_\theta - [p + (1 - R(\theta))P]\dfrac{\mathrm{d}m_\theta}{\mathrm{d}\theta}}{m_\theta^2} = 0$$

which, noting that $\dfrac{\mathrm{d}m_\theta}{\mathrm{d}\theta} = R(\theta)$, gives

$$\frac{-\dfrac{\mathrm{d}R}{\mathrm{d}\theta}(\theta)}{R(\theta)}\,m_\theta + R(\theta) = \frac{p + P}{P}. \tag{51}$$

Note that if the failure rate is constant and equal to λ, the left-hand term of Eqn (51) is:

$$\lambda \frac{1 - e^{-\lambda\theta}}{\lambda} + e^{-\lambda\theta} = 1$$

and Eqn (51) will not balance. It makes sense not to carry out preventive maintenance in the case of an exponential distribution. Let us now examine the case where the distribution of the uptime of the fragile component is lognormal. The probability density of the uptime is therefore:

$$\frac{1}{\sigma\sqrt{2\pi}t} \exp\left[-\frac{(\log t - M)^2}{2\sigma^2}\right].$$

Putting $u = (\log t - M)/\sigma$ and denoting by $F(u)$ the distribution function of the standard normal distribution, we obtain:

$$F(u) = \text{Prob}\,(U \leq u).$$

and (cf. Ch. 1), $m = \exp(M + \sigma^2/2)$.

We can show, cf. Exercise 2, that

$$m_\theta = mF(\nu - \sigma) + \theta[1 - F(\nu)] \tag{52}$$

where

$$\nu = \frac{\log\theta - M}{\sigma}.$$

and it follows that:

$$\frac{C(\theta)}{C_\infty} = \frac{F(\nu) + \frac{p}{P}}{\left(1 + \frac{p}{P}\right)\left(F(\nu - \sigma) + \frac{\theta}{m}[1 - F(\nu)]\right)}. \tag{53}$$

Putting $\alpha = \dfrac{\theta}{m} = \theta \exp\left(-M - \dfrac{\sigma^2}{2}\right)$ (use factor), we notice that $v = \dfrac{\sigma}{2} + \dfrac{\log\alpha}{\sigma}$ and therefore that $\dfrac{C(\theta)}{C_\infty}$ is a function of v and σ only. Given the distribution function of the standard normal distribution, we can plot the curves in Fig. 2 for $\sigma = 0.3$.

In a practical problem, we draw the curve $\dfrac{C(\theta)}{C_\infty}$. We can see graphically whether preventive maintenance is necessary and, if so $\left(\min\dfrac{C(\theta)}{C_\infty} < 1\right)$, the value of the optimum frequency.

It is important to mention the fact that *wearing* parts, for which σ is between 0.3 and 0.6, give clearly defined minima of the ratio $\dfrac{C(\theta)}{C_\infty}$, and even more so as p/P becomes smaller, cf. Fig. 2. For this reason, it is generally advisable to perform preventive maintenance on wearing parts (belts, brake linings, bearings, chains, etc.) which are, in most cases, inexpensive compared with the cost of a failure.

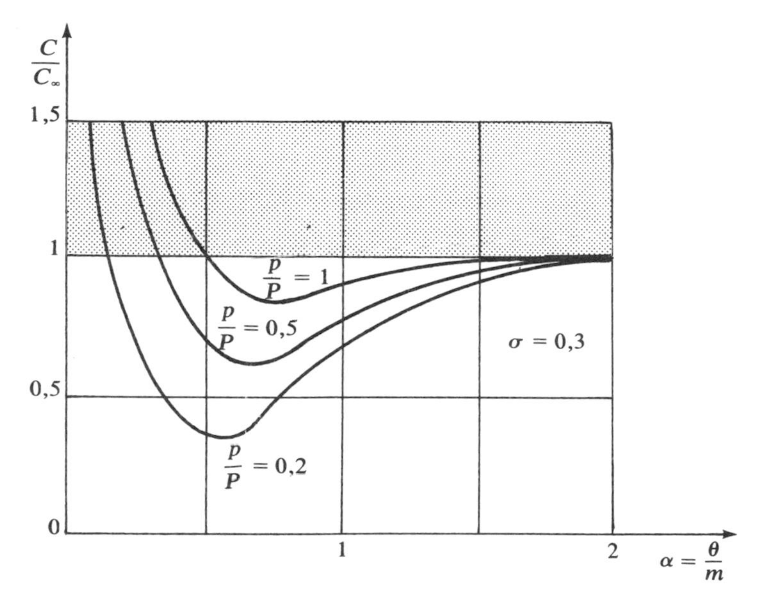

Fig. 2.

On the other hand, for *fatigue parts* ($\sigma \simeq 0.9$), clearly defined minima of $\frac{C(\theta)}{C_\infty}$ only occur for extremely low values of p/P (of the order of 0.05). Now, fatigue parts are relatively expensive and so there is generally no advantage in replacing them preventively.

A large number of exercises and useful problems relating to equipment wear and replacement can be found in [Desbazeille, 1972].

EXERCISES

Exercise 1 — Effect of staggered testing

Assuming $H4$ and $H5$ (section 4.3), compute the probability $P_1(t)$, as a function of the test stagger, of the standby system being required but both standby units are unavailable or fail to start.

Exercise 2 — Prove Eqn 52

Exercise 3 [Desbazeille, 1972]

For a given type of equipment, we isolate a mean service time T and a mean time to repair τ; the availability ratio is therefore:

$$A = \frac{T}{T + \tau}.$$

If preventive maintenance is performed (assuming the equipment is restored to the 'as new' condition) at time θ and if the mean time allocated to this maintenance e is less than τ, we obtain a new availability ratio A^*.

(1) If the service distribution of the equipment is $R(t)$, express this ratio as a function of T, τ, e and $s(\theta) = \dfrac{1}{R(\theta)}\displaystyle\int_0^\infty R(t)\,dt$.

(2) Is it worth doing this if the failure rate is constant? Or if it is proportional to the time? In the latter case, take $R(t) = \exp(-\frac{1}{2}\lambda t^2)$; $\lambda = 10^{-2}$; $e/\tau = 0.5$.

SOLUTIONS TO EXERCISES

Exercise 1 [Gondran and Pagès, 1975]

For $t = NT$, $P_1(t) = \dfrac{\lambda_S t}{T}\displaystyle\int_0^T I_1(t)I_2(t)\,dt + \lambda_S t[2yID_m + y^2]$.

Let x be the stagger between the two tests.

Compute the integral $\dfrac{1}{T}\displaystyle\int_0^T I_1(t)I_2(t)\,dt$.

The curves $I_1(t)$ and $I_2(t)$ have the following form:

(1) When neither of the components is under repair.

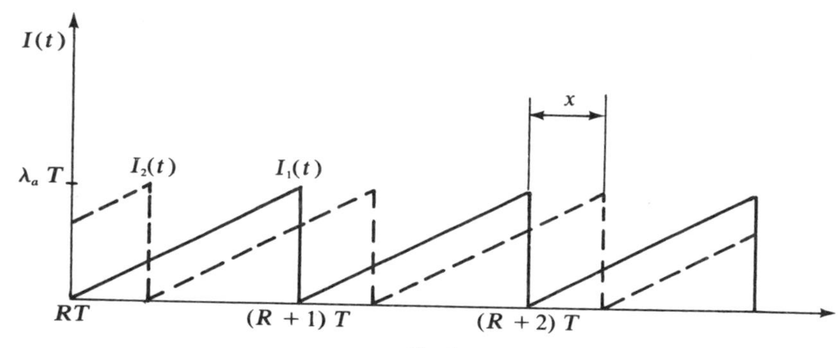

Fig. 3.

(2) When component 1 is under repair.

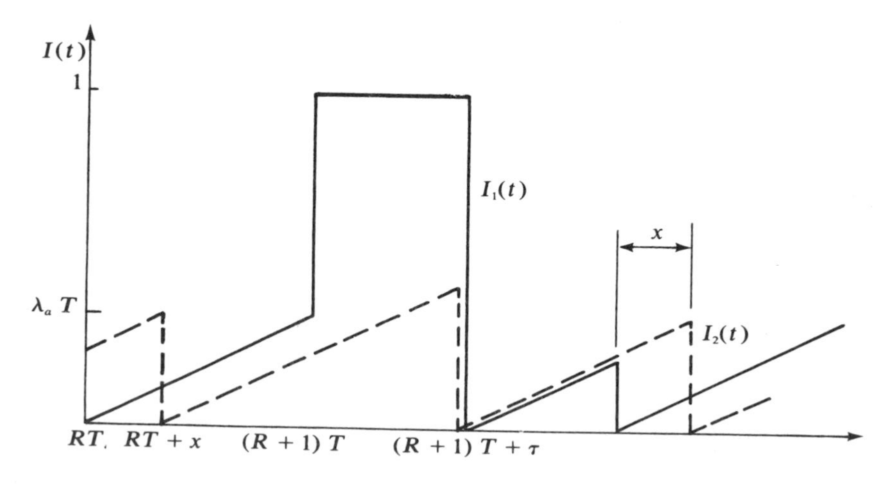

Fig. 4.

(3) When component 2 is under repair.

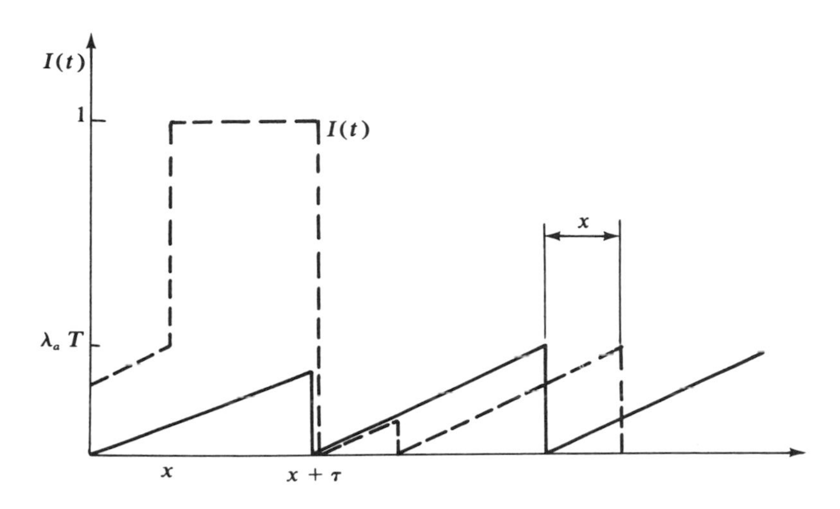

Fig. 5.

Over N cycles, a component fails $N(\gamma + \lambda_a T)$ times on average.

The mean unavailability time due to uncertainties about the state of the components is (cf. Fig. 1):

$$\int_0^x \lambda_a t(\lambda_a t + \lambda_a T - \lambda_a x)\,\mathrm{d}t + \int_x^T \lambda_a t(\lambda_a t - \lambda_a x)\,\mathrm{d}t$$

or

$$\lambda_a^2 T\left(\frac{x^2}{2} - \frac{xT}{2} + \frac{T^2}{3}\right).$$

The mean unavailability time due to repair is (cf. Figs 2 and 3)

$$(\lambda_a T + \gamma) \int_{T-x}^{T-x+\tau} \lambda_a t\,\mathrm{d}t + (\lambda_a T + \gamma) \int_x^{x+\tau} \lambda_a t\,\mathrm{d}t.$$

Let $\lambda_a^2 T\left(1 + \frac{\gamma}{\lambda_a T}\right)(\tau + T)\tau$

hence:

$$I(x) = \frac{1}{T}\int_0^T I_1(t)\,I_2(t)\,\mathrm{d}t = \lambda_a^2\left[\frac{x^2}{2} - \frac{xT}{2} + \frac{T^2}{3} + \tau\left(1 + \frac{\gamma}{\lambda_a T}\right)(\tau + T)\right].$$

$I(x)$ is minimum for $x = T/2$ and maximum for $x = 0$ or T.

$$I\left(\frac{T}{2}\right) = \frac{5}{24}\lambda_a^2 T^2 + \lambda_a^2 \tau\left(1 + \frac{\gamma}{\lambda_a T}\right)(\tau + T)$$

and

$$I(T) = I(0) = I\left(\frac{T}{2}\right) + \frac{\lambda_a^2 T^2}{8}.$$

Exercise 2

$$m_\theta = \int_0^\theta R(t)\,\mathrm{d}t = \int_0^\theta t\frac{\mathrm{d}R}{\mathrm{d}t}(t)\,\mathrm{d}t + \theta R(\theta).$$

Putting $v = (\log\theta - M)/\sigma$, we obtain:

$$m_\theta = \frac{1}{\sqrt{2\pi}} \int_{-\infty}^{\nu} e^{M+u\sigma} e^{-u^2/2}\, du + \theta[1 - F(\nu)]$$

$$m_\theta = e^{M+\sigma^2/2} \frac{1}{\sqrt{2\pi}} \int_{-\infty}^{\nu} \exp\left(-\frac{(u-\sigma)^2}{2}\right) du + \theta[1 - F(\nu)]$$

$$m_\theta = mF(\nu - \sigma) + \theta[1 - F(\nu)].$$

Exercise 3 [Desbazeille, 1972]

(1)
$$A^* = \frac{1}{1 + \dfrac{\tau - R(\theta)(\tau - e)}{T - R(\theta)\, s(\theta)}}$$

$$A^* > A \Leftrightarrow \int_0^{\theta} R(t)\, dt > [1 - kR(\theta)] \int_0^{\infty} R(t)\, dt$$

with $k = 1 - e/\tau$.

(2) Case 1 $R(t) = e^{-\lambda t}$ $A^* < A$

Case 2 $R(t) = \exp\left(-\frac{1}{2}\lambda t^2\right)$.

$$A^* > A \Leftrightarrow \frac{1}{\sqrt{2\pi}} \int_0^{\mu_0} \exp\left(-\frac{1}{2}\mu^2\right) d\mu > \frac{1 - k\exp\left(-\frac{1}{2}\mu_0^2\right)}{2}$$

with $\mu_0 = \lambda^{1/2}\theta$

for $k = \frac{1}{2}$, $A^* > A$ for $\mu_0 \simeq 1.1 \Rightarrow \theta_0 = 11$

for $k = \frac{1}{2}$, $A^* - A$ is maximum for the solution of

$$\int_0^{\mu_0} \exp\left(-\frac{1}{2}\mu^2\right) d\mu = \frac{2}{\mu_0}\left(1 - \frac{1}{2}\exp\left(-\frac{1}{2}\mu_0^2\right)\right)$$

or $\mu_0 \simeq 1.6 \Rightarrow \theta_0 = 16$

but the gain is slight.

REFERENCES

BUCHAN J. and KOENIGSBERG E. (1965): *Gestion scientifique des stocks*; Eyrolles, Paris.

CARNINO A. and GACHOT B. (1976): Défaillances d'un système de protection d'un réacteur considérées en tant qu'événements rares, ISPRA Seminar, *Task Force on Problems of Rare Events in the Reliability Analysis of Nuclear Power Plants.*

CORAZZA M. (1975): *Techniques mathématiques de la fiabilité prévisionnelle des systèmes*; Capadues Editions, Toulouse.

DESBAZEILLE G. (1972): *Exercices et problèmes de recherche opérationnelle*; Dunod, Paris.

FAURE R. (1971): *Eléments de la recherche opérationnelle*; Gauthier Villars, Paris.

GACHOT B. (1977): Les défauts de cause commune; Jouy-en-Josas Conference on *Fiabilité et disponibilité des systèmes mécaniques et de leurs composants*, 1978.

GONDRAN M. (1978): *Les modes communs, une tentative de classification*, Report of the working party on statistics and decision theory applied to rare events; Committee on the Safety of Nuclear Installations, OECD Nuclear Energy Agency.

GONDRAN M. and MINOUX M. (1979): *Graphes et Algorithmes*; Collection des Etudes et Recherches EDF, Eyrolles, Paris.

GONDRAN M. and PAGES A. (1975): *Une politique de test pour les éléments en secours*; EDF Note, HI 1965/02.

JACOBS I. M. and MARRIOT P. W. (1969): *Guideline for determining safe test intervals and repair times for engineered safeguards*; General Electric n° APED 5736.

KELLY J. (1959): L'entretien préventif est-il justifié? *Revue de la SOFRO*, no. 10.

KETELLE J. D. Jr (1962): Least-cost allocation of reliability investment; *Operations Research*, **10**, pp. 249–265.

MEISTER (1964): Methods of Predicting Human Reliability in Man Machine Systems; *Human Factors*, pp. 621–646.

MONTMAYEUL R. (1977): *Erreur humaine et fiabilité*; EDF-DER Note, Report no. 77, p. 44/264.

N.C.S.R., U.K.E.A. (1977): C.S.N.I. Task Force on *Rare Events Research Sub-group on Common Mode Failures*; Interim Report Summary, August 1977.

PAGES A. (1977): *Proposition d'une méthodologie de calcul du niveau optimal du stock de sécurité*; EDF Note, HI 2435/02.

PAGES A. (1979): *L e programme ESOPS (Evaluation du stock optimal des pièces de sécurité)*; EDF Note, HI 3010-02.

POLLYAK Y. G. (1963): Errors in reliability prediction caused by statistical relation between component failures; *Telecomm. Rad. Eng.*, no. 4, pp. 1–8.

ROY, C. (1979): *Une modélisation des modes communs*; EDF Note, HI/3120-02.

SWAIN A. (1976): *Sandia Human Factors Program for Weapon Development*; SAND 76-0326.

SWAIN A. (1977): Estimating Human Error Rates and Their Effects on System Reliability; Jouy-en-Josas Conference on *Fiabilité et disponibilité des systèmes mécaniques et de leurs composants*, 1978.

U.S.A.E.C. (1975): *Reactor Safety Study. An Assessment of Accident Risks in U.S. Commercial Nuclear Power Plants*; WASH-1400, NUREC-75/014.

CHAPTER 9

CASE STUDY (TRAVELLING BRIDGE CRANE)

J.-F. Barbet
Research Engineer
E.D.F. Research and Development Department

1. AIM OF THE EXERCISE

Qualitative and quantitative reliability analysis can be applied to any type of engineered system, whether electrical, pneumatic, hydraulic or mechanical. This example is drawn from a probability analysis of a materials-handling crane [Barbet, 1980]. Although deliberately simplified it nevertheless highlights the main aspects of the probabilistic approach.

A travelling crane is a system comprising two distinct sections: a mechanical section consisting of components such as ropes, gears, pulleys, etc. and an electrical (electromechanical, electronic) section forming the equipment's supervisory control system.

We shall assume here that the supervisory control system's contribution to the risk of the load falling is negligible. Consequently, we propose basically to analyse the risks attributable to the mechanical part of the system, i.e. the hoisting winch ('hoisting' refers equally to raising or lowering the load).

To do this, we have a diagram of the kinematic chain of the crane's hoisting gear, details of the nature and role of the main components as well as the failure rates of these components.

First, we analyse the system failure modes and how they affect that system (FMEA) (*). We then construct the reliability block diagram or the fault tree. From this, we determine the minimal cut sets and can therefore calculate the probability of overall system failure, i.e. the probability of the load falling.

2. DESCRIPTION OF THE KINEMATIC HOISTING CHAIN

We shall first describe the operation of the kinematic chain, and then its constituent components.

(*) Commonly used acronym for Failure Modes and Effects Analysis.

Figure 1 gives an exploded view of the chain. Its logic diagram is shown in Fig. 2. A schematic representation of this kind presupposes a thorough and detailed examination of the system plans.

2.1. Operation

The hoisting gear may be broken down into three main subassemblies (see Figs 1 and 2):

—the hoisting subsystem proper: the load is raised or lowered by an electric motor (2) (**). The motor shaft transmits the torque to the high-speed input shaft in the reduction gearing (7) via a flexible coupling (5). The reduction gearing imparts movement to two shafts (low-speed output) (8) each driving a drum (9) by means of a pinion and wheel coupling. Coiled around each drum is a rope (10) capable of supporting the total load by itself. Each rope passes down to the pulley block and back up to the crane bridge where it passes round a return pulley (11), back down to a second pulley in the block and finally to a fixed point (12) integral with the bridge. The forces at the two fixed points are balanced by a balancing bar (17).

—the braking devices: braking is provided by two disc brakes. A main service brake (3) on the motor shaft above the flexible coupling provides routine braking for each stop-go sequence. An emergency auxiliary brake (6) on the high-speed shaft of the reduction gearing is applied with delayed action after the main brake. Both of these brakes are applied in the event of a voltage drop.

—the safety devices. There is an overspeed detector (16) on the shaft of each drum. If the drum rotation speed exceeds a certain threshold, the overspeed detector disconnects the crane's power supply, causing both brakes to be applied simultaneously. Both overspeed detectors operate on a redundancy basis. An unbalance detector (15) mounted on the balancing bar cuts off the current and causes both brakes to be applied simultaneously if there is excessive imbalance of forces between the two fixed points.

2.2. Description of components

In this section we shall describe the components comprising the chain, together with the assumptions underlying their design.

(1) Power supply — This system supplies the electric hoist motor, the electric motors (of the hydraulic units) of the brakes (see (3) and (6)) and the crane's supervisory control transmission system. Supply failure results in the loss of hoist motor torque but also shuts down the motors of the hydraulic braking units, causing the brakes to be applied. Failure of the supervisory control power supply has the same effect. We shall therefore only consider hoist motor supply failure.

(**) The numbers in brackets refer to Figs 1 and 2 and section 2.2.

(2) Electric hoist motor

This component has the following failure modes:
— breakage of the motor shaft,
— loss of torque due to rotor or stator failures.
The overall failure rate is that of the data item 'motor failure' (cf. Table 1).

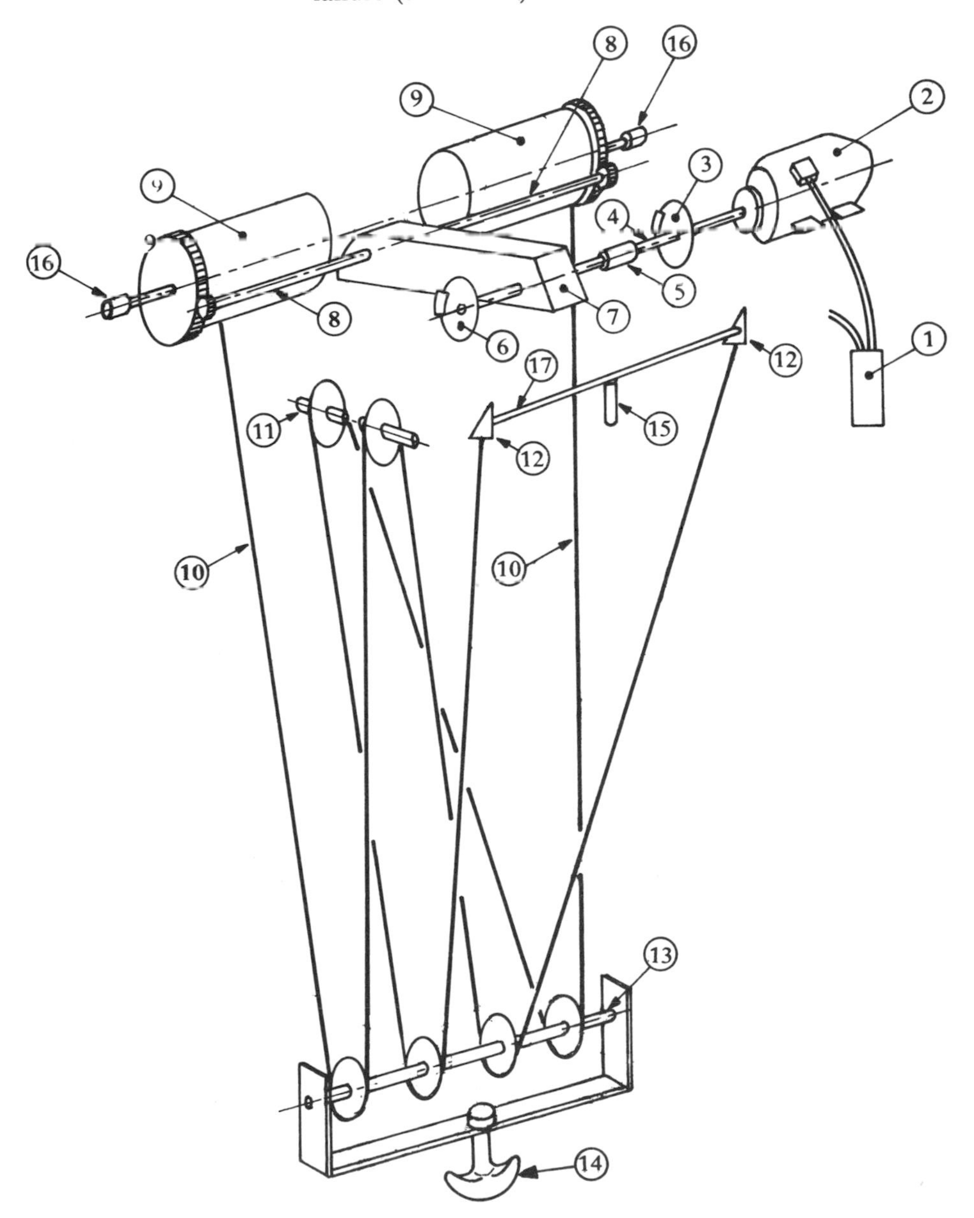

Fig. 1. *Kinematic hoisting chain.*

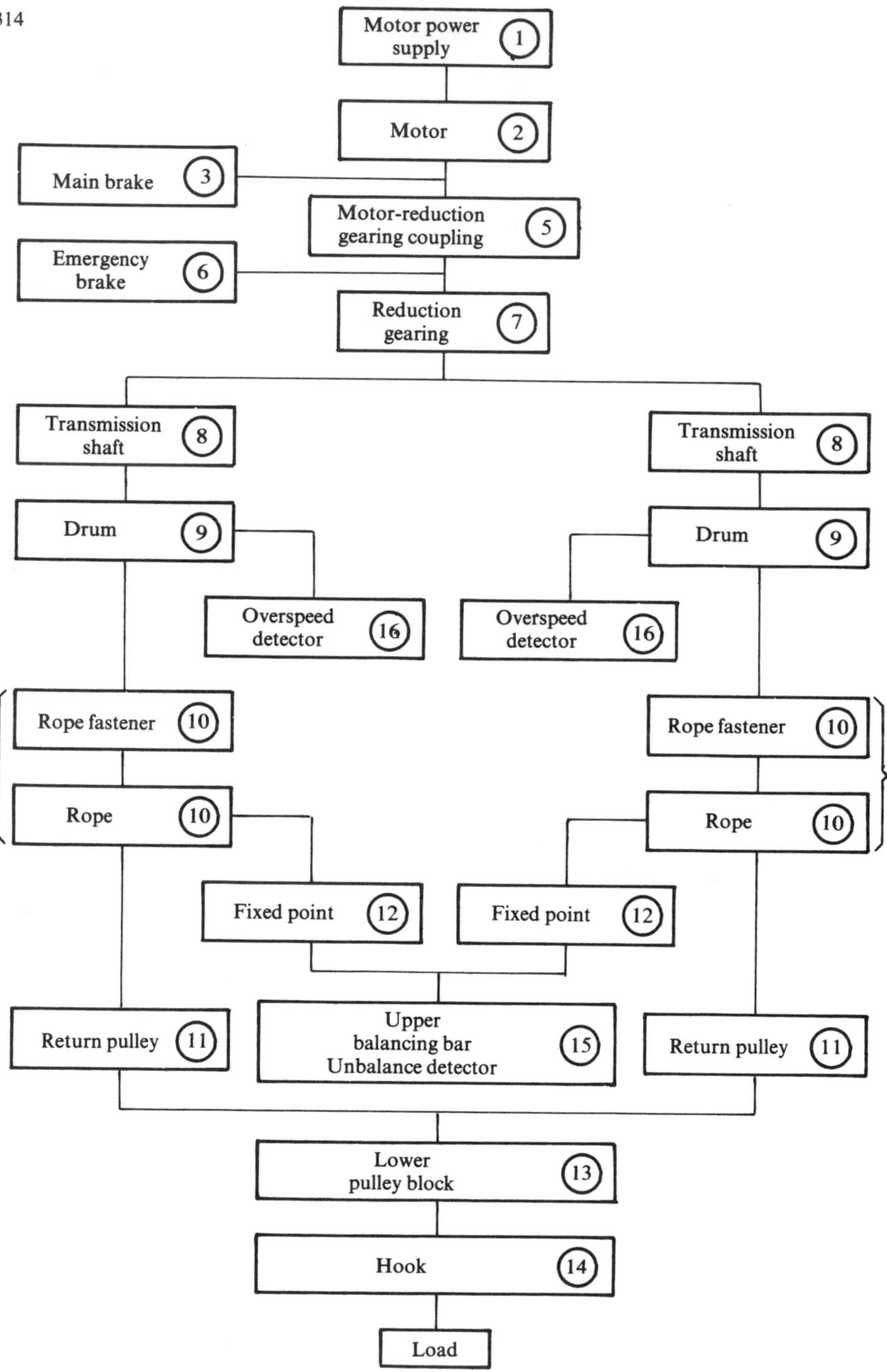

Fig. 2. *Logic diagram of kinematic chain.*

(3) Main service brake	This is a disc brake whose jaws are kept apart by a hydraulic set whose pressure is generated by an electric motor. In the event of a pressure drop, the jaws close on the disc. This brake responds to each motor stop signal, thereby ensuring that the load is secured.
(4) Motor shaft	This component will be assumed to be integral with the motor. Its failure is included in the possible failure modes of the motor.
(5) Coupling	This coupling is of the flexible type. It transmits motor torque to the high-speed input shaft in the reduction gearing. Its flexibility enables it to absorb dynamic torsional effects.
(6) Emergency brake	This brake is identical to the main brake. Under normal operating conditions it is only applied after a time-lag following main brake application: if the latter has operated properly, then the emergency brake has no further effect. On the other hand, on receiving an overspeed signal (see (16)) or load unbalance signal (see (15)), the emergency brake is applied simultaneously with the main brake.
(7) Reduction gearing	This system reduces the speed of the hoist motor to a lower one for driving the drums. The reduction is obtained through gears enclosed in an oil-tight case. The continuous presence of oil enhances gear reliability.
(8) Low-speed output shafts of the reduction gearing	Each shaft imparts movement to a drum via a pinion driving a toothed wheel integral with the drum.
(9) Drum	
(10) Ropes	Each rope has one end anchored to the drum by a fastener. The failure rates of the rope and fastener are merged to form one item of data.
(11) Return pulleys	These are mounted on the crane bridge. Each has an independent shaft.
(12) Fixed point	The other end of the rope is secured by a wedge box: this system is such that greater rope traction increases the grip.
(13) Pulley block	This consists of four pulleys on a common shaft, two pulleys per rope. We shall assume that failure of a pulley (rope breakage or derailment) results in pulley block failure (risk of shaft breakage).

(14) Lifting beam	This serves as intermediary between pulley block and load.
(15) Unbalance detector	The forces exerted at the two fixed points are balanced by a balancing bar (see (17)) at bridge level. If the force at one of the fixed points increases, the balancing bar tips and above a certain threshold activates the detector (limit-of-travel type). The latter then disconnnects the power, thereby causing the brakes to be applied.
(16) Overspeed detector	This detector is driven by the drum. If the drum rotation speed exceeds a certain limit (about 10% above the nominal speed), the detector cuts off the power and applies the brakes.
(17) Balancing bar	This component is similar to the arm of a balance and equalizes the forces exerted by the ropes at their fixed ends. The probability of this component failing will be disregarded.

3. RELIABILITY DATA AND OPERATING CONDITIONS

In addition to qualitative analysis of this system which presupposes detailed knowledge of each component (structure, function, performance), quantitative analysis requires knowledge of two types of numerical data: reliability data, and the crane operating conditions.

3.1. Reliability data

The reliability data required are listed in Table 1.

Provided the equipment has been tested over a sufficiently long period, has been maintained and has only operated a limited number of hours per year, it may be regarded as having passed the early failure period but not having reached the wear-out failure period. The failure rates are therefore taken as constant.

These data were obtained from surveys [Barbet, 1980] conducted in industries using similar equipment, particularly the iron and steel industry, or from the technical literature [U.S.N.R.C., 1975; IEEE, 1977; Marcovici and Ligeron, 1974; N.T.I.S., 1975].

Note: Reliability data should be used with the utmost caution. For a given failure rate λ, we need to know the failure modes concerned, and the equipment for which we intend using this value must be very similar if not identical to that listed.

Take, for example, the case of the pulley, a component for which very different failure rate values are to be found.

(1) In the American R.A.D.C. handbook [N.T.I.S., 1975], for pulleys used in aviation (on-board equipment), without any further details on the manufacture or models of these pulleys, we find a failure rate $\lambda = 5 \times 10^{-6}$/h.

Table 1 – *Reliability data*

Events	Failure rate per component and per hour
Loss of power supply to motor	10^{-6}
Failure of a motor	7.7×10^{-6}
Breakage of a flexible coupling	9×10^{-9}
Breakage of a reduction gearing	5.7×10^{-7}
Breakage of a transmission shaft	1.1×10^{-6}
Breakage of a pinion-wheel coupling	1.1×10^{-5}
Bursting of a drum, failure of welds or of bearings	9×10^{-9}
Failure of a rope which had been checked and changed in accordance with relevant standards	ε
Breakage of a pulley shaft	9×10^{-9}
Breakage of a pulley	6×10^{-9}
Derailment of a rope from a pulley	9×10^{-7}
Derailment of a rope from a pulley fitted with anti-derailment systems	ε
Failure of a 'wedge box' type fixed point	9×10^{-9}
Hook breakage	2×10^{-8}
Overspeed detector failure	2×10^{-8}
Failure of a disc brake subjected to some thirty loadings per hour	4.8×10^{-7}
Failure of a limit-of-travel device (unbalance detector)	per component and by demand 3×10^{-4}

(2) Using a USINOR survey compiled from five years' observations of 200 materials-handling cranes, each comprising on average six pulleys, a statistical method developed by the T.R.W. Systems Group [Barbet et al. 1979; Welker and Lipow, 1974] has made it possible to work out a failure rate $\lambda = 6 \times 10^{-9}$/h for breakage of these pulleys (These are steel, centrifugally-moulded pulleys whose surface finish meets specified requirements.) This failure rate is comparable with that given in the Rasmussen report [U.S.N.R.C., 1975] for weld failure, namely 3×10^{-9}/h.

3.2. Crane operating conditions

Remember that, in order to simplify the exercise, we are only considering the hoisting phases – with no distinction between raising and lowering. The crane is assumed to be operating for 100 hours per year in this mode. Consequently, the result obtained is incomplete. For a complete system analysis, one would have to incorporate the results for the crane's other operating modes (start-stop sequences, traversing, etc.).

4. QUALITATIVE ANALYSIS

It must be stressed at the outset that these methods are only used to complement all the other design procedures. In particular, they presuppose that all the components have been suitably dimensioned using recognized mathematical rules (cf., for example [F.E.M., 1970]) and that manufacturing checks, tests and controls have been properly implemented. The failures taken into account are assumed to be catastrophic (total and sudden) and random in nature.

This example describes complementary qualitative reliability analysis methods. We first carry out FMEA (Failure Modes and Effects Analysis), which is basically an inductive method. Two graphical representation methods are then used: the fault tree, a deductive method, and the reliability block diagram.

Finally, we determine the minimal cut sets.

4.1. FMEA

This method consists of recording all the failure modes of every system component and examining the possible effects of each of these modes on the system.

For this purpose and also to make the analysis as complete as possible, we organize the information under various headings, e.g.:

— component identification: item reference, designation, type, location;

— function, states;

— failure modes;

— possible causes of failure (internal, external causes): this column is very useful for trying to find possible common mode failures (e.g. human error);

— effects on the system;

— means of detection: if a failure does not have an immediate effect on the risk of the specified undesirable event occurring (in this case, fall of the load), it is important to know whether the human operator has some means of detecting this failure and of taking appropriate action. In this case, the human operator plays a part in the reliability (or availability) of the system;

— frequency of inspection or testing: as we saw previously, it is necessary to know whether or not the system is tested and under which conditions. In the case of the materials-handling crane in question, the crane is tested prior to commencing its 100 operating hours and not subsequently: this column will therefore be left blank;

— observations.

FMEA analysis of the winch is listed in the table on the following pages. It is of course not an exhaustive list but gives an example of what can be achieved.

PROJECT:
SYSTEM:

ANALYSIS OF COMPONENT FAILURES AND THEIR EFFECTS ON THE SYSTEM (FMEA)

REFERENCE DOCUMENT

COMPONENT IDENTIFICATION (reference, designation, type, location)	FUNCTIONS AND STATES	FAILURE MODES	POSSIBLE CAUSES OF FAILURE (internal, external causes)	EFFECTS ON THE SYSTEM	MEANS OF DETECTION	INSPECTION OR TEST FREQUENCY	OBSERVATIONS
1. Power supply unit.	Supplies: the hoist motor, the motor of the hydraulic unit to keep the brakes open during hoisting, the supervisory control system.	Loss of supply.	Mains failure. Tripout. Breakage of power cable within the system.	Motor shutdown. Brake application.	Motor shutdown.		
2. Motor.	Drives kinematic chain. Only supplies torque when power on.	Burnt-out winding. Supply cut off. Breakage of drive shaft.	Overcurrent. Mechanical fatigue. Moisture.	Loss of lifting torque. Overspeed start-up. When overspeed detected: brake application.	Overspeed.		
3. Service brake.	Provides normal stopping of hoisting. Also applied in event of overspeed or unbalance between ropes.	No braking torque.	Braking instruction not transmitted. Faulty braking characteristics of brake (disc, linings, pads, etc.).	If other brake operating: no effect on system.			
4. High-speed transmission shaft.	Transmission of motor torque or braking torque.	Breakage.		Breakage of the kinematic chain rendering motor torque and service-brake torque ineffective. Overspeed start-up and application of emergency brake.			

PROJECT:
SYSTEM:

ANALYSIS OF COMPONENT FAILURES AND THEIR EFFECTS ON THE SYSTEM (FMEA)

REFERENCE DOCUMENT

COMPONENT IDENTIFICATION (reference, designation, type, location)	FUNCTIONS AND STATES	FAILURE MODES	POSSIBLE CAUSES OF FAILURE (internal, external causes)	EFFECTS ON THE SYSTEM	MEANS OF DETECTION	INSPECTION OF TEST FREQUENCY	OBSERVATIONS
5. Flexible coupling.	Links the motor output shaft to the reduction-gearing input shaft. Consists of a flexible part which damps the dynamic torsional forces.	Breakage of flexible part.		Loss of coupling elasticity.	Noise.		Regular inspection essential.
		Total breakage.		Breakage of kinematic chain similar to 4.	Overspeed.		
6. Emergency brake	Applied after a time-lag to support main brake for normal stopping and simultaneously in event of overspeed or rope unbalance. Acts on high-speed input shaft to reduction gearing.	No braking torque.	Similar to 3.	No effect during normal operation. In event of overspeed and if main brake ineffective, fall of load.			Regular inspection essential.
7. Reduction gearing.	Reduces the nominal motor speed to the drum driving speed.	Breakage.	Defective lubrication.	Breakage of kinematic chain: fall of load.			
8. Pinion/wheel type reduction gearing-drum couplings.	Driving the drums.	Breakage.		Breakage of one of the subchains (drum-rope): load transferred to other subchain causes unbalance at fixed points. Handling halted by unbalance detection and brake application.	Handling stopped.		

PROJECT:
SYSTEM:

ANALYSIS OF COMPONENT FAILURES AND THEIR EFFECTS ON THE SYSTEM (FMEA)

REFERENCE DOCUMENT

COMPONENT IDENTIFICATION (reference, designation, type, location)	FUNCTIONS AND STATES	FAILURE MODES	POSSIBLE CAUSES OF FAILURE (internal, external causes)	EFFECTS ON THE SYSTEM	MEANS OF DETECTION	INSPECTION OF TEST FREQUENCY	OBSERVATIONS
9. Drums.		Bursting.		Probable loss of overspeed detection on failed drum. Effects similar to 8.	Handling stopped.		
10. Ropes.		Breakage.	Coming unfastened from drum.	Similar effects to 9.	Handling stop. Banging and whipping of rope (not necessarily).		Negligible probability of failure if ropes inspected as per regulations.
11. Return pulleys.		Shaft breakage.	Impact on shaft caused by rope leaving pulley groove.	Effects similar to 9.	Handling stop. Possible rope breakage.		
		Pulley breakage. Rope leaves pulley groove.		Possible rope or shaft breakage by effects similar to 9.			
12. Fixed points.	Wedge boxes. Used to anchor rope to frame.	Coming loose.		Same effects as 9.	Handling stop. Fall of rope.		
13. Lower pulley block.	Shaft: common to the four pulleys and both subchains. Pulleys.	Breakage.	Impact on shaft caused by rope leaving pulley groove.	Probable fall of load.	Handling stop. Possible fall of rope. Fall of load.		
		Rope leaves pulley groove. Pulley breakage.		Possible rope breakage, therefore same effects as 9. Possible shaft breakage.			

PROJECT:
SYSTEM:

ANALYSIS OF COMPONENT FAILURES AND THEIR EFFECTS ON THE SYSTEM (FMEA)

REFERENCE DOCUMENT

COMPONENT IDENTIFICATION (reference, designation, type, location)	FUNCTIONS AND STATES	FAILURE MODES	POSSIBLE CAUSES OF FAILURE (internal, external causes)	EFFECTS ON THE SYSTEM	MEANS OF DETECTION	INSPECTION OF TEST FREQUENCY	OBSERVATIONS
14. Hook.	Used to suspend load.	Breakage.		Fall of load.			
15. Upper balancing bar.	Mechanical system. Balances the two fixed points.	Fails to detect unbalance.		No immediate effect. If one subchain is out of balance, the fault will go unnoticed.	The crane continues to operate after failure of a subchain.		
	Detects any slack in either of the subchains and halts handling by applying the brakes.	Emits a false unbalance signal.		Handling stop.			
16. Overspeed detectors.	Terminate handling when the drum rotation speed exceeds a given threshold.	Threshold set too high. Fails to detect a threshold exceedance.	Human error. Failure of control system transmitting the stop instruction.	No effect during normal operation. If either detector is operating: no effect. If both failed: fall of load.			
		Unwanted operation.		Handling stop.			

4.2. Fault tree

It will be recalled that a fault tree is a diagrammatic representation of the various possible combinations of events liable to produce the undesirable event. It is made up of successive levels such that each event is generated from lower-level events via various logic operators (or gates). This process is continued until we arrive at basic events which are mutually independent and whose probabilities can be evaluated. These basic events may be failures, human errors, external conditions, etc.

In order to facilitate the analysis, the winch's kinematic chain can be divided up into a number of subchains (cf. Fig. 3) as follows:

—subchain *A*: this subchain comprises components between the ropes and the load and which are a single-failure source;

—subchain *B*: this is made up of the two subchains *B*1 and *B*2 which are completely identical and redundant, together with the unbalance detector. Subchains *B*1 and *B*2 are each made up of the transmission shaft from reduction gearing to drum, the drum itself, the rope/fastener unit, the return pulley and the fixed point. In the event of one of the subchains failing, the unbalance detector must detect the load transfer to the other subchain and terminate handling. If this fails to happen, handling may continue with the other subchain;

—subchain *C*: this consists solely of the reduction gearing which is also a single-failure source;

—subchain *D*: this consists of the motor-reduction gearing coupling and the emergency brake which is applied on receipt of an overspeed signal if the coupling fails;

—subchain *E*: this consists of the motor unit backed up by the main brake and the emergency brake.

The fault tree is shown in detail in Fig. 4.

4.3. Reliability block diagram

It will be recalled that in this representation the components whose failure is of itself sufficient to cause system failure are arranged in series, and those whose failure does not cause system failure, except in combination with the failure of other components, are arranged in parallel with the latter.

The case of the winch in question therefore poses the problem of incorporating subchains *B*1, *B*2 and the unbalance detector into the reliability block diagram. Subchain *B* must therefore be entered globally in the diagram as a supercomponent to be analysed and computed separately (cf. section 5.1). In fact, the unbalance detector does not have a load maintaining role but solely one of ensuring system safety in the event of either of subchains *B*1 or *B*2 failing (brake application). It cannot therefore be entered separately in the reliability block diagram. This difficulty also arises in connection with the fault tree, when determining the minimal cut sets (cf. section 4.4).

The corresponding reliability block diagram is shown in Fig. 5. It comprises four 'inputs' (motor power supply, overspeed detector 1, overspeed detector 2, auxiliary brake) and one 'output' (hook). For the load to be maintained, it is necessary that at

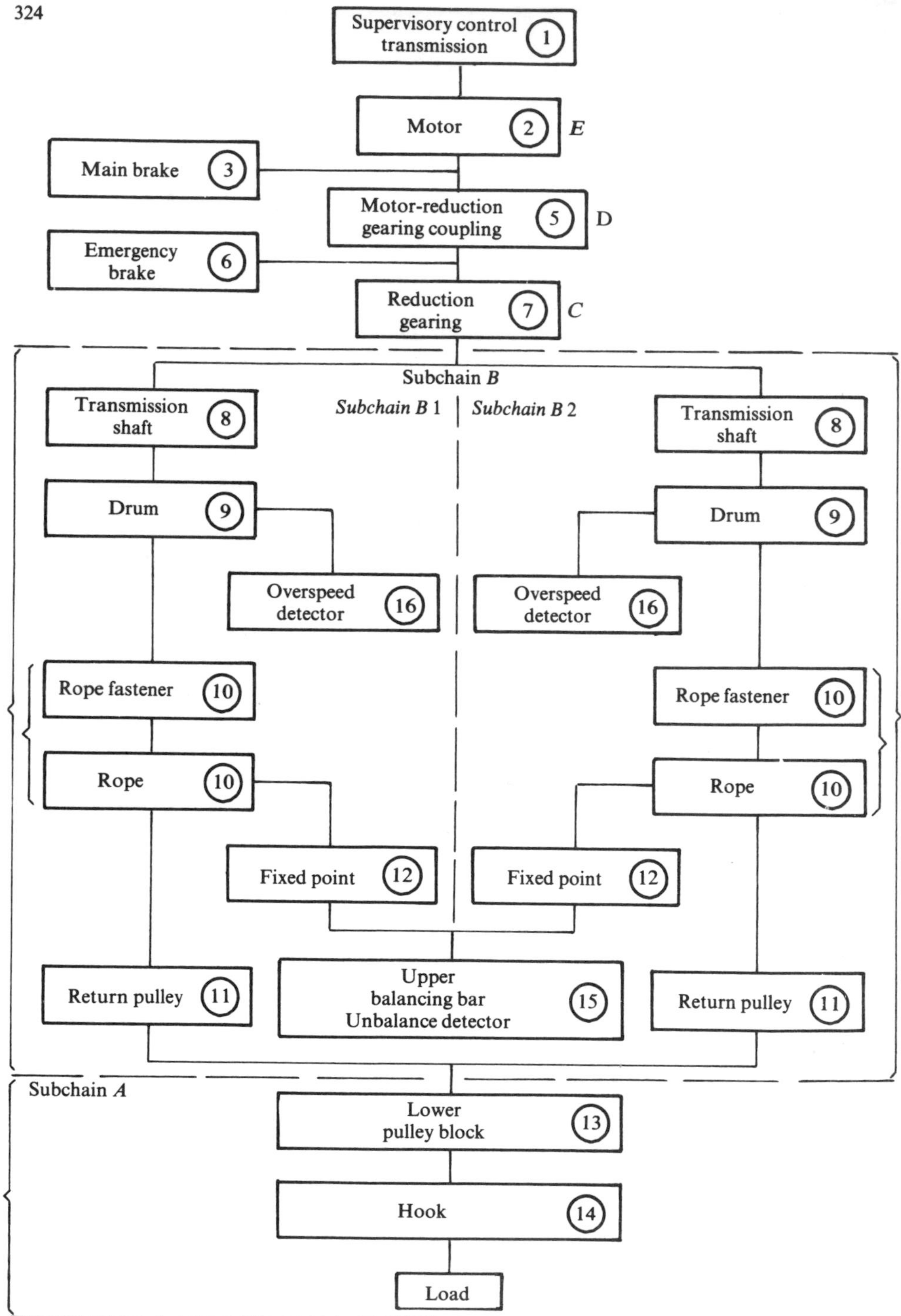

Fig. 3. *Logic diagram of kinematic chain.*

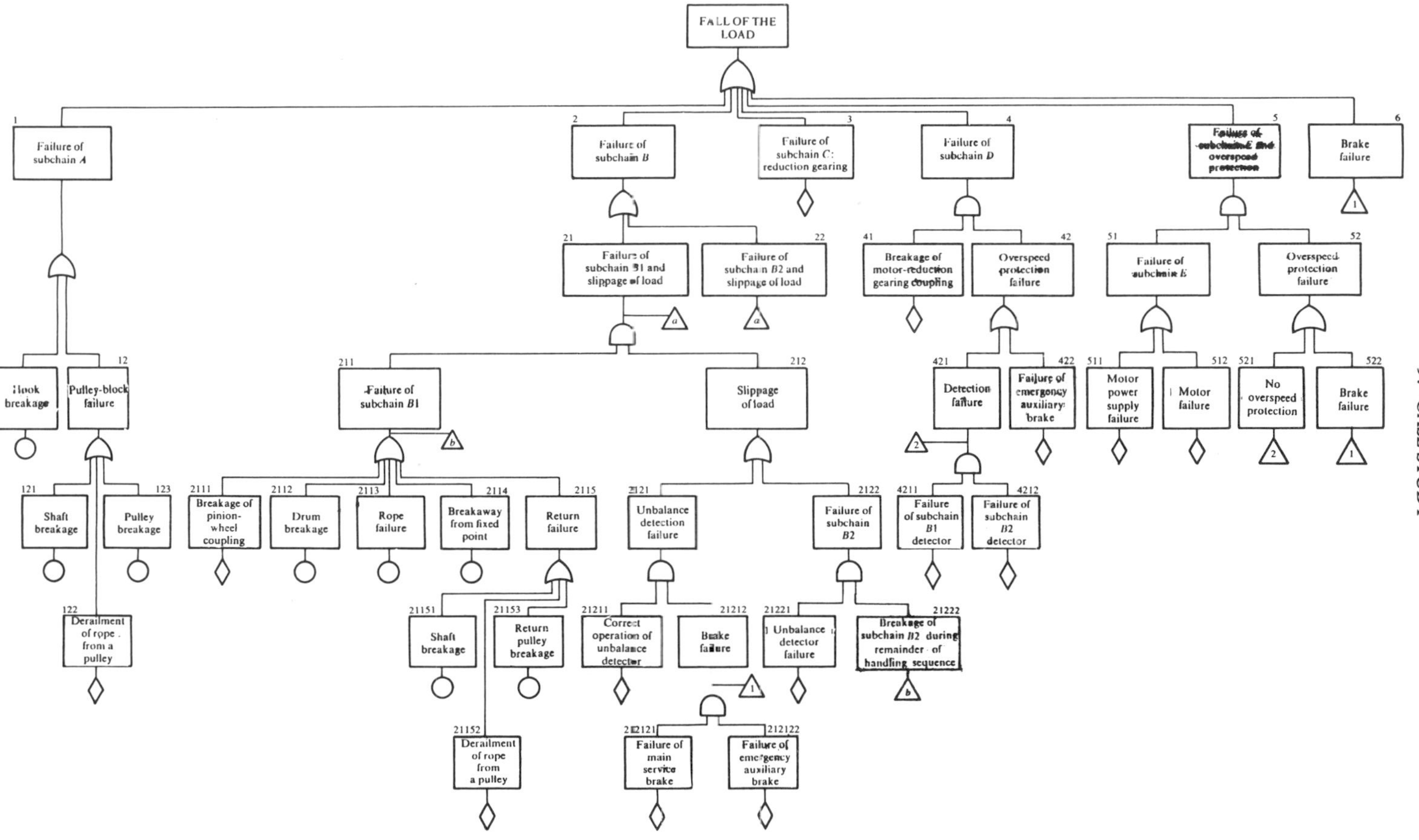

Fig. 4. *Fault tree.*

5. QUANTITATIVE ANALYSIS

Based on the results of qualitative analysis, this is now a matter of calculating the probability of the winch releasing the load, given the failure rates of the winch components (cf. Table 1) and its duty time.

First, we establish the equivalent failure rate of the supercomponent comprising subchains $B1$ and $B2$ and the unbalance detector. We then hand-calculate the probability of the load falling. Finally, we examine how to handle this problem using the DEFAIL and FIABC computing codes.

5.1. Calculating the unavailability of the supercomponent comprising subchains $B1$, $B2$ and the unbalance detector

This supercomponent can be represented by Fig. 6.

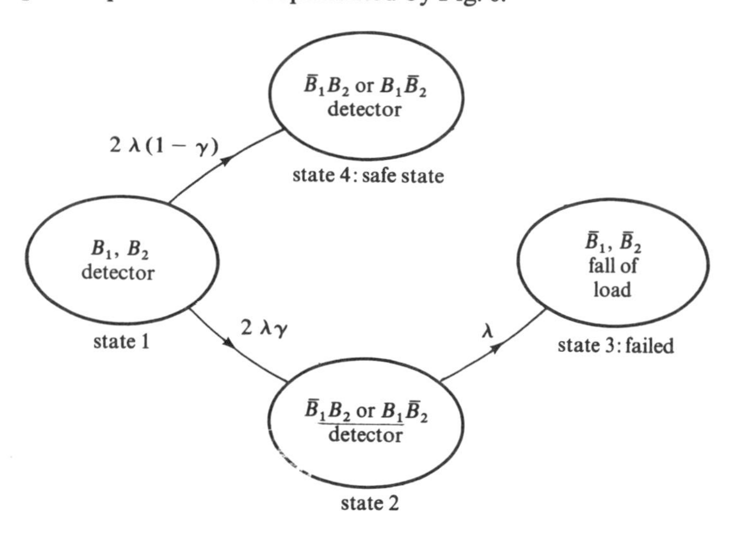

Fig. 6. *State transition diagram.*

with:

B_i: operating state of subchain B_i,
$\bar{B}_i$: failed state of subchain B_i,
λ: hourly failure rate of subchain $B1$ and $B2$,
γ: failure rate under load of the unbalance detector.

The supercomponent can be in one of the four following states:

—state 1: both subchains $B1$ and $B2$ and the unbalance detector are operating properly.

—state 2: either of subchains $B1$ or $B2$ is failed and the unbalance detector has not operated.

—state 3: subchains $B1$ and $B2$ are both down. This is the failed state of the supercomponent: the load is dropped.

—state 4: one of subchains $B1$ or $B2$ is down, the unbalance detector has operated by issuing the instruction to terminate handling and apply the brakes. This is a safe state.

Solving the set of differential equations associated with this diagram, we obtain (cf. Exercise):

$$P_3(t) = \gamma(1 - 2\,e^{-\lambda t} + e^{-2\lambda t}).$$

This probability represents the system unavailability, since state 4 is a safe state. As the quantity $2\lambda T$ (T = duty time) was assumed to be very small, expansion of the exponentials gives:

$$P_3(T) \simeq \lambda^2 \gamma T^2 .$$

5.2. Hand-calculating the probability of the load falling

The calculation uses results from the determination of the minimal cut sets. Initially, we calculate the value of the probability of the supercomponent failing, then the value of the failure probabilities of the other components present in the minimal cut sets and, finally, the end result.

5.2.1. *Calculating the probability of the supercomponent failing*

The supercomponent corresponds to subchain B comprising the two subchains $B1$ and $B2$ (failure rate λ) together with the unbalance detector (failure rate under load γ)

- $\gamma = 3 \times 10^{-4}$/d (cf. Table 1)
- calculation of the λ of $B1$ or $B2$; this is equal to the sum of the failure rates of their constituent components (cf. Fig. 5).

Components	Failure modes	λ/h
reduction gearing—drum transmission	breakage of transmission shaft	1.1×10^{-6}
	breakage of pinion-wheel coupling	1.1×10^{-5}
drum	breakage	9×10^{-9}
rope	breakage	ε
return pulley	derailment from return pulley	9×10^{-7}
	pulley breakage	6×10^{-9}
	breakage of pulley shaft	9×10^{-9}
fixed point	release of rope	9×10^{-9}
	total for sub-chains $B1$ or $B2$	1.3×10^{-5}

We then calculate the failure probability of the supercomponent:

$$\begin{aligned} P_\alpha &= \lambda^2 \gamma \tau^2 \qquad \text{with } \tau = 100 \\ &= (1.3 \times 10^{-5})^2 . 3 \times 10^{-4} . 10^4 \\ &= 5.1 \times 10^{-10}. \end{aligned}$$

5.2.2. *Calculating the failure probabilities of the other components*

As the values of the λt are small compared to 1, all the following calculations are performed using the approximation $P = 1 - e^{-\lambda t} \simeq \lambda t$.

Probability of failure of subchain A

Components	Failure modes	λ/h
Hook	Breakage	2×10^{-8}
Lower pulley block	Breakage of one of the four pulleys	$4 \times 9 \times 10^{-9} = 3.6 \times 10^{-8}$
	Rope derailment from one of the four pulleys	$4 \times 9 \times 10^{-7} = 3.6 \times 10^{-6}$
	Breakage of pulley shaft	9×10^{-9}
	Total for subchain *A*	3.7×10^{-6}

hence the failure probability of subchain *A* is:

$$\begin{aligned} P_\beta &= 3.7 \times 10^{-6} \times 100 \\ &= 3.7 \times 10^{-4}. \end{aligned}$$

Probability of failure of reduction gearing

The failure rate is: 5.7×10^{-7}/h.
The failure probability is therefore:

$$P_\gamma = 5.7 \times 10^{-7} \times 100 = 5.7 \times 10^{-5}.$$

Probability of failure of the brakes and of the motor-reduction gearing coupling

The second order cut sets, [3, 6] and [5, 6], can be written in Boolean form as follows:

$$[6] \times ([3] + [5])$$

Components	Failure modes	λ/h
[3] main service brake	application failure	4.8×10^{-7}
[6] emergency auxiliary brake	application failure	4.8×10^{-7}
[5] motor-reduction gearing coupling	breakage	9×10^{-9}

hence, the failure probability of the brakes and the coupling is:

$$P_\delta = 4.8 \times 10^{-7} \times 100 \times [4.8 \times 10^{-7} \times 100 + 9 \times 10^{-9} \times 100]$$
$$= 2.4 \times 10^{-9}.$$

Probability of failure of the overspeed detectors and the components which they protect

The calculation corresponds to the last three cut sets [5, (16-1), (16-2)], [2, (16-1), (16-2)] and [1, (16-1), (16-2)] which can be written in Boolean form as follows:

$$([5] + [2] + [1]) \,.\, (16\text{-}1) \,.\, (16\text{-}2)$$

Components	Failure modes	λ/h
[1] motor power supply	loss of supply	10^{-6}
[2] motor	failure: loss of torque	7.7×10^{-6}
[5] motor-reduction gearing coupling	breakage	9×10^{-9}
(16-1) or (16-2) overspeed detector	failure to detect an overspeed	2×10^{-8}

hence, the overall failure probability is:

$$P_\varepsilon = (10^{-6} \times 100 + 7.7 \times 10^{-6} \times 100 + 9 \times 10^{-9} \times 100) \cdot (2 \times 10^{-8} \times 100)^2$$
$$= 3.5 \times 10^{-15}.$$

Note: Component [5] (motor-reduction gearing coupling) appears twice in the calculation, but this is found to have no numerical effect.

5.2.3. *Computing the final probability*

$$P = P_\alpha + P_\beta + P_\gamma + P_\delta + P_\varepsilon$$
$$= 5.1 \times 10^{-10} + 3.7 \times 10^{-4} + 5.7 \times 10^{-5} + 2.4 \times 10^{-9} + 3.5 \times 10^{-15}$$
$$= 4.3 \times 10^{-4}.$$

5.3. Using the DEFAIL and FIABC programs

The DEFAIL program [Mulet Marquis and Dubreuil-Chambardel, 1984] enables us to analyse a fault tree and determine from it the operating paths and the minimal cut sets. Using the FIABC program [Pagès, 1979], we can then compute the system availability and reliability.

The numerical data required by the latter are the operating times of the system and the failure rates of the components with their respective error factors (error factor = ratio of the upper bound of the confidence interval of the data item to the median value). It therefore provides the following results:

—the successful paths,
—the sets of cuts,
—the critical operating states,
—the system unavailability at the end of the operating time, with the associated confidence intervals,
—the importance factors of the various components.

A typical application of the FIABC program appears in Fig. 7.

6. CONCLUSION

We summarize the results obtained in the above example, work out the corrective measures necessary and subsequently make a few comments on the methods employed.

6.1. Synthesis of results obtained and possible action to be taken

Quantitative analysis highlights the overwhelming importance of subchain A which in itself accounts for 88% of the probability value obtained. In fact, the components of subchain A are single-mode failure sources.

Of all the possible failures of the components of subchain A, the greatest risk is that of rope derailment from the pulleys on the lower pulley block. One improvement would be to provide the pulleys with anti-derailment systems ensuring that the rope remained in the pulley groove. If this is done, the probability of derailment becomes negligible and the probability of subchain A failing is reduced to 6.5×10^{-6}. The probability of the load falling then becomes 5.6×10^{-5} and it is then the probability of reduction gearing failure, also a single-mode failure, which becomes predominant.

A second possible improvement might be to eliminate the emergency brake on the high-speed input shaft of the reduction gearing and to replace it by two emergency brakes mounted directly on the drums. Failure of the reduction gearing would therefore be backed up by the overspeed detection system and these emergency brakes on the drums. Calculations would show that the probability of the load falling would then become practically the same as the probability of subchain A failing, i.e. $P = 6.5 \times 10^{-6}$.

The design could be improved still further by duplicating subchain A. For this purpose it would be necessary to separate the pulleys of the lower pulley block corresponding to each rope, each pulley having its own shaft, and to duplicate the load suspension devices.

COMPONENT 1 : MOTOR + POWER SUPPLY
COMPONENT 2 : MOTOR-REDUCTION GEARING COUPLING
COMPONENT 3 : REDUCTION GEARING
COMPONENT 4 : BRAKE
COMPONENT 5 : AUXILIARY BRAKE
COMPONENT 6 : OVERSPEED DETECTOR 1
COMPONENT 7 : OVERSPEED DETECTOR 2
COMPONENT 8 : SUBCHAINS *B*1 AND *B*2 + UNBALANCE DETECTOR
COMPONENT 9 : PULLEY BLOCK
COMPONENT 10 : LIFTING BEAM

THE SYSTEM COMPRISES 10 SUPERCOMPONENTS

```
&LCARAC
IN=          10,ICOX=          5,DELTAT=  10.0000000   ,NBPAS=        10,TMAX=  100.000000  ,NIV=        1,ATTENT=  .0
IMAXC=        4,NEGLI=      1000,ICOURB=              0,ISEED=     50000,TINFI=  100.000000  ,ISIMU=      0,PNEG=  1.00000000
IFIC=         5,KPRINT=        1,NREPET=              3,PUNCH=  .0
&END
```

COMPONENTS	TYPE	PARAMETERS			
1	3	0.87000E-06	0.0	0.67000E-05	
2	1	0.80000E-08			
3	1	0.50000E-06			
4	2	0.16000E-06	0.0	0.16000E-07	
5	2	0.16000E-09	0.0	0.16000E-07	
6	1	0.17000E-07			
7	1	0.17000E-07			
8	10	0.13000E-04	0.0	0.30000E-03	
9	3	0.90000E-08	0.0	0.90000E-08	0.0
		0.90000E-08	0.0	0.90000E-08	0.0
		0.90000E-08	0.0	0.90000E-06	0.0
		0.90000E-06	0.0	0.90000E-06	0.0
		0.90000E-06			
10	1	0.20000E-07			
11	EQUIVALENT TO 5				
12	EQUIVALENT TO 5				
13	EQUIVALENT TO 7				

Fig. 7.

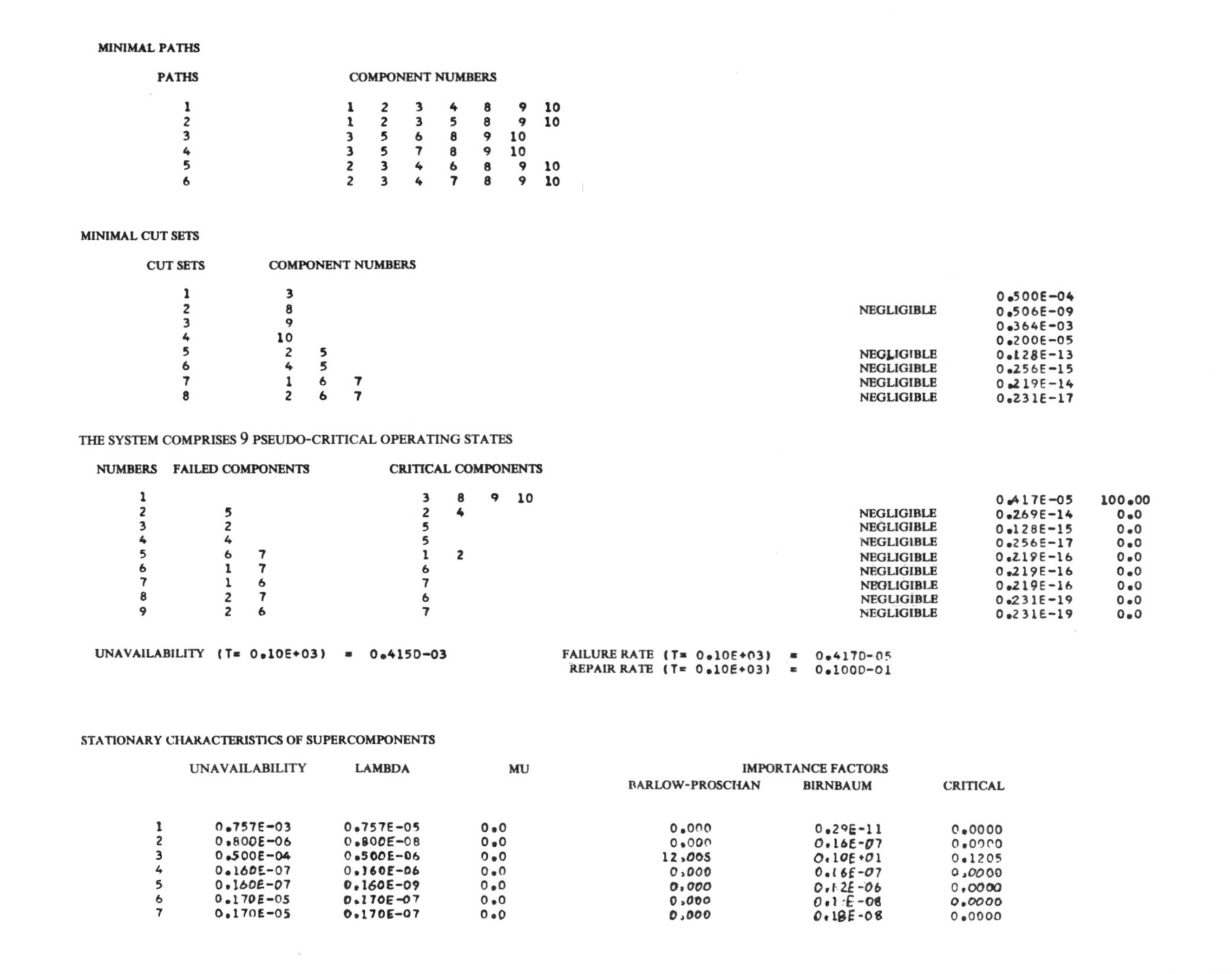

MINIMAL PATHS

PATHS	COMPONENT NUMBERS						
1	1	2	3	4	8	9	10
2	1	2	3	5	8	9	10
3	3	5	6	8	9	10	
4	3	5	7	8	9	10	
5	2	3	4	6	8	9	10
6	2	3	4	7	8	9	10

MINIMAL CUT SETS

CUT SETS	COMPONENT NUMBERS				
1	3				0.500E-04
2	8			NEGLIGIBLE	0.506E-09
3	9				0.364E-03
4	10				0.200E-05
5	2	5		NEGLIGIBLE	0.128E-13
6	4	5		NEGLIGIBLE	0.256E-15
7	1	6	7	NEGLIGIBLE	0.219E-14
8	2	6	7	NEGLIGIBLE	0.231E-17

THE SYSTEM COMPRISES 9 PSEUDO-CRITICAL OPERATING STATES

NUMBERS	FAILED COMPONENTS		CRITICAL COMPONENTS						
1			3	8	9	10		0.417E-05	100.00
2	5		2	4			NEGLIGIBLE	0.269E-14	0.0
3	2		5				NEGLIGIBLE	0.128E-15	0.0
4	4		5				NEGLIGIBLE	0.256E-17	0.0
5	6	7	1	2			NEGLIGIBLE	0.219E-16	0.0
6	1	7	6				NEGLIGIBLE	0.219E-16	0.0
7	1	6	7				NEGLIGIBLE	0.219E-16	0.0
8	2	7	6				NEGLIGIBLE	0.231E-19	0.0
9	2	6	7				NEGLIGIBLE	0.231E-19	0.0

UNAVAILABILITY (T= 0.10E+03) = 0.415D-03

FAILURE RATE (T= 0.10E+03) = 0.417D-05
REPAIR RATE (T= 0.10E+03) = 0.100D-01

STATIONARY CHARACTERISTICS OF SUPERCOMPONENTS

	UNAVAILABILITY	LAMBDA	MU	IMPORTANCE FACTORS		
				BARLOW-PROSCHAN	BIRNBAUM	CRITICAL
1	0.757E-03	0.757E-05	0.0	0.000	0.29E-11	0.0000
2	0.800E-06	0.800E-08	0.0	0.000	0.16E-07	0.0000
3	0.500E-04	0.500E-06	0.0	12.005	0.10E+01	0.1205
4	0.160E-07	0.160E-06	0.0	0.000	0.16E-07	0.0000
5	0.160E-07	0.160E-09	0.0	0.000	0.12E-06	0.0000
6	0.170E-05	0.170E-07	0.0	0.000	0.1[illegible]E-08	0.0000
7	0.170E-05	0.170E-07	0.0	0.000	0.18E-08	0.0000

Fig. 7. (*cont.*)

8	0.506E-09	0.101E-10	0.0	0.000	0.10E+01	0.0000
9	0.364E-03	0.364E-05	0.0	87.515	0.10E+01	0.8781
10	0.200E-05	0.200E-07	0.0	0.480	0.10E+01	0.0048

TIME	UNAVAILABILITY	FAILURE RATE	UNRELIABILITY
0.0	0.916D-15	0.416D-05	0.0
10.00	0.416D-04	0.417D-05	0.416D-04
20.00	0.832D-04	0.417D-05	0.833D-04
30.00	0.125D-03	0.417D-05	0.125D-03
40.00	0.166D-03	0.417D-05	0.167D-03
50.00	0.208D-03	0.417D-05	0.208D-03
60.00	0.249D-03	0.417D-05	0.250D-03
70.00	0.291D-03	0.417D-05	0.292D-03
80.00	0.332D-03	0.417D-05	0.333D-03
90.00	0.374D-03	0.417D-05	0.375D-03
100.00	0.415D-03	0.417D-05	0.416D-03

Fig. 7. (*cont.*)

WITH ANTI-DERAILMENT SYSTEM

IDENTICAL SYSTEM WITH NEW PARAMETERS

```
&LCARAC
IN=          -1,ICOX=          5,DELTAT=  1.00000000    ,NBPAS=            0,TMAX=  .0          ,NIV=          0,ATTENT=  .0
IMAXC=        4,NEGLI=      1000,ICOURB=               0,ISEED=       50000,TINFI=  100.000000   ,ISIMU=          0,PNEG=  .0
IFIC=         5,KPRINT=        1,NREPET=               0,PUNCH=  .0
&END
```

COMPONENTS	TYPE	PARAMETERS			
1	3	0.87000E-06	0.0	0.67000E-05	
2	1	0.80000E-08			
3	1	0.50000E-06			
4	2	0.16000E-06	0.0	0.16000E-07	
5	2	0.16000E-09	0.0	0.16000E-07	
6	1	0.17000E-07			
7	1	0.17000E-07			
8	10	0.13000E-04	0.0	0.30000E-03	
9	3	0.90000E-08	0.0	0.90000E-08	0.0
		0.90000E-08	0.0	0.90000E-08	0.0
		0.90000E-08			
10	1	0.20000E-07			

MINIMAL CUT SETS

CUT SETS	COMPONENT NUMBERS				
1	3				0.500E-04
2	8			NEGLIGIBLE	0.506E-09
3	9				0.450E-05
4	10				0.200E-05
5	2	5		NEGLIGIBLE	0.128E-13
6	4	5		NEGLIGIBLE	0.256E-15
7	1	6	7	NEGLIGIBLE	0.219E-14
8	2	6	7	NEGLIGIBLE	0.231E-17

THE SYSTEM COMPRISES 9 PSEUDO-CRITICAL OPERATING STATES

NUMBERS FAILED COMPONENTS   CRITICAL COMPONENTS

Fig. 7. (*cont.*)

```
1                                   3   8   9  10                                             0.565E-06   100.00

UNAVAILABILITY (T= 0.10E+03) = 0.563D-04                FAILURE RATE (T= 0.10E+03) = 0.565D-06
                                                         REPAIR RATE (T= 0.10E+03) = 0.100D-01
```

STATIONARY CHARACTERISTICS OF SUPERCOMPONENTS

	UNAVAILABILITY	LAMBDA	MU	IMPORTANCE FACTORS BARLOW-PROSCHAN	BIRNBAUM	CRITICAL
1	0.757E-03	0.757E-05	0.0	0.000	0.29E-11	0.0000
2	0.800E-06	0.800E-08	0.0	0.000	0.16E-07	0.0000
3	0.500E-04	0.500E-06	0.0	88.494	0.10E+01	0.8880
4	0.160E-07	0.160E-06	0.0	0.000	0.16E-07	0.0000
5	0.160E-07	0.160E-09	0.0	0.000	0.82E-06	0.0000
6	0.170E-05	0.170E-07	0.0	0.000	0.13E-08	0.0000
7	0.170E-05	0.170E-07	0.0	0.000	0.13E-08	0.0000
8	0.506E-09	0.101E-10	0.0	0.002	0.10E+01	0.0000
9	0.450E-05	0.450E-07	0.0	7.964	0.10E+01	0.0799
10	0.200E-05	0.200E-07	0.0	3.540	0.10E+01	0.0355

Fig. 7. (*cont.*)

A diagram of a kinematic chain, modified accordingly, is shown in Fig. 8.

Complete and detailed analyses using the methods described in this example have been performed for a number of materials-handling cranes [Barbet, 1976, 1977, 1980]. The results have revealed weaknesses and have led to a number of

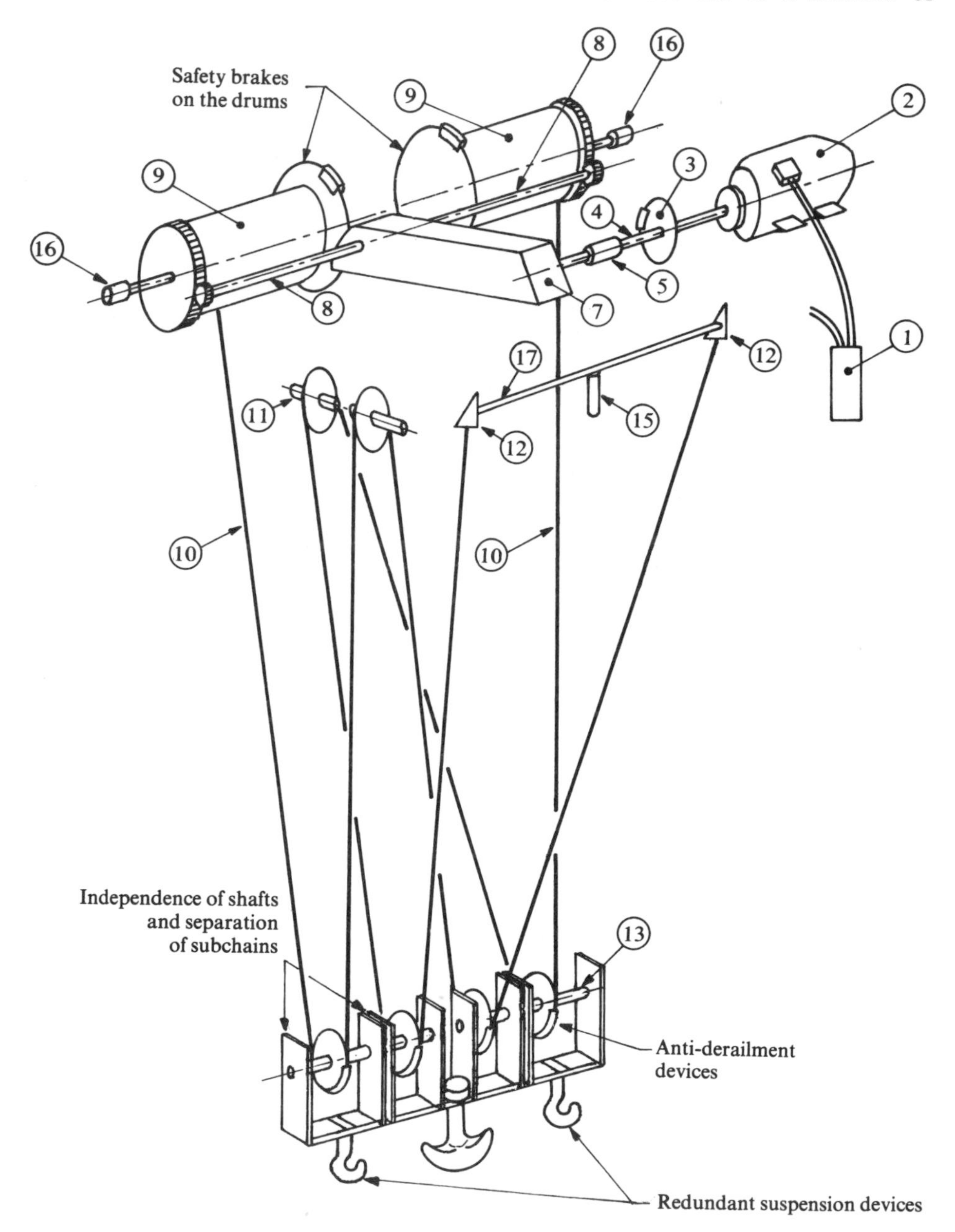

Fig. 8. *Possible improvement to kinematic hoisting chain.*

modifications. This clearly shows that this type of analysis must be carried out in parallel with design studies in order to ensure that any resulting modifications are not excessively difficult and costly to implement.

6.2. Comments on methods used

The methods used in this example raise certain points:

—the figures obtained assume that the failures likely to occur are independent and random, and do not therefore take account of possible common mode failures. It must be remembered that a common mode failure is a failure which may affect several components either simultaneously or successively. It is difficult to incorporate this type of failure quantitatively, but any reliability study must also examine the possibility of common mode failures.

—the FMEA is a very useful method which is virtually indispensable if we want to make sure that we have examined all possible component malfunctions in the system under test: it is a guarantee of exhaustiveness and proper understanding of the system.

—the fault tree is a practical tool but cannot be used in every case. It provides a representation of the system logic and of the possible failure combinations. The reliability engineer can use this representation not only for his own purposes but also to explain the problems involved to persons not versed in reliability analysis techniques. On the other hand, when the size of the tree increases, it soon becomes unwieldy and reducing it to a more manageable size may be difficult.

—the reliability block diagram can be constructed directly without constructing the fault tree. It is a good overall representation of the operating possibilities of the system, provides a quick way of determining the cut sets and facilitates quantitative analysis.

—availability can often be hand-calculated, as Poincaré's formula often only needs to be applied to the first term. If the system is non-repairable, reliability equals availability. In the case of repairable systems, hand-calculating the reliability soon proves difficult despite the existence of approximate formulae (cf. Chapter 6, section 2.3.2). In any case, when the number of components becomes large, the use of programs is necessary and yields additional results (confidence intervals, importance factors, etc.).

EXERCISE

Calculate the unavailability of the supercomponent made up of subchains $B1$ and $B2$ and the unbalance detector (section 5.1, Fig. 6).

SOLUTION TO EXERCISE

This unavailability (which is also the unreliability, as the system is non-repairable during its mission) is equal to $P_3(t)$.

We first have to calculate $P_1(t)$, then $P_2(t)$.

— Probability P_1 of being in state 1, with both subchains $B1$ and $B2$ and the unbalance detector operating.

We have:

$$P_1(t+\Delta t) = P_1(t).[1-(2\lambda.(1-\gamma)).\Delta t + 2\lambda.\gamma.\Delta t + 0(\Delta t)]$$

and letting Δt tend to 0, we find:

$$\frac{dP_1(t)}{dt} = -2\lambda.P_1(t).$$

with $P_1(0) = 1$

hence $$\overline{P}_1(s) = \frac{1}{s+2\lambda}$$

or $$P_1(t) = e^{-2\lambda t}.$$

— Probability P_2 of being in state 2, either of subchains $B1$ or $B2$ being failed and the unbalance detector not having functioned.

We have:

$$P_2(t+\Delta t) = P_1(t).(2\lambda.\gamma.\Delta t + 0(\Delta t)) + P_2(t).(1-\lambda.\Delta t + 0(\Delta t))$$

and letting Δt tend to 0, we find:

$$\frac{dP_2(t)}{dt} = 2\lambda\gamma P_1(t) - \lambda.P_2(t).$$

Using the Laplace transform, we obtain:

$$s\overline{P}_2(s) = 2\lambda.\gamma.\overline{P}_1(s) - \lambda.\overline{P}_2(s) \qquad \text{with } P_2(0) = 0$$

or $$\overline{P}_2(s) = \frac{2\lambda.\gamma}{s+\lambda}\cdot\overline{P}_1(s)$$

$$= \frac{2\lambda.\gamma}{(s+\lambda)(s+2\lambda)} = 2\gamma\left(\frac{1}{1+\lambda} - \frac{1}{s+2\lambda}\right)$$

hence $$P_2(t) = 2\gamma(e^{-\lambda t} - e^{-2\lambda t}).$$

— Probability P_3 of being in state 3, both $B1$ and $B2$ having failed (fall of load):

We have:

$$P_3(t+\Delta t) = P_2(t).(\lambda.\Delta t + 0(\Delta t)) + P_3(t)$$

and letting Δt tend to 0, we find:

$$\frac{dP_3(t)}{dt} = \lambda P_2(t)$$

hence $s\cdot\overline{P}_3(s) = \lambda\cdot\overline{P}_2(s)$ with $P_3(0) = 0$ and therefore

$$\overline{P}_3(s) = \frac{\lambda}{s}\overline{P}_2(s)$$

$$= \frac{2\lambda^2\gamma}{s(s+\lambda)(s+2\lambda)} = \gamma\left(\frac{1}{s} - \frac{2}{s+\lambda} + \frac{1}{s+2\lambda}\right)$$

or:

$$P_3(t) = \gamma(1 - 2e^{-\lambda t} + e^{-2\lambda t}).$$

The equivalent failure rate of the supercomponent can then be calculated by writing:

$$\begin{aligned}\Lambda(t) &= \frac{-F'(t)}{F(t)} = \frac{P_3'(t)}{1-P_3(t)} \\ &= \frac{2\lambda\gamma\, e^{-\lambda t} - 2\lambda\gamma\, e^{-2\lambda t}}{(1-\gamma) + 2\gamma\, e^{-\lambda t} - \gamma\, e^{-2\lambda t}} \\ &= 2\lambda\gamma \cdot \frac{e^{-\lambda t} - e^{-2\lambda t}}{(1-\gamma) + 2\gamma\, e^{-\lambda t} - \gamma\, e^{-2\lambda t}}.\end{aligned}$$

As the values of $\lambda \cdot t$ and γ remain small compared to 1, we obtain the approximate value of $\Lambda(t)$:

$$\Lambda(t) \simeq 2\lambda \,.\, \gamma \,.\, \lambda \,.\, t = 2\lambda^2 \gamma t \,.$$

REFERENCES

Barbet J. F. (1976): *Fessenheim* 1 et 2. *Etude probabiliste du pont de manutention du conteneur d'assemblages combustibles usés*; EDF Note HT 13/41/76.

Barbet J. F. (1977): — CP1 — *Etude probabiliste du pont de manutention du conteneur d'assemblages combustibles usés*; EDF Note HT/13/11/77.

Barbet J. F. (1980): *Une méthode d'étude probabiliste de la sûreté des ponts de manutention de centrales nucléaires*, European Seminar on System Reliability, 17–19 June, Bordeaux.

Barbet J. F., Charton P. and Hovnanian P. (1979): *Saint-Laurent-des Eaux B. Etude probabiliste du pont polaire du bâtiment réacteur*; EDF Note HT 13/21/79.

F.E.M. (1970): *Règles pour le calcul des appareils de levage*; Fédération Européenne de la Manutention.

IEEE (1977): *Guide to the Collection and Presentation of Electrical, Electronic, and Sensing Component Reliability Data for Nuclear Power Generating Stations*; IEEE Std 500-1977.

Marcovici C. and Ligeron J. C. (1974): *Utilisation des techniques de fiabilité en mécanique*; Technique et Documentation, Collection PSI.

Mulet Marquis D. and Dubreuil-Chambardel A. (1984): DEFAIL 84, FIABC84: un ensemble de logiciels pour l'évaluation qualitative et quantitative de la fiabilité d'un système; Note EDF HI 4737-02.

N.T.I.S. (1975): *Non-electronic Reliability Notebook*, RADC-TR 75-22; distributed by N.T.I.S. (US Department of Commerce).

U.S.N.R.C. (1975): *Reactor Safety Study. An Assessment of Accident Risks in U.S. Commercial Nuclear Power Plants*; WASH-1400 (NUREG 75/015).

Welker E. L. and Lipow M. (1974): Estimating the exponential failure rate from data with no failure events. TRW Systems Group — Redondo Beach, California; *Annual Reliability and Maintainability Symposium*, Los Angeles.

APPENDIX 1

QUANTILES OF THE STANDARDIZED NORMAL DISTRIBUTION

Values of u_α such that:

$$\int_{u_\alpha}^{\infty} \frac{e^{-t^2/2}}{\sqrt{2\pi}} dt = \alpha . \qquad (1)$$

There are numerous tables giving the values of u_α as a function of α. A large bibiliography may be found in [Johnson and Kotz, 1970]

There are also several programs for evaluating u_α. The table given below was computed using the XFROMP algorithm [Cunningham, 1959].

α	0.00	0.01	0.02	0.03	0.04	0.05	0.06	0.07	0.08	0.09
0.0	INFINI	2.3263	2.0537	1.8808	1.7507	1.6449	1.5548	1.4758	1.4051	1.3408
0.1	1.2816	1.2265	1.1750	1.1264	1.0803	1.0364	0.9945	0.9542	0.9154	0.8779
0.2	0.8416	0.8064	0.7722	0.7388	0.7063	0.6745	0.6433	0.6128	0.5828	0.5534
0.3	0.5244	0.4959	0.4677	0.4399	0.4125	0.3853	0.3585	0.3319	0.3055	0.2793
0.4	0.2533	0.2275	0.2019	0.1764	0.1510	0.1257	0.1004	0.0753	0.0502	0.0251

When α is greater than 0.5, Eqn (1) shows that:

$$u_\alpha = - u_{1-\alpha} .$$

APPENDIX 2
QUANTILES OF THE CHI-SQUARED DISTRIBUTION

Values of $\chi^2_\alpha(v)$ such that:

$$\int_{\chi^2_\alpha(\nu)}^{\infty} \frac{t^{(\nu-2)/2}\,\mathrm{e}^{-t/2}}{2^{\nu/2}\,\Gamma\left(\frac{\nu}{2}\right)}\,\mathrm{d}t = \alpha . \tag{1}$$

There are numerous tables giving the value of $\chi^2_\alpha(v)$ (cf. [Johnson and Kotz, 1970]). Sometimes these tables give the value of $\bar{\chi}^2_\alpha(v)$ such that:

$$\int_0^{\chi^2_\alpha(\nu)} \frac{t^{(\nu-2)/2}\,\mathrm{e}^{-t/2}}{2^{\nu/2}\,\Gamma\left(\frac{\nu}{2}\right)}\,\mathrm{d}t = \alpha .$$

These two values are linked by the relationship:

$$\bar{\chi}^2_\alpha(v) = \chi^2_{1-\alpha}(v).$$

Computing $\chi^2_\alpha(v)$ using a program

The CHISQD algorithm [Goldstein, 1973] proposes two series expansions for evaluating $\chi^2_\alpha(v)$ depending on whether v is greater than or less than $2+4|u_\alpha|$. Note that these series expansions are applicable for non-integer values of v, which makes it possible to determine the quantiles of the gamma distribution. In fact, by putting $t/2 = \lambda x$ and $v/2 = \beta$, Eqn (1) can be written as:

$$\int_{\chi^2_\alpha(2\beta)/2\lambda}^{\infty} \frac{\lambda^\beta x^{\beta-1}\,\mathrm{e}^{-\lambda x}}{\Gamma(\beta)}\,\mathrm{d}x = \alpha \tag{2}$$

hence
$$X(\beta, \lambda, \alpha) = \frac{\chi^2_\alpha(2\beta)}{2\lambda} .$$

However, the errors are very significant for small values of v. We can then proceed as as follows if we have a program for computing the distribution function F of the gamma distribution:

— write Eqn (2) in the form $1 - F(X) = \alpha$,

— solve this equation by Newton's method,

— take as starting point the value $X = \lambda(1-\alpha)^{1/\beta}$ which is the asymptotic value when β tends to zero (cf. [Waller et al., 1977]).

This method gives good results for small values of β; by combining it with the series expansion suggested in [Goldstein, 1973] for large values of β, we obtain a program for compiling the following chi-squared table, or the gamma distribution quantiles table given in section 6.7 of Chapter 7.

DEGREES OF FREEDOM	$\alpha = 0.95$	$\alpha = 0.90$	$\alpha = 0.80$	$\alpha = 0.20$	$\alpha = 0.10$	$\alpha = 0.05$
1	.393E-02	.158E-01	.642E-01	1.642	2.706	3.841
2	.1026	.2107	.4463	3.219	4.605	5.991
3	.3518	.5844	1.005	4.642	6.251	7.815
4	.7107	1.064	1.649	5.989	7.779	9.488
5	1.145	1.610	2.343	7.289	9.236	11.07
6	1.635	2.204	3.070	8.558	10.64	12.59
7	2.167	2.833	3.822	9.803	12.02	14.07
8	2.733	3.490	4.594	11.03	13.36	15.51
9	3.325	4.168	5.380	12.24	14.68	16.92
10	3.940	4.865	6.179	13.44	15.99	18.31
11	4.575	5.578	6.989	14.63	17.27	19.68
12	5.226	6.304	7.807	15.81	18.55	21.03
13	5.892	7.041	8.634	16.98	19.81	22.36
14	6.571	7.790	9.467	18.15	21.06	23.68
15	7.261	8.547	10.31	19.31	22.31	25.00
16	7.962	9.312	11.15	20.47	23.54	26.30
17	8.672	10.09	12.00	21.61	24.77	27.59
18	9.390	10.86	12.86	22.76	25.99	28.87
19	10.12	11.65	13.72	23.90	27.20	30.14
20	10.85	12.44	14.58	25.04	28.41	31.41
21	11.59	13.24	15.44	26.17	29.61	32.67
22	12.34	14.04	16.31	27.30	30.81	33.92
23	13.09	14.85	17.19	28.43	32.01	35.17
24	13.85	15.66	18.06	29.55	33.20	36.41
25	14.61	16.47	18.94	30.68	34.38	37.65
26	15.38	17.29	19.82	31.79	35.56	38.89
27	16.15	18.11	20.70	32.91	36.74	40.11
28	16.93	18.94	21.59	34.03	37.92	41.34
29	17.71	19.77	22.48	35.14	39.09	42.56
30	18.49	20.60	23.36	36.25	40.26	43.77
31	19.28	21.43	24.26	37.36	41.42	44.99
32	20.07	22.27	25.15	38.47	42.58	46.19
33	20.87	23.11	26.04	39.57	43.75	47.40
34	21.66	23.95	26.94	40.68	44.90	48.60
35	22.47	24.80	27.84	41.78	46.06	49.80
36	23.27	25.64	28.73	42.88	47.21	51.00
37	24.07	26.49	29.64	43.98	48.36	52.19
38	24.88	27.34	30.54	45.08	49.51	53.38
39	25.70	28.20	31.44	46.17	50.66	54.57
40	26.51	29.05	32.34	47.27	51.80	55.76
41	27.33	29.91	33.25	48.36	52.95	56.94
42	28.14	30.77	34.16	49.46	54.09	58.12
43	28.96	31.63	35.07	50.55	55.23	59.30
44	29.79	32.49	35.97	51.64	56.37	60.48
45	30.61	33.35	36.88	52.73	57.51	61.66
46	31.44	34.22	37.80	53.82	58.64	62.83
47	32.27	35.08	38.71	54.91	59.77	64.00
48	33.10	35.95	39.62	55.99	60.91	65.17
49	33.93	36.82	40.53	57.08	62.04	66.34
50	34.76	37.69	41.45	58.16	63.17	67.50

REFERENCES

BHATTACHARJEE G. P. (1970): The incomplete gamma integral. Algorithm A532; *Applied Statistics*, **19** (3).

CUNNINGHAM S. (1959): From normal integral to deviate. Algorithm AS24; *Applied Statistics*, **18** (3).

GOLDSTEIN R. B. (1973): Chi-square quantiles. Algorithm 451; *Communications of the ACM*, **16** (8).

JOHNSON N. L. and KOTZ S. (1970): *Continuous univariate distributions* 1; Houghton Mifflin, Boston.

WALLER R. A., JOHNSON M. M., WATERMAN M. S. and MARTZ H. F. Jr. (1977): *Gamma Prior Distribution Selection for Bayesian Analysis of Failure Rate and Reliability*; International Conference on Nuclear Systems Reliability Engineering and Risk Assessment, Gatlinburg, Tenn., U.S.A.

INDEX